BIOLOGISTS AND THE PROMISE OF AMERICAN LIFE

FROM MERIWETHER LEWIS TO ALFRED KINSEY

PHILIP J. PAULY

PRINCETON UNIVERSITY PRESS

PRINCETON AND OXFORD

Library of Congress Cataloging-in-Publication Data

Pauly, Philip J.
Biologists and the promise of American life: from Meriwether Lewis to
Alfred Kinsey / Philip J. Pauly
p. cm
Includes bibliographical referneces (p.).
ISBN 0-691-04977-7 (cl: alk. paper)
1. Biology—United States—History. 2. United States—Civilization. I. Title.
QH305.2.U6 P38 2000
570′.973—dc21   00-020896

This book has been composed in Electra

The paper used in this publication meets the minimum
requirements of ANSI/NISO Z39.48-1992 (R1997)
(*Permanence of Paper*)

www.pup.princeton.edu

Printed in the United States of America

10 9 8 7 6 5 4 3 2 1

This book is for Nick,
a more cultured child
than I was

CONTENTS

This is the rest of the story. More years ago than I care to admit, in writing *Controlling Life*, a biography of the experimental biologist Jacques Loeb, I examined what kind of scientific world this German Jewish immigrant encountered on his arrival in the United States in 1891.[1] I emphasized the differences between the biotechnological impulses that motivated Loeb and the academic evolutionism of most life scientists at the new American research universities. Such a contrast was sufficient for the purposes of a biography, but my characterization of the scope and interests of American biologists left a nagging dissatisfaction in my mind and in the minds of some otherwise sympathetic readers. My awareness that there was a larger American scientific landscape extending beyond Harvard, the University of Chicago, and the Marine Biological Laboratory, and that its inhabitants shared purposes with Loeb, was epitomized in the figure of the horticultural wizard Luther Burbank. Administrators at the University of California at Berkeley, who hired the cosmopolitan Loeb in 1902, worked enthusiastically, if unsuccessfully, at the same time to make that self-educated Yankee nurseryman a member of their faculty.

My aim in this book is to sketch the contours of the landscape of American biological science, to see how the people who occupied that territory worked to change it, and to understand the degree to which they used their science to shape the country in which they lived. In contrast to my earlier book, I devote little attention to particular discoveries, experiments, or theories. In order to comprehend the influence of a heterogeneous collection of individuals, working from the beginning of the nineteenth century to the middle of the twentieth century, I move from the outside in—from settings and aims to communities, programs, and visions. As a former biographer I am deeply aware of both the difficulty and the importance of reconstructing the thoughts and actions of individual scientists. But grasping the larger picture entails a different perspective.

Ultimately, however, the view of American biology presented here is consistent with that presented in *Controlling Life*. One can follow Michel Foucault in articulating the shackling structure of power/knowledge. More benignly, using vocabulary associated with John Dewey, it is possible to describe these enterprises as the actions of intelligence in the modern world. What follows is a concrete study of how a particular group of Americans, who increasingly called themselves biologists, gained a communal identity, forged a set of tools for dealing with the natural world and with other people, and shaped the future in ways that were sometimes — if not always — precisely as they wished.

The same, I hope, could be said about this book. When I was younger I did not understand why serious scholars waxed effusive in their acknowledgments; with age and mishap I have become wiser, or perhaps just self-indulgent. I began to think seriously about the issues addressed here about ten years ago when Dorothy Ross and Olivier Zunz asked me to participate in a conference in Bellagio on the murky problem of "modernism" in the human sciences. Listening to David Hollinger, James Kloppenberg, and Richard Rorty were, in retrospect, formative experiences.[2] Work on that paper was possible because, fortuitously, I then held a Rockefeller Foundation Humanities Fellowship at the Wood Institute of the College of Physicians of Philadelphia. I am deeply grateful for the freedom that fellowship provided and for the interchanges it made possible with the institute's director that year, Nancy Tomes. I refined my ideas in 1992 at Rutgers' Center for the Critical Analysis of Contemporary Culture. George Levine provided participants a lively scholarly home, and it was made more lively still by Norman Levitt's energetic criticisms.

These skirmishes, which ultimately formed one of the streams leading to the so-called science wars, were intellectually interesting but a scholarly dead end. A diagnosis of lymphoma reinforced my conviction in the positive value of a noncombatant status in these debates. Three historians of American science reminded me of the existence and potential of more constructive approaches. John C. Burnham has been a mentor and friend for two decades. Never a slave to fashion, he has combined intellectual openness with a vocal faith in civilization. Nathan Reingold emphasized that empirical inquiry has led historians to more and more interesting questions than those dreamt of by philosophers. Robert Kohler's writings, especially *Lords of the Fly*, have been for me, as for others, some of the best exemplars of such inquiries.[3]

Support for this project came from the National Science Foundation (SBR-9421880) and the National Endowment for the Humanities (RH-21244-95); the value of these grants was enhanced through the enlightened policies that Rutgers, the State University of New Jersey, has implemented to support its faculty's research. The libraries and archives essential to this work are evident, in part, in the citations. The Internet has both increased their work and made the allocation of credit more difficult. I am particularly grateful to Susan Fraser (New York Botanical Garden), Jean Cargill (Botany Libraries, Harvard University), Jean Monahan (Marine Biological Laboratory), Jon Roberts (Bancroft Library, University of California), Deborah Day (Scripps Institution of

Oceanography), Janice Goldblum (National Academy of Sciences), and Bertram Zuckerman (Fairchild Tropical Gardens). At the Smithsonian Institution Archives, Pamela Henson and Paul Theerman provided all the guidance an amateur of that unique American organization could wish. Finally, Susan Witzell of the Woods Hole Historical Collection and Museum responded with enthusiasm and kindness to innumerable queries and speculations about her village's history.

Numerous individuals answered questions, provided information, or shared their own unpublished writings. Garland Allen, Toby Appel, Keith Benson, Adele Clarke, Nathaniel Comfort, Kathy Cooke, Kevin Dann, Patricia Gossell, Sally Kohlstedt, Jane Maienschein, Lynn Nyhart, Naomi Oreskes, Ronald Rainger, David Rhees, John Servos, Michael Sokal, and Leila Zenderland were particularly helpful. Kathryn and Robert Jacob offered hospitality and wonderful conversation. Mary P. Winsor provided details of Harvard history, encouragement at conventions and via email, and a careful reading of the entire text. One couldn't ask more of a professional colleague. At Rutgers I received particular support from Jack Cargill, James Reed, Thomas Slaughter, Mark Wasserman, and Virginia Yans. Paul Israel and Susan Schrepfer took the time to read and comment on the entire manuscript. Mike Siegel of the Rutgers geography department's Cartographic Laboratory took me through the intricacies of mapmaking. Jackson Lears provided the marvelous service of suggesting, perhaps with irony, the book's title.

I have been lucky to have Sam Elworthy as an editor. He responded to this author's sometimes overstated anxieties with unflagging good cheer; his unique combination of a New Zealand background and enthusiasm for science and American life provided the ideal kind of ideal audience.

Now for the more personal. One of the minor fantasies that sustained me through chemotherapy was the hope that I would be able to acknowledge in print the people who kept me alive. It's taken longer than anticipated, but I'm glad finally to be able in this medium to thank Drs. Edward Abramowitz, Michael Geraghty, and Philip Friscia for their effective professional services and realistically reassuring bedside manners. The overburdened staff of the Long Island College Hospital kept me going. The Rutgers University history department—in the persons of its then leaders, Ziva Galili, Paul Clemens, and Mark Wasserman—acted as an incredibly humane institution. Gerald Grob showed me concretely the meaning of collegial concern. Lily Kay reminded me that therapeutics was a more advanced—and hopeful—science than statistics. I also want to remember the heroic people a generation earlier who tested, with their bodies, the mixtures of cell poisons that have become an often effective, if hardly optimal, treatment.

Closer to home, I am grateful for the continuing love of my parents, Edyth and Vincent Pauly, and for the support provided by my siblings. Agnes and Leo Bogart were on the scene with the kind of day-to-day support that made a profound difference. John Maggiotto, Cathy Mulhern, and Louise Aiello were true friends. Jackie Williams brought an empathy to our family that made a real difference.

My son, Nick, had to absorb some difficult lessons at an early age. He has slowed completion of this book, and for that I'm grateful. My wife, Michele Bogart, kept our family together. Her devotion and determination have been extraordinary. I feel extremely lucky to be her partner. Death is a natural process, but I, like the individuals in this book, continue to find the alternative much more appealing.

Brooklyn, New York
12 October 1999

BIOLOGISTS AND THE PROMISE OF AMERICAN LIFE

Toward a Cultural History of American Biology

Newspaper magnate Edward W. Scripps and biologist William E. Ritter had much to discuss. One Saturday morning in June 1919, Scripps drove the few miles from his southern California estate to the Scripps Institution for Biological Research, the starkly modern laboratory he and his sister Ellen had built on a hillside overlooking the Pacific, just north of the village of La Jolla. Ritter, Scripps's longtime friend and the institution's founding director, took his visitor to his large second-floor office. The two men—one a self-described "damned old crank," the other a gentlemanly scientist-philosopher—talked into the afternoon; as the coastal haze lifted, and they could see up and down the beach from the institution's balcony, they agreed to collaborate on a project in what they called "socio-biology."[1]

Scripps and Ritter's socio-biology was not the same as the "sociobiology" developed and promoted, some six decades later, by Harvard entomologist Edward O. Wilson. While the latter was fundamentally an interpretation of the continuities between animal and human nature, the former was above all a blueprint for culture—for the improvement of modern American life. It combined scientific understanding, philosophical convictions, organizational initiatives, and a national vision.

The two Californians' project originated from a particular concern. Only a few days earlier, the German government had accepted the terms of the treaty that President Woodrow Wilson and the Allies had crafted to shape the world after World War I. Scripps and Ritter anticipated that the global preeminence the United States was now claiming would oblige Americans to "carry responsibilities such as no other people in the history of the world have ever had to bear." In a twist on Wilson's famous phrase, Scripps argued that now that the world was "safe for democracy," the problem was "to make democracy safe for the world."[2]

Scripps and Ritter viewed this sociopolitical problem, however, from a broad biological perspective. They took seriously the fact that people were animals, and they were convinced that all human activities—including voting, newspaper publishing, and scientific investigation—ought to be comprehensible within a properly conceived science of biology. Improving democracy depended on making the human organisms residing in the United States

Figure I–1

Scripps Institution for Biological Research, early 1910s. Ritter and Scripps are figures 8–1 and 8–2. Photograph courtesy of the Scripps Institution of Oceanography Archives, University of California, San Diego.

"more intelligent." Breeding for higher IQ was not what they had in mind; both men considered eugenics politically retrograde and probably futile. Rather, they conceived their socio-biological problem to be the search for a way to improve the breadth of knowledge and depth of understanding of millions of people.

Scripps and Ritter believed that, in the twentieth century, "more intelligence must consist largely in more intelligence about science." They understood that newspapers were the primary avenue through which adult Americans learned new things extending beyond the personal. They thus drew from their discussion a specific proposal: to create an "American Association for the Dissemination of Science," soon to be renamed Science Service. In the next months, Scripps would endow, and Ritter preside over, a news agency whose mandate was to increase and improve the presentation of science in the nation's press.

Science Service's originators anticipated that much of their organization's work would be expository in content and incremental in impact. But they expected that certain moments could be crucial. Thus, in 1925 Ritter was ready to preside over Science Service's special program to send experts to Dayton, Tennessee, to shape press coverage of the trial of the biology teacher John T. Scopes. From Ritter's perspective, this paradigmatic American media event provided biologists an unparalleled opportunity to explain evolution to millions, to advocate scientific attitudes, and thus, Ritter hoped, to make people "more intelligent." It was one step in biologists' efforts to improve American life.

Three generations later, Scripps and Ritter's convictions—that the promotion of democracy and the reinforcement of American global power were biological problems, that a biological solution to these problems should be the creation of a newspaper bureau devoted to science popularization and advocacy, and, more broadly, that such national planning was part of the mission of an institution for "biological research"—seem strange. The aim of this book is to make those activities understandable. Certain aspects of their thinking—for example, their emphases on newspapers and on Wilsonian rhetoric—were peculiar to themselves and to the unique circumstances of 1919. But their deeper conviction—that biology provided a foundation for improving American life—had long motivated American biologists and their supporters. I want to recover the history of that conviction.

Throughout the nineteenth century, and well into the twentieth century, American biologists sought to influence national development. Their interests included determining what the extent and boundaries of the United States ought to be, exploiting the resources of the North American continent in the interest of the Euro-American population, defining American identity, and creating a sensibility among Americans appropriate to their position in the world—a sensibility that would be liberal, secular, and humanistic. Biologists' engagement with these problems was an important element in gaining support from national leaders in the development of biology, in shaping a national scientific network, and in orienting scientific work in particular directions. Although scientists' hopes were often more grandiose than their accomplishments, their accomplishments were not negligible. Interest in improving American life existed, however, in tension with other scientific desires, which were more individual and purely intellectual. At particular moments this tension broke biologists into openly competing groups.

The story of these efforts can be divided into three parts. The first interprets the development of natural history in the United States in the nineteenth century. There was initially a close identity among national development, efforts to organize naturalists across the nation, and the ability to think about the continent's natural history and biological future. The earlier efforts, prior to the Civil War, centered on a private university—Harvard—and its two leading naturalists, Asa Gray and Louis Agassiz. After the war the center of activity shifted to the newly prominent federal government—to a group of naturalists who had come from the provinces to work under the creator of the National Museum, Spencer Baird. Their efforts peaked in the 1880s in an integrated republican program, centered on the aptly named Cosmos Club, in which natural resources, national development, and political advancement reinforced each other. This program continued into the twentieth century in a more technocratic form within the Department of Agriculture and the Bureau of Immigration, where scientific bureaucrats sought to shape the biotic and demographic future of the country.

Part 2 examines the appearance of a new academic culture of biology in the last quarter of the nineteenth century. Federal naturalists overreached in the 1880s: their organizational base was too limited, their intellectual aspirations too grand, and their scientific practice too tedious. The alternative was a multitude of endeavors that were specialized both geographically and intellectually. Attempts to organize the life sciences nationally and to integrate disciplines were decoupled from state power and national goals. Although this period has been characterized as a "search for order," searches for order in the life sciences were at most partly successful; both geographic dispersion and interdisciplinary competition were against them. Academic biologists, however, had one great success that was specific to their science: at Woods Hole, Massachusetts, they took advantage of the unique characteristics of their domain to project a single science of biology with national scope. By focusing exclusively on a cosmopolitan, universalizing quest for basic knowledge defined by academic standards, however, they separated themselves from American society and its problems.

Part 3 deals with the efforts of this national academic elite, organized around the science of biology, to reengage with American problems. Whereas in the nineteenth century this engagement occurred on the level of organisms, in the twentieth century it was focused on the more abstract level of biological functions. One place where the proponents of biology were notably successful—but in a way that is retrospectively so obvious that both scientists and historians have ignored its importance—was in schools. Biology, designed as a science that would bring children into self-conscious adulthood, became a near-universal experience for American middle-class adolescents. Grander efforts—epitomized in the Scripps-Ritter partnership—conveyed a scientific philosophy to the American people that was fundamentally progressive. In addition, biologists sought to direct and alter the "breeding" of American people. Grand schemes for eugenic improvement failed, largely because they were incompatible with the new framework for academic research. Sexual biology, by contrast, was an area of triumph. The high school textbook writer and evolutionary biologist Alfred Kinsey profoundly influenced what Americans did in their most private lives.

What follows is, in essence, a cultural history of American biology, extending from the beginning of the nineteenth century to the middle of the twentieth century. However, it is neither antiquarian nor encyclopedic. The key terms here — biology, American, and cultural history — need to be understood in both the multiplicity of their meanings and in the exclusions they entail.

The word "biology" has, from its coinage in the early nineteenth century to the present, often referred to a broad sector of learning, synonymous with "life science." More specialized disciplines, ranging from cytology to ecology and from bacteriology to ornithology, have been included under such a biological umbrella. Simultaneously, however, biology has been a word around which particular research groups have rallied, in the belief that the seemingly chaotic agglomeration of scientific collectives dealing with organisms and their properties needed a coherent, essentially academic, core. The nature of that core has varied, with priority given to adaptation, evolution, the cell, or protoplasm, and, more recently, DNA and information. But advocates of academic biology have sought to conceive a subject that linked basic instruction with the most advanced research. One of the most important events in the history of American biology was the self-conscious effort in the late nineteenth century to make biology an academic nucleus. By preferentially following this academic group, I craft a narrative that reaches into the 1950s; at that point biology, even more than other American sciences, became so large and complex that the conceptual tools used here wear out. But, as chapter 3 indicates, the story is quite selective even earlier. I can only gesture toward the range of work done in agriculture, ecology, and medicine in the early twentieth century in the hope that others will explore the dimensions and significance of these enterprises.

"American" is not a word about which I can say anything new; I can only scratch the surface of reflections that other scholars have made. The linguistic presumption that residents of the United States exercised in appropriating a term dealing with the entire western hemisphere preferentially to themselves is widely recognized. The more focused issues involve its pairing with biology: "American biology" can mean a number of distinct but potentially linked entities. It can refer to the New World setting, the organisms either evolutionarily native to, or presently resident in, the American continents, or in the continental United States. Alternatively, it can describe the work of scientists trained and employed in the United States, in contrast to those in Europe and elsewhere. Less obviously, it has meant a scientific network that extended beyond particular localities and regions to include the entire nation, or continent. American biology also refers to a national style of science, or to a particular set of problems identified with U.S. scientists. Finally, it could mean some explicitly national, and even nationalistic, American bioscientific project. This book tries to encompass all these meanings, but, as I have indicated, ultimately focuses most fully on the last. I will say the least about style. Some historians have used this notion to characterize national scientific communities to good effect, but the concept is limited in its static quality and its deemphasis on change and on outcomes. As such it obscures what I am trying to accomplish.

This book is a "cultural" history in a straightforward, although hardly undisputed, sense. My methodological starting point is that science is an activity that can be understood, to a significant degree, with the same interpretive tools historians apply to such domains as painting, philosophy, or political reform. My examination of American biologists moves from their geographic and social environments to their particular activities. It emphasizes the formation and interactions of small but communally structured groups, rather than the theories and experimental discoveries made by individuals. It seeks to interpret their plans and visions as much as possible in light of their immediate settings, and it emphasizes the degree of overlap—sometimes conscious, as in the case of Ritter, but sometimes not—between the ways they thought about their objects of study and the ways they thought about themselves.

I am particularly interested in the history of "culture," however, because the word expresses precisely what was most important about American biology in the nineteenth and early twentieth centuries, and what distinguished biology, broadly conceived, from every other human activity, including other scientific disciplines. If we put aside the mostly twentieth-century disputes over the anthropological and literary meanings of culture and return, with historicist sensitivity, to the usage accepted among educated Americans in the 1800s, we find that culture referred, above all, to the intersection of the biological and the technological. *The Century Dictionary*, one of the major American intellectual products of the late nineteenth century, defined culture first as "tillage," second as "the act of promoting growth in animals or plants, . . . specifically the process of raising plants with a view to the production of improved varieties," and, third, as a central element in the new science of bacteriology. It only then noted the recent extension of the word to encompass "the systematic improvement and refinement of the mind," and finally, citing British anthropologist E. B. Tylor and American biologist W. K. Brooks, equated culture with learning and civilization.[3]

I sought to capture this sequence and prioritization earlier, in using "culture" to characterize the aim of Scripps and Ritter's socio-biology. American biology, in sum, was an ongoing effort on the part of scientists in the United States to "culture" the western hemisphere and its organisms—to influence the distribution, reproduction, and growth of plants, animals, and humans, and to improve them. From this perspective, scientists' plans and actions become more important than theories and discoveries. We can interpret discussions about development and evolution not only as arguments about truth, but also—and sometimes most centrally—as a search for tools.

The meanings embedded in such a cultural history of American biology come together through the allusion in my title. Herbert Croly's *The Promise of American Life* (1909) was the central manifesto of the Progressive Era. This oracular book identified the promise of American life as the general realization of prosperity, freedom, "individual and social excellence," and, ultimately, "a larger amount of vitality."[4] The core of the text was a narrative of national history, from the end of the eighteenth century to the present, organized around the argument that progressive development was not providential or natural, but depended on intelligence and planning. Its conclusion emphasized that the fundamental problem for educated Americans in the twentieth century was to reach beyond the narrow desire for technical excellence and professional status to influence the public and improve the nation.[5]

American biologists were significant participants in the processes Croly sketched. They embodied clearly the tension he outlined between the ideals of professional competence and national engagement. More important, they were involved, in senses that were both broad and deep, in movements to understand "American life" and to realize its "promise," or possibilities. Biologists provided the scientific facts that formed the bases for secular thinking about organisms, including humans. More pointedly, they were identified with evolutionism, the theoretical complex that stood at the base of progressive philosophy. They also came to provide guidance concerning the human future. They could grasp (as we will see in part 3) Croly's call to improve "the methods whereby men and women are bred."[6]

My interest in linking biologists to this well-known statement of Progressive Era reform is, to a considerable extent, a matter of integrating history of science and American history, and, more specifically, of showing that biologists were important figures in American development. In recent decades, historians of biology, myself included, have produced detailed studies that have illuminated a variety of scientific changes. This work has connected scientists to other Americans' activities in only fragmentary ways. While granting the extent of diversity and disagreement among biological scientists, I focus on the simple, continuing commitments of national leaders and spotlight the broader influences that particular groups of scientists had. If I persuade readers that a continuing network of biologists were significant participants in national movements that had consequences, I will be satisfied.

But I would like to do more. Historical writing on American biologists has involved an awkward juxtaposition. While monographers have pointed to significant scientific achievements in areas ranging from paleontology and taxonomy to embryology, cytology, physiology, and genetics, the overall image of the place of life science in American culture has been decidedly more negative. The most fully developed, and certainly best known, narrative of the history of American biology prior to the 1920s is a story, extending from Thomas Jefferson, through Samuel Morton and Louis Agassiz, to Nathaniel Shaler and Charles Davenport, that combines racism, sexism, social darwinism, and eugenics. My reading of American biology's past, however, is different. On the one hand, great American discoverers and clear observational or experimental achievements were few and far between. On the other, as I have indicated, the general narrative is one that includes limits and tragedy, but is ultimately a story of progress.

I link this thinking about history of biology with the larger reassessment, over the last decade, of the Progressive Era and progressive values more generally. Historians' distrust of American nationalism, elite visions, and progress—in their Progressive Era manifestations and more broadly—surfaced in the 1960s in conjunction with bipolar racial tensions and fights over American intervention in Vietnam. That era, however, is now as far in the past as the Depression was then. A decade after the end of the cold war, in a country that is rapidly returning to an ethnic complexity analogous to the situation that existed for a hundred years prior to the mid-twentieth century, it is not surprising to find efforts to reconnect in a positive way with the intellectual leaders of that time.

After an academic generation in which the Progressives were criticized as elitist, naive, and hypocritical, historians have returned to what these people in fact aimed for and achieved. Pragmatist philosophers such as William James and John Dewey have been the focus of efforts by such historians as David Hollinger, Robert Westbrook, and James Livingston to recover a vision that was sensitive to science, the complexity of modern experience, and the openness of the future.[7] Philosopher Richard Rorty and attorney Thomas Geoghegan have recently returned explicitly to Croly for a progressive vision applicable to the post–cold war, post-welfarist situation. Lastly, Michael Lind's *Next American Nation* can be seen as a sequel to *The Promise of American Life*—a polemical tract that forcefully reasserts the positive value of American nationalism, provides a realistic yet progressive historical overview of the meaning of American national identity, and points toward a future that could be fundamentally egalitarian, free, and racially amalgamated.[8] My story of biologists fits within this framework. It describes efforts of small groups of Americans to build a prosperous, liberal, secular, and humane nation, composed in part of organisms that would reason and experiment.

PART I

NATURALISTS AND NATIONAL DEVELOPMENT
IN THE NINETEENTH CENTURY

Natural History and Manifest Destiny, 1800–1865

LEWIS TO BARTON TO PURSH: THE LACK OF TEAMWORK AMONG AMERICAN NATURALISTS

On 23 September 1806, after more than two years' absence, Meriwether Lewis, William Clark, and their Corps of Discovery returned to St. Louis from the West. The largest part of their cargo, carefully collected and then transported over thousands of miles of mountains, prairies, and rivers, consisted of zoological and botanical specimens. Charles Willson Peale put the most exotic objects, including a live prairie dog and the skin of a bighorn sheep, on display for the public at his museum in Philadelphia's Independence Hall. Horticulturists William Hamilton and Bernard McMahon received the expedition's seeds and cuttings, including the optimistically named Osage orange, and made them commercially available within a few years.[1] Realization of the expedition's larger aims, however—contributing to international science, providing an overview of the resources of the Louisiana Territory, and displaying to the world that the leaders of the new republic were determined to learn about and exploit the North American West—depended on publishing the expedition's discoveries. This part of the project was a disaster.

Lewis's largest and most important collection consisted of hundreds of carefully pressed and dried plants, many of them unknown species. The task of identifying, naming, and describing these specimens was taken up, even before Lewis's return, by Benjamin Smith Barton, professor of botany at the University of Pennsylvania and vice-president of the American Philosophical Society, the country's most important scientific organization. Barton published nothing on the plants, however, prior to his death in 1815. He was distracted by his duties as a physician, professor, and scientific leader, as well as by good living and gout. Yet even had these burdens not overwhelmed him, he probably would never have completed his *Flora* of the Lewis and Clark expedition. Barton simply did not know enough to describe such a large collection of unknown plants properly. His promise to complete this national service had been unrealistic from the beginning.[2]

Some species did make it into print, but only because of the actions of a second dubious figure. In 1806 Barton hired Frederick Pursh, a German immigrant of indeterminate origins, to assist him in preparing his *Flora*.

Figure 1–1

Herbarium specimen of buffalo berry (*Shepherdia argentea*), collected by Meriwether Lewis. Arrow points to the branch removed by Pursh. On this specimen, see Cutright, *Lewis and Clark*, 361. Photograph courtesy of the Ewell Sale Stewart Library, the Academy of Natural Sciences of Philadelphia.

Unbeknownst to Barton or anyone else, Pursh appropriated pieces of numerous Lewis and Clark specimens for his own private plant collection, or herbarium. He left for England with these plants, and with the descriptions he had prepared, just before the War of 1812. He quickly obtained access to the extensive herbarium and botanical library owned by aristocratic botanical amateur Aylmer Lambert, and in 1814 he published his *Flora Americae Septentrionalis* (Flora of North America), which included descriptions of the new species collected by Lewis, along with other American plants.[3]

Were it not for Pursh's theft, Lewis and Clark's botanical discoveries probably would have been lost to science.[4] But the work of describing had been done by a European, not an American, and the publication appeared in London, rather than in Philadelphia. These issues mattered little to most Americans, but they had real consequences both for the status of American practitioners of natural history and—more important—for the international stature of the United States. Pursh's book, published in London a few months before the British burned the White House, was one further indication to educated Europeans that the leaders of the United States did not have effective control over the territory they claimed—that the political future of North America, especially the area west of the Mississippi, was still open.

The immediate reasons for the strange fate of the Lewis and Clark plants were Barton's misfeasance, Pursh's duplicity, and Lewis's inability to supervise (he became governor of Louisiana Territory in 1807 and died by suicide two years later, his own report unfinished). But deeper problems underlay these individual failings. This chapter examines why American naturalists in the early 1800s were able to produce major works, such as Alexander Wilson's nine-volume *American Ornithology*, yet were unable to execute projects, like the Lewis and Clark *Flora*, that involved substantial taxonomic or geographic scope. Next it describes the efforts by a few American naturalists to construct scientific domains on the scale of the continent. The central figure in this effort was Harvard botanist Asa Gray. Between 1835 and 1855 he organized, on the model of the mercantile proprietorships pioneered by his Boston patrons, a far-flung, long-term venture to comprehend the botany of North America. Swiss zoologist Louis Agassiz, who joined Gray at Harvard in 1847, was implicitly a partner in Gray's effort to make Harvard the center for studying the "nature" of America. Their efforts would aid Americans in asserting hegemony over "their" continent.

The chapter concludes with a reexamination of the well-known battles between Gray and Agassiz in the years around 1860. The configuration of American natural history was bound up with the political compromises that kept the Union afloat from 1820 to 1860. The sectional crisis, which increasingly threatened to break up the American empire in the 1850s, heightened the importance of differing visions of America's natural "history" and its future and highlighted controversy over the roles of naturalists on the national scene. The enthusiastic embrace by Gray, a conservative Presbyterian, of Charles Darwin's intellectually radical theory of evolution in 1859 makes sense if we see the degree to which he saw it as a way to understand America—both its history and its looming struggle for existence.

NATURE IN THE EARLY REPUBLIC

Between the Revolution and the 1820s, the United States changed in at least three fundamental ways. First, American nationalism became an established ideology. This included both the belief that the United States were independent of England and, in a triumph of sentiment over grammar, the idea that the United States was a single entity. Second, the country's effective boundaries moved, dramatically, from southern Georgia and the Appalachians to the Caribbean and the Rockies. Finally, a society based on patrician values and leadership became one in which democratic individualism and social equality were celebrated, even if not generally realized.

These changes substantially affected the study of American plants and animals. The natural history of the United States was not, of course, a new enterprise in 1776. The first colonists had acquired both practical information and useful plants from the Indians they had encountered. European botanical and zoological writers described many North American species in the course of the 1600s, and by the mid-1700s, a number of scientific travelers had visited British America in their search for novelties. Colonials such as Philadelphia nurseryman John Bartram collected specimens and provided information for the small but growing number of Europeans interested in the world's animals and plants. Still, the nationalism, territorial expansion, and democratic ideology associated with the Revolution and its aftermath both accelerated interest in the plants and animals of "America"—ambiguously, both the nation and the continent—and made certain kinds of study of them more difficult.[5]

Revolutionary concern to justify the existence of the new United States stimulated interest in natural history. In 1785, two years after the Peace of Paris, Thomas Jefferson published his only book, *Notes on the State of Virginia*. By far the largest part of this work discussed the plants, animals, and humans living in the indefinitely large territory Virginia claimed. Some of this material was a straightforward survey of the region's natural resources, but Jefferson's real concern was to convey the excellence of American organisms. Europeans skeptical about the Revolution had argued that the New World climate caused animals to "degenerate," or decline over the course of generations, in vitality and intelligence. They implied that European Americans were doomed to revert to the uncivilized mediocrity displayed by the continent's aborigines. Jefferson attacked the degeneration hypothesis by systematically comparing the weights of American animals to comparable European species and by defending the intelligence, morals, and sexual vitality of the Indians. He then pointed to the contributions that Benjamin Franklin, astronomer David Rittenhouse, and, by implication, he himself were making to civilized learning to show that the United States was not a nation of dunces. Both the facts of natural history and the pursuit of this subject would, from Jefferson's perspective, support his deepest hope: that the United States should succeed as an independent, civilized nation.[6]

This interest in describing organisms specific to America and, by doing so, displaying that at least some of the white population of the United States were participants in civilization, continued into the nineteenth century. At his museum, opened in Philadelphia in 1786, Charles Willson Peale displayed together the first reconstructed skeleton of the "American mastodon," stuffed specimens of such nationally symbolic native animals as the bald eagle and wild turkey, and the portraits he had painted of Washington, Jefferson, and other Revolutionary leaders.[7] In the 1800s, Alexander Wilson's *American Ornithology*, the first large-format illustrated natural history book published in the United States, combined descriptions of American birds with accounts of Wilson's own activity as a collector and observer. John James Audubon's self-presentation as an American frontiersman was an integral part of *Birds of America* and its companion text, the *Ornithological Biography*. Even a more severely technical work like Thomas Say's *American Entomology* was "American" in both its animals and its author.[8]

The spread of Anglo-Americans across the continent had obvious consequences for the study of natural history. On the practical level, the establishment of roads, river routes, and pioneer settlements made travel by naturalists possible. Within five years of the logistically unprecedented Lewis and Clark expedition, the young English immigrant Thomas Nuttall was able to collect plants as far west as the future North Dakota by joining John Jacob Astor's private expedition to Oregon. Seven years later he was able to wander, essentially alone, through what would become Arkansas and Oklahoma.[9] American commercial and political expansion also gradually raised the level of ambition among naturalists. While Wilson and Say, like Pursh, produced compilations that were limited to some of the organisms from some of the regions of the continent, by the late 1810s a few naturalists were discussing what it would mean to produce a truly comprehensive work on the plants or animals of North America. The northern and southern boundaries of that geographic entity were not clear, but it certainly extended from the Atlantic to the Pacific.[10]

The democratization of American society affected both the population and the behavior of those seriously interested in natural history. In Europe, inquiry in this area had long been guided by the wealthy. In the new republic, by contrast, it was pursued by a heterogeneous mix of individuals. A small number of republican gentlemen, including Jefferson, Barton, Columbia College natural history professor and U.S. senator Samuel L. Mitchill, and Scottish-American philanthropist William Maclure, sought to lead American efforts. A variety of self-absorbed individualists, well-bred but uninterested in accepted markers of social standing, worked alongside them. John Bartram's son William, for example, gave up the pursuit of economic success and scientific fame in the 1780s. Nuttall and the Philadelphia gentleman Say were both indifferent to money and status.[11] The most interesting and unusual group, however, came from undistinguished, or checkered, backgrounds; they saw in natural history a path out of obscurity and perhaps toward prosperity. Peale had moved from saddler's apprentice to portrait painter to museum proprietor.[12] Alexander Wilson was a weaver, convicted in Scotland of libel and extortion, who hoped that his illustrated volumes on American birds would redeem his name in his new country (they did) and provide financial security for his family (they did not).[13] The shipwrecked Franco-Italian merchant-turned-naturalist Constantine Rafinesque pursued one scheme after another for twenty years in the hope of gaining fame and fortune. The illegitimate ne'er-do-well Audubon ultimately became wealthy from *Birds of America* and its successor volumes. Amos Eaton, a New York real estate speculator convicted of forgery in 1811, was able to get out of jail, reestablish his reputation, and return to prosperity by teaching botany.[14]

This miscellaneous collection of American naturalists produced remarkable work, through some truly heroic labors. Wilson completed eight illustrated volumes before his death, attributed to overexertion, in 1813. Nuttall published, and probably printed with his own hands, the pioneering *Genera of North American Plants*, and he crossed the continent on foot when he was nearly fifty. Audubon's greatness as an artist-naturalist is well known.

Yet this science had obvious and important limitations. Production was slow and failures frequent. The fate of Lewis's plants was in fact not exceptional. Nuttall never published many of his discoveries, Say's *American Entomology* ended prematurely, and Wilson left his project unfinished. Audubon took nearly twenty years to complete the study, drawing, and printing of his bird folios. The converse flaw of rushed and sloppy work was also common, most prominently in the case of Rafinesque: he published so many new species in such obscure places that his work was essentially ignored by his contemporaries. Succeeding generations of taxonomists struggled to sort out what he had in fact done.[15] Moreover, projects on American natural history were limited in both geographic and taxonomic scope. Nearly all plant studies were circumscribed regionally, and both botanists and zoologists focused on a few groups. Flowering plants were of much greater interest than mosses, lichens, and fungi; birds and mollusks received a great deal of attention, but little was done with fish or most arthropods.

American naturalists commonly ascribed their difficulties to inadequacies in patronage and infrastructure. Federal support for natural history lapsed for nearly a decade on Jefferson's retirement in 1809. Travel, transport, and communication were difficult. Reference collections of books and specimens, necessary to determine whether an object was in fact new to science, were incomplete. The country lacked the resources and the audience for both scientific journals and the elaborately illustrated volumes that at that time seemed essential to natural history.

Focus on economic and institutional underdevelopment, however, diverts attention from the more fundamental problem: a lack of shared values and mutual trust. In England, natural history was guided by cosmopolitan aristocrats. Men such as Sir Joseph Banks controlled a material, intellectual, and moral "economy" in which objects and information were exchanged according to the informal but elaborate rules long established within aristocratic networks. Gifts of specimens were sent through personal connections, and gestures of reciprocity—whether in the form of rare objects or the scientific immortality of a name, such as *Banksia* or *Franklinia*—were carefully calculated to individuals' positions within both natural history and genteel society.[16] The Revolution had detached American naturalists, as it did Americans more generally, from this system. The geographically dispersed, fluctuating, and socially miscellaneous population of enthusiasts for knowledge of plants and animals in the United States was unable, or unwilling, to behave in ways associated with pre-Revolutionary gentility.[17]

Barton was most notorious. In addition to his neglect of Lewis's collection, he stole specimens from Peale's museum while on its board of trustees. More routinely, he shortchanged associates exchanging specimens: as the German-American pastor and botanist Henry Muhlenberg commented, Barton's rule was that "it is more blessed to receive than to give."[18] But he was hardly alone. Audubon hoaxed Rafinesque with drawings of fictitious fish, began his project with scant regard for Wilson's claims to priority, and based some of his drawings on unacknowledged tracings of Wilson's work.[19] Rafinesque fabricated the *Walam Olum*, a supposed Indian creation epic.[20] The first secretary of Philadelphia's Academy of Natural Sciences absconded with the only copy of its minutes after he was expelled for nonpayment of dues.[21] Members of New York's Lyceum of Natural History embezzled money that was to fund excavation of a mastodon skeleton. On a more mundane level, fewer than half of the people who subscribed to the new *American Journal of Science* in the late 1810s paid their bills.[22]

It was this lack of trust among American naturalists that limited projects to those that could be completed by individuals or, in the usual mode of premodern production, by a family. As a consequence, most ventures were specialized and, in spite of nationalistic rhetoric, were determined more by the personal interests of naturalists and their local audiences than by perceived national needs. Birds were beautiful and interesting to a select upper-class audience. Hard-shelled mollusks were ideal collectibles—colorful, pocket-sized, and, after cleaning and drying, odorless and permanent. They were thus major foci for work and publication, in spite of the fact that neither was particularly useful, from either an economic or nationalistic perspective.

The science that was useful, at least in the long run, was botany. As Jefferson and Lewis understood, collecting and identifying the flora of an unexplored territory was one of the simplest and most effective ways for a nation to signify its disinterested interest in that territory's natural resources and thereby stake a preliminary claim to political primacy. Learning about the plants indigenous to a region also provided a rough index of its climate and agricultural potential. Lastly, botanical explorers hoped to find new useful species—trees to claim and cut, and plants that could be propagated for agricultural, horticultural, or medicinal purposes.

Attaining these goals, however, required the collection of large numbers of specimens over extensive territories and their placement within a taxonomic system that, by 1800, included thousands of species and was rapidly moving beyond comprehension by any individual. Thus, in spite of the portability of seeds and dried plant fragments, botanical exploration was not really feasible as an individual project. While the French-Canadian voyageurs who transported Nuttall up the Missouri were harsh in calling him a "madman" because he appeared to them obsessed about locating seemingly useless plants, there is something pathetic in his largely futile effort, extending over three decades, to collect and describe essentially all the flowering plants of North America on his own. Yet no one in the United States was in an intellectual or social position to mount any larger, more coordinated effort.[23]

James Fenimore Cooper highlighted the narrowness, petty competitiveness, and ineffectuality of American naturalists through the comic figure of Dr. Obediah Bat in *The Prairie* (1827). Bat entered the novel, set in the Nebraska wilderness, complaining that he was unable to find "even a blade of grass that is not already enumerated and classed." He then claimed that he had seen a new carnivore, which he immediately named "*Vespertilio; Horribilis, Americanus.*" He projected "a Historia Naturalis, Americana, that would put the sneering imitators of the Frenchmen De Buffon to shame!" until the young girl to whom he was bragging explained that the animal was in fact his own ass. Cooper's illiterate hero, Natty Bumppo, easily showed that Bat's "bookish larning and hard words" were totally irrelevant to either practical or spiritual appreciation of "the creatur's of the Lord" inhabiting North America.[24]

THE EDUCATION OF JOHN TORREY

The possibilities and problems that confronted young Americans interested in becoming naturalists in the first quarter of the nineteenth century can be seen in miniature in the early career of botanist John Torrey. His struggles

Figure 1–2

John Torrey, about 1830.
Photograph courtesy of the
LuEsther T. Mertz Library of
the New York Botanical Garden.

also clarify the consequences for American national power that followed from botanists' limitations during these decades.

Torrey was born in New York City in 1796. His father, William, was an imposing figure who had done the dirty work of the Revolution, commanding the troops who hanged Benedict Arnold's British contact, Major John André, as a spy. William Torrey married into a wealthy New York family and became a successful merchant and local political leader—the kind of gentleman who could command the attention and respect of the rough crowds of artisans who increasingly swarmed the Manhattan streets. John, by contrast, was shy and weak. He found in natural history (and especially botany) a subject that took him away from the increasingly intense competition of life in the nascent American metropolis.[25] Still, learning botany required peculiar initiatives in New York City in the early 1810s. John Torrey took his first lessons in the subject, from the incarcerated Amos Eaton, in the secluded surroundings of Newgate Prison in Greenwich Village; he could do this because his politically connected father was the prison's business manager .[26]

By the time he was thirty, Torrey had become the most prominent botanist in the United States. In 1817, while still a student at New York's College of Physicians and Surgeons, he helped to organize the Lyceum of Natural History. He made numerous botanical excursions to Long Island, New Jersey, and upstate, and in 1819 he published *A Catalogue of Plants Growing Spontaneously within Thirty Miles of the City of New York*.[27] Living with his family, he divided his time between a desultory medical practice and his real interest in plants. His resources, social position, and knowledge enabled him to establish exchanges with a significant number of both American and foreign botanists, and he accumulated a substantial herbarium.

As early as 1818, Torrey asserted that the most important American botanical project would be a new *Flora* to supplant the work of that "notorious liar & plagiarist" Pursh.[28] Many new plants had been found in the last decade, and it was important for Americans to gain control over the botany of their continent. By the early 1820s Torrey expected to lead other American botanists in a cooperative effort to produce a comprehensive work. Yet he was unable to get the project off the ground.

Part of the problem was socioeconomic. His father, while comfortably well-off, was not wealthy enough to provide him an independent income. In 1824 Torrey married Eliza Shaw, "a woman of . . . great independence, extremely benevolent, and with a capacity for government and control." The henpecked Torrey struggled for the next three decades to maintain his family in the style to which they were accustomed. He found work as a professor of chemistry, first at West Point and then, more permanently, at the College of Physicians and Surgeons and as an adjunct at Princeton. His correspondence was peppered with complaints about his need for money and lack of time. Botany became a part-time occupation.[29]

The more serious issue, however, was Torrey's inability to project convincingly that American botanists, working under his leadership, were in control of North America's plants. In 1823, William Jackson Hooker, the University of Glasgow professor who was the unofficial coordinator of British imperial botany, learned of Torrey's scheme. He immediately reminded the young American that he should "of course include the recent discoveries of our arctic travelers."[30] A year later he notified Torrey that he was sending his own collectors to the "N W coast of America" and was planning a "Flora of the British possessions & of the Arctic regions of N America."[31] On learning that Torrey was unhappy that he might "anticipate some of [Torrey's] descriptions," Hooker put the son of the killer of Major André (who had recently been reburied, as an imperial hero, in Westminster Abbey) in his place:

> Of the propriety of a British Botanist undertaking a description of the
> Plants belonging to British America, or to those parts of N. America not be-
> longing to the United States, & where so much has been done in investigat-

ing its vegetable production, & at such a heavy expense, with so great risk & even loss of life, by people of our own country, there can be *no question*. Indeed it would be a disgrace to this nation were the thing not to be done here.

Pointing to the collections acquired by recent and planned British expeditions to the Northwest Territories, the Pacific coast, and the Columbia River region, he asserted that "the Government of this country & the public bodies" [the Royal Horticultural Society and the Linnean Society] who had sent out naturalists "will, for their own credit's sake, have them published *first in this country*."[32]

Both Torrey and Hooker understood that this dispute was not merely botanical. Hooker's informant regarding Torrey's initial plans was a young plant collector named David Douglas, who had been sent to the United States to scour nurseries for interesting fruit tree varieties. A few years later, Hooker and the Royal Horticultural Society sent Douglas to the Columbia River region, where he collected specimens of the valuable fir later named in his honor. On his return he provided W. R. Hay of the British Colonial Office with a report outlining the reasons why Britain should push to establish the Columbia River as the boundary between British America and the United States. Hay decided that the best way to press Britain's diplomatic advantage in the northwest would be to contrast "British scientific enterprise and concern for the Columbia region" with "American ignorance and neglect." He therefore arranged to have Douglas sent back to America to collect plants, to provide exact geographic data on the Pacific region as far south as San Francisco and also to establish a presence in Hawaii. Douglas died suddenly in Hawaii, but Hooker obtained his collections and incorporated them into his *Flora Boreali-Americana*. Douglas's geographic records went to the Colonial Office.[33]

Torrey was left with Hooker's bland reassurance that "we are both serving the cause of science, & each in a way that the other cannot do."[34] For the next decade, Torrey continued to accumulate specimens, but otherwise came no closer to achieving his scientific and national goal of an American *Flora of North America*.

Asa Gray, American Botanical Entrepreneur

In the two decades after Torrey struggled and failed, Asa Gray succeeded. In the 1830s and 1840s he built a national network of botanists moving toward an inventory of American plants. He pushed aside foreign naturalists and gained for Americans the power to conceptualize the natural territory of North

America. Gray's achievement was partly a matter of timing—he was the beneficiary of a major phase of American economic development and of a resurgence on the part of the American upper class, as well as of the efforts of Torrey and other botanists who had come before him. But personality also made a difference. While Torrey was a diffident gentleman struggling to maintain his old-fashioned ideals in an uncongenial environment, Gray was a single-minded entrepreneur. He understood how to use the resources available in the United States in the 1840s to organize and maintain botanical work on a national scale, and did what he had to do to achieve that goal.

Gray firmly recalled that it had been an accident that he was born in the Methodist, rather than the Presbyterian, section of the Mohawk River valley village of Saquoit in 1810.[35] In the early nineteenth century, upstate New York and other rapidly developing western areas were sharply divided between populist religious enthusiasts and conservative adherents of learning and methodical accumulation.[36] Gray's Presbyterian father, who began in the dirty work of tanning but prospered through farming and careful investment in real estate, was clearly in the latter category. He supported his oldest son's education, first at a classical academy, and then at a local medical school. In 1831, however, Asa Gray rejected a career as a doctor. He believed that success would require too much accommodation to the prejudices of his uneducated patients and would be too limited in scope. Botany, which he studied in medical school, offered a clearer sense of intellectual and social order, as well as a larger field of action. Gray was not deterred by the fact that no American at this time made a living as a botanist. That would come. He pursued the subject for the rest of his life with a Presbyterian combination of sobriety and intensity. His career is best seen as the long-term venture of a Yankee entrepreneur.

Gray found his initial opportunity at home, in his access to the still underexplored regions of upstate New York. In 1830 he began collecting local plants seriously. A year later, by teaching a summer botany course at his medical school, he was able to get students to do some of that routine work for him. He avoided wasting specimens on local friends not in a position to "repay" him, but he sent unsolicited material to recognized American and European collectors, noting accurately that "botanical presents are like gifts among the Indians. They always presuppose a return." Gray's aggressive marketing brought him into contact with more established botanists, most notably John Torrey. In 1833 Torrey invited Gray to New York City to become his personal assistant.[37]

Torrey saw in Gray the young man he himself once hoped to be and supported him with platonic passion. He gave Gray a room in his Greenwich Village townhouse and put him to work organizing plants according to the newly predominant natural system of classification. He introduced Gray to other American naturalists and induced them to show Gray their collections. Gray made the most of these opportunities. He carefully studied the plants he saw and began to publish descriptions of certain groups. He produced an elementary textbook designed to supplant that of Torrey's old teacher, Amos Eaton; on meeting Eaton, Gray arrogantly informed him that he was behind the times.[38] Gray was confident about his future: in May 1836 he wrote a friend that his work thus far had "prepared me most thoroughly for future progress, and if I happen to pursue Botany undividedly for a little time I shall (entre nous) be soon the best botanist in this country."[39]

A few months later he cashed in. In 1836, Torrey decided that he finally had the income, library, herbarium, connections, and, in Gray, the trained assistant necessary to push his *Flora* project forward. He offered Gray $300 per year to write the formulaic Latin descriptions. Gray turned him down. Torrey countered by asking Gray to propose a salary, promising him a share in the profits and suggesting that he could write monographs on the most

Figure 1–3

Asa Gray, about 1861. Photograph courtesy of the Missouri Botanical Garden Archives.

interesting material. What Gray wanted, however, was neither money nor access, but status. He held out for coauthorship, and ultimately Torrey agreed.[40]

Having established a partnership, Torrey and Gray turned to other American botanists. Success depended on extracting descriptions of as many unpublished specimens as possible from those knowledgeable about the plants of different regions; these materials would then be situated within the taxonomic framework Torrey and Gray were developing. The genteel Torrey was the front man in these delicate negotiations, persuading Thomas Nuttall, for example, to agree to share the results of his Western travels. Gray worked as the enforcer, pushing contributors to send in specimens and descriptions of plants they considered new, and then, with samples and publications before him, denying most claims to priority (and the names associated with them) in order to make nomenclature unambiguous and consistent. Seventy-eight individuals contributed material in some form to volume 1 of the *Flora*; their contributions were managed by Gray.[41]

Torrey and Gray justified their project on a number of grounds. They argued that Linnaeus's artificial system (based on counting pistils and stamens) was outmoded and that American plants needed reclassification according to the "natural system," based on deeper structural affinities, which European botanists had developed over the previous four decades. In addition, a general flora would bring together, and thereby strengthen, the standing of names Americans had given to "their" plants; such a text would thus enable American botanists to pursue the work of naming without having to defer continually to Europeans. The primary importance of the *Flora of North America*, however, as Torrey had known, was implicit in its title and proposed scope. In contrast to earlier works, it would enable botanists in the United States to comprehend together all the flowering plants of all of North America. Moreover, the boundaries of Torrey and Gray's "North America" were strikingly similar to those of what the United States would soon become. As Gray later recalled, their *Flora* "took the initiative in annexing Texas, ten years before its political incorporation into the Union," and "California was also annexed at the same time."[42] It is not surprising, then, that they included the disputed Oregon region. On the other hand, they did not consider either the Arctic or Mexico south of Texas and California to be part of North America. Their *Flora* was an instrument for imagining that a continental United States was a single natural biogeographic unit; it asserted the existence of something that, a few years later, would be seen as Anglo-Americans' "Manifest Destiny" to realize politically and militarily.

Torrey and Gray prepared and published two sections of their work in order to display their methods and publicize their intent.[43] Gray then left for a year in Europe, where he examined major herbaria containing American plants. His larger aim in this trip was to show European botanists, especially Hooker, that he and Torrey were so knowledgeable and productive that non-Americans should withdraw from the competition to process the plants of North America. On his return to the United States, Gray began to deliver on these promises. He completed new sections of the *Flora* annually between 1840 and 1843, and he joined the *American Journal of Science* as botanical editor. He became, as he had predicted, the "best botanist" in the country.

In 1842 Gray's status was recognized, and his ventures were underwritten, through an appointment as professor of natural history at Harvard University. This was not primarily a job to teach undergraduates. A number of the Boston businessmen who ran Harvard (most notably Benjamin Greene, John Amory Lowell, and Charles Loring) were serious amateur botanists. They wanted someone who could advise them about their collections, develop the Harvard Botanic Garden as a source of rare plants, and popularize botany nationwide. More broadly, they appreciated scholars who combined efficiency, interest in accumulation, and a nationalistic project. They had recently named Jared Sparks, the individual most deeply engaged in tracking down the documents essential to an understanding of the nation's political origins, to a new chair in American history. They realized that Gray was pursuing a parallel project in American natural history, and they wanted to support it.[44]

Gray's perception of his work changed substantially as he began to view it from the perspective of Cambridge. The Torrey-Gray *Flora* suddenly seemed premature. A truly worthwhile inventory of the continent's plant species would require a much longer period of work and a broader comparative search into South America and eastern Asia. It would also entail a different set of professional relations. Gray abandoned his formal partnership with Torrey and began to construct his own network of collectors and collaborators.

In his strategies and values, Gray emulated the Brahmin businessmen with whom he associated—Greene, Lowell, and, most immediately, Loring, whose daughter, Jane, Gray married in 1848. Loring, a member of an established mercantile family, had become prominent as a lawyer in marine insurance cases—that is, in enforcing long-range, long-term business relations. In 1854 he became head of the Massachusetts Hospital Life Insurance Company, the primary caretaker of the estates of upper-class widows and orphans, the major source of capital for Brahmin enterprises, and the financial base for the Massachusetts General Hospital. The Boston elite saw in Loring the combination of enterprise, discretion, and judgment necessary to reconcile doing well and doing good. He was, in a meaningful sense, the most trusted man in Massachusetts. He provided a clear model for his son-in-law.[45]

The key to Gray's botanical proprietorship, as with those of other Bostonians, was the recognition that with strategic vision, institutional stability, and a reputation for probity, a single individual and a few helpers could maintain a far-flung venture on relatively little capital. Gray's central office—the Harvard Botanic Garden and Herbarium, also his home—was always small enough so that he could maintain close control of the work. On arriving in Cambridge he replaced the longtime gardener, who had his own local connections, with more tractable Europeans. In the herbarium his only full-time assistants for many years were his wife, the illustrator Isaac Sprague, and a succession of "gluers" (laborers who prepared specimens for preservation and exchange). When he finally appointed a permanent assistant in 1872, he chose the pathologically shy Sereno Watson.[46]

Gray's vision was directed outward—to the network of agents he developed and maintained. These ranged from such genteel provincial collaborators as William Sullivant in Ohio and George Engelmann in Missouri, to rougher collectors on the frontier, such as Charles Wright, August Fendler, and Charles Parry. These men regularly fed specimens and descriptions to Cambridge, where priority was certified, names assigned, credit given, and texts and duplicates marketed internationally. Gray controlled this national network through a variety of means. The textbooks he wrote and rewrote and his practical *Manual* broadcast botanical learning, affirmed his unique standing, and provided a taxonomic framework within which collectors could work.[47] Through extensive correspondence he maintained personal contact with his widely scattered helpers, providing both general information and specific directions, as well as the encouragement that someone of consequence was sympathetic to their work and was utilizing it. Lastly, he provided material support that kept his collectors going—books, supplies, expense money, and, in a few cases, cash wages.[48]

The increased scope of activity by the federal government after 1840 was crucial to Gray's ability to maintain this network. One change was mundane, yet essential: the establishment of national mail service. Whereas in the early 1840s Gray and Torrey struggled to send letters the few hundred miles between Cambridge and New York via a courier service, by 1851 the post office was carrying Gray's letters (with millions of others) routinely up to three thousand miles for three cents each.[49] Renewed federal involvement in exploration had a more specific impact. Beginning with the United States Exploring Expedition of 1838–42, which investigated the Pacific Northwest (along with Polynesia), the government mounted a series of expeditions to the West. Many of these, including John C. Fremont's treks to the Great Basin and beyond, the Mexican Boundary Survey, and the Pacific Railroad Survey, included substantial natural history work. The expeditions furnished, explicitly or surreptitiously, jobs and logistical support for collectors; equally important, they provided funds for the preparation and publication of elaborate descriptive accounts. In the case of the Exploring Expedition, for example, Gray was able to demand in 1848 a $7,200 fee (when his Harvard salary was only $1,000), plus expenses for a year-long trip to Europe, to write up the expedition's botany; the government subsidized the resulting illustrated large-format volumes and distributed them throughout the world.[50]

The trick was to maximize federal assistance to botany while minimizing the ability of government employees to interfere with the transfer of specimens and information to Gray's control. Gray negotiated long and hard with Charles Wilkes, commander of the Exploring Expedition, to avoid the professionally embarrassing requirement that Gray's standard botanical Latin be accompanied by English descriptions, which Wilkes argued would make the work more accessible to American readers.[51] When Gray began publishing material from the Mexican Boundary Survey without government approval in 1853, Survey leader William Emory threatened legal action to have the specimens returned, but to no effect.[52] Gray's longstanding friendship with Joseph Henry, another upstate New Yorker who rose to national scientific leadership from obscure origins, was crucial here. As head of the Smithsonian Institution from its creation in 1846 to his death in 1878, Henry was in Washington but ostentatiously not of it. He supported Gray's and other botanists' requests for government resources and collections and defended their efforts to set their own standards.

The final element in Gray's mercantile strategy was the maintenance of understandings about the division of territory with his potential chief competitors: Hooker and Torrey. He was able to induce the former to cede North America after completion of the *Flora Boreali-Americana* in 1840. In return, Gray tried to keep clear of other regions in British intellectual and political spheres, such as Australia, South Asia, and Polynesia. When his job with the Exploring Expedition required him to work on this last region, he conspicuously displayed his deference to Hooker and his son Joseph by providing miscellaneous services to them and their friend, Charles Darwin.[53]

Gray's dealings with Torrey were of course more intimate and delicate. In the mid-1840s Torrey was unhappy with Gray's declining interest in the *Flora* project. Gray's sudden announcement of his engagement to Jane Loring upset him more deeply, in ways he only partly understood, because it destroyed his dream that Gray would transform their partnership into an enduring botanical union by marrying one of his daughters.[54] In the aftermath of this breakup the two men competed directly, each placing a collector on the staff of the Mexican Boundary Survey and then publishing separately.[55] Torrey, however, ultimately recognized the futility of competition with his former protégé, and he resumed collaboration with the man he now acknowledged as the leader of American botany. Their relationship stabilized in 1853 when Torrey obtained, through the influence of his friends, the position of chief assayer of gold and silver for the federal government. The job required great rectitude and some knowledge of chemistry, but only routine attention; as a consequence, Torrey became a more productive botanist and Gray's valued partner.

Gray's enterprise worked because, in spite of his complaints about overwork and inadequate funding, he was sufficiently efficient in identifying and publishing materials to forestall competing efforts from other Americans. At the same time he maintained a reputation as an honest taxonomic broker. By comparison with his predecessors, he did not flood the international botanical market with cheap goods—with large numbers of claims for new American species, made without assessment of comparable material. He thus added a great deal of value to the species he did certify.

By the late 1840s, Gray had established "American botany." His reach extended outward from Cambridge into schools, homes, and frontier outposts throughout the country. The continent's geographically and taxonomically scattered plant enthusiasts had become unified as parts of a network at whose center he sat. He defined the botanical extent of North America and had placed control over the study of that region's plants within the United States.

As Gray reached for national botanical leadership, the Mexican War and brinkmanship with the British over Oregon were bringing the political boundaries of the United States into line with the natural boundaries he and Torrey had established earlier. Gray took the opportunity to refine his thinking about America. In an 1848 lecture on the as yet undeveloped subject of biogeography, Gray argued that "America" was a single natural domain whose flora was distinct from that of Europe and Asia. Within that domain, boundaries were minimized. Gray described a "North N. American Kingdom" that extended from Canada to the southern tip of the Alleghenies and from the Atlantic Ocean to Oregon. While recognizing that "from a particular view" this kingdom could be divided into "3 pretty distinct provinces"—"western wooded," "middle unwooded," and "eastern wooded"—he argued that some species were common to east and west, and, "as annexation is the order of the day,—we unite them into one."[56] Transcontinental travelers might perceive dramatic changes in the course of their passage from east to west, but, Gray assured them, they were still living within a fundamentally unitary region of the natural world.

Gray, like the nationalistic landscape painters Asher Durand and Frederick Church, imagined a continental empire that was essentially an extension of their native Empire State, New York. From Gray's biogeographic perspective, the South in particular was an afterthought. His "Southern North-American Kingdom" was a minor enclave along the Gulf of Mexico. In contrast to the expansive northern kingdom, it was quite distinct from the "Californian & New Spain" region in the Southwest. Neither region was important for defining the nature of America.[57]

GRAY, AGASSIZ, AND THE IMPENDING CRISIS

From the time they clashed over Darwin's theory of evolution in 1859, Asa Gray and the zoologist Louis Agassiz have been seen in contrast to each other. Their modern biographers, respectively, Hunter Dupree and Edward Lurie, showed clearly that their public disagreement was a consequence of deep differences between them in philosophy and in personality. While Agassiz was a Platonist and a domineering crowd-pleaser, Gray was an empiricist whose careful advocacy for Darwin ultimately established his intellectual primacy among American naturalists.[58] Interest in the controversies between Gray and Agassiz, which extended from 1858 to 1866, has, however, deemphasized the degree to which the scientific projects and strategies of the two men were parallel. For more than a decade prior to 1859, Gray and Agassiz formed a union to make Harvard, in the words of Agassiz's brother-in-law Cornelius Felton, "the acknowledged center of science in the United States." It was the hub for a national network of scientific workers and supporters, and a clearinghouse for the inventory of the flora and fauna—the "natural productions"—of North America.[59]

Recognizing the basis and extent of the Gray-Agassiz union in fact illuminates their ultimate conflict. Specifically, we can see that the collapse of their relationship occurred in conjunction with the collapse of that larger Union in whose Nature they were both interested. At the center of their argument over evolution lay the subject that most prominently straddled politics and science during this period—American geography. An important part of the reason Gray brought Darwin across the Atlantic so rapidly and so dramatically was that he needed intellectual reinforcement in his fight to keep the United States synonymous with America.

Louis Agassiz came to the United States from his native Switzerland in 1846 at the age of thirty-nine. He was already internationally known as an embryologist, the author of a massive treatise on fossil fishes, and the chief proponent of the Ice Age concept. As a consequence, he was uniquely situated to become a leader of science in America. Yet when he was appointed professor of geology and zoology at Harvard, he was unsure how to proceed. In matters small and large, he emulated the colleague closest to him, Asa Gray.

Figure 1–4

Louis Agassiz, about 1857. Photograph courtesy of Stanford University Archives.

Like Gray, Agassiz linked himself personally to the Brahmin elite. In 1850 he married Elizabeth Cabot Cary, a descendant of a number of prominent old Boston families. He gained daily visibility throughout the country by immediately publishing a textbook, *Principles of Zoölogy*, for use in colleges and high schools.[60] Most important, he began to do for animals what Gray was doing for plants: to establish at Harvard a center for a long-term project to comprehend the continent's fauna. This project developed rapidly in the course of the 1850s. In 1853 he launched an effort to gather a comprehensive collection of American fishes, mollusks, and crustaceans. Next he initiated a ten-volume series titled *Contributions to the Natural History of the United States*.[61] Finally, between 1857 and 1859, he established the foundations for a national zoological museum.

The structure of Agassiz's enterprise was the same as Gray's. He established a central operation largely made up of subordinates he had gathered during his years in Switzerland; these immigrant workers, in what one assistant called Agassiz's "scientific factory," had little standing in the United States apart from their master.[62] He also began to build a nationwide network of deferential collaborators and collectors. As early as 1848 he was maneuvering to make younger American zoologists, such as Dickinson College professor Spencer Baird, into junior partners in his activities. In 1853 he distributed six thousand copies of a circular throughout the country, requesting preserved specimens of fishes, mollusks, and crustaceans.

In addition, Agassiz sought to establish a modus vivendi with the federal government, both as a source of funds and specimens and as a potential competitor. He quickly formed close ties with the powerful leader of the Coast Survey, Alexander Dallas Bache, obtaining a Survey vessel for an exploratory cruise off Nantucket in 1847, and then devoting part of the winter of 1851 to a Survey-sponsored study of reef formation in the Florida Keys.[63] At the Smithsonian, Agassiz's admirer Joseph Henry pressured Baird, who became Henry's assistant for natural history in 1850, to lend material to Agassiz freely.

Finally, Agassiz had to deal with the problem of showing that his enterprise was producing results—was processing organisms into knowledge. Like Gray twenty years earlier with *Flora*, he made sure that some proposed publications appeared rapidly. The first two volumes of *Contributions*, made up of an introductory overview of the principles of classification and a technical monograph on turtles, appeared within four years of the series' prospectus. Agassiz anticipated that this evidence of productivity would give him breathing space to proceed more broadly and deliberately.

Agassiz's project, like Gray's, was national in its subject and organization, and nationalistic in its intent. Both the fish circular and the *Contributions* prospectus took as their domain the "natural history of the United States" or, equivalently, "the animal creation of this continent."[64] Agassiz advertised that in both authorship and patronage, his work would be "an American contribution to science." When he addressed a committee of the Massachusetts legislature in early 1859, requesting funds to establish a "museum of natural history and comparative zoology," he emphasized that the prosperity of the nation depended on knowledge of its resources, including "the animals and plants which inhabit its surface." He declaimed that for more than a decade he and his "collaborators" had been collecting objects "from all parts of the country; from the Pacific as well as the Atlantic; from the Northern and Southern extremities of the continent." A museum in Cambridge would make this information available. Within twenty years, he predicted, his museum would rival, "and I hope stand above," the national museums of France and England.[65]

Appreciation of the extent to which Gray and Agassiz were involved in a common project leads to a fuller understanding of what distinguished them from each other. In part their antagonism stemmed from differences in their objects of study and timing of their ventures. In part it derived from their different attitudes about nature and the nation.

By mid-century, a system for processing American herbarium specimens was in place. Preservation technology consisted of absorbent paper, portable presses, and vermin-proof tin boxes. Express companies could ship packages of dried plants reliably and cheaply. A herbarium of thousands of specimens could be maintained in a few rooms. Gray was thus able to manage the nation's botany, essentially from his home, on a few thousand dollars per year drawn from Harvard and a few friendly patrons. His primacy was acknowledged by a network that had grown slowly over the quarter century during which he had built his career.

Agassiz, by contrast, came into an arena that was significantly more difficult to manage. A cohort of zoologists was already established in the United States when he appeared in the latter 1840s. Killing, preserving, and transporting animal specimens were messy and expensive enterprises. Fish and frogs could be packed in barrels of whiskey, but this preservative was notoriously subject to pilfering in the field before—and sometimes after—specimens were placed in it. Mammals required careful treatment with arsenic or other dangerous poisons. Specimens needed considerable storage space, and regular maintenance was necessary to protect most from insects and rot.

In order to succeed, therefore, Agassiz needed to be much more aggressive than Gray both in his pursuit of money and in his efforts to dominate his colleagues. These activities of course generated jealousy and opposition. Agassiz used his international reputation and charisma to raise unheard-of sums: subscribers to *Contributions* pledged more than $300,000, and by 1859 private and government patrons had donated $150,000 for his museum.[66] His constant public presence put Gray in the shade, and his thirst for support threatened, from Gray's perspective, to drain the well for science in Boston.

Agassiz's drive to dominate was directed especially at Spencer Baird. Although young and without a public reputation, Baird had the potential, through his Smithsonian position, to control all the federal resources that could be turned to zoological work. Initially Agassiz sought to draw Baird under his wing. But when he realized that Baird intended to act independently, he began to criticize him, both privately and publicly, for incompetence. He hoped that Smithsonian secretary Henry would block Baird's plan for a national museum in Washington. The most likely alternative site for the government's collections would, of course, be Cambridge. As late as 1860 Agassiz was able to get Henry to agree in principle that he could have everything he wanted from the Smithsonian.[67] Baird worked hard to stay on good terms with Agassiz while maintaining his independence; increasingly he resented Agassiz's patronizing interference.[68]

The most important difference between Gray and Agassiz, however, lay in their different experiences of the "nature" of America. Gray was an archetypal Yankee who traveled little but was certain he knew his country. His complacent faith in the natural unity of the American expanse was evident in his easy jokes in the late 1840s about the "annexation" of the continent's three botanical "provinces" into one. His taxonomic work was an effort to inventory the specific contents of this presumed geographical unit.

Agassiz, by contrast, was deeply interested in the diversity he saw in the United States. He had grown up in a part of Europe in which ethnic, religious, and political distinctions were both fine-grained and enduring, and his early training had imbued him with the idea that the earth contained numerous well-defined "zoological provinces." As a foreigner and an enthusiastic scientific traveler (he was able to boast in 1859 that he had visited all but two of the states east of the Rocky Mountains), he devoted considerable attention to the differences he saw among the organisms of America. The premise of his study of freshwater animals was that the United States was a terrestrial "ocean" of land in which isolated "islands" of water existed, each with its own characteristic fauna. Agassiz pushed aside decades of American intellectual evasions to assert firmly that Africans, because they had come from a different zoological province than Europeans, belonged to a different species. He was widely viewed as the articulator of "ethnological laws" according to which Africans were primordially suited to both the climate and social arrangements peculiar to the southern half of the United States.[69]

Agassiz was interested in diversity both for its own sake and as the necessary background for his deeper message, that life contained ideal unities. In his public lectures, he demonstrated that the welter of phenomena hid fundamental symmetries that were crucial evidence of the thought God had invested in his Creation. Only a zoologist with broad knowledge and philosophical genius could uncover and display these principles. The cultural purpose of his collections and his museum, he explained to Massachusetts legislators, was "to exhibit the thoughts of the Creator as manifested in the visible world"—to enlighten the public regarding the wonderful existence of transcendent unity amid diversity.[70] From Agassiz's perspective, his support for polygenism—the idea that African and European Americans were zoologically distinct—could easily be reconciled with Christian belief in the unity of mankind and with recognition that Europeans had definite moral duties toward Africans.[71]

In presenting his paradoxical juxtapositions of manifest diversity and fundamental unity, Agassiz was participating in a form of cultural work that was
characteristic of America in the 1850s. Following the Mexican War and the
ramshackle Compromise of 1850, mainstream political leaders, most notably
Stephen Douglas, struggled to reconcile the divergent visions that southern
slaveholders, northern capitalists, and midwestern farmers had for the future
of their new continental empire. Would the West be populated by free whites
or by a mixture of races related through slavery; or would the sharp eastern
division between North and South be extended? Douglas and his allies argued
that the commitment to local self-determination and the acceptance of uncertainty embodied in the concept of "popular sovereignty" were compatible
with belief in a single, transcendent national identity. Yet, the possibility of
such compatibility in fact depended during this period on the existence of a
small group of national leaders who could maintain the tattered fabric of
compromise they had woven; it depended, ultimately, on the breadth of their
personalities.

Louis Agassiz represented this same kind of leadership operating on a scientific stage adjacent to the main political drama. He had taught in Massachusetts and South Carolina, and had done fieldwork in both northern Michigan
and southern Florida. His lectures, presented throughout the country, reconciled theism and freethinking, professional elitism and cultural democracy, a
cosmopolitan identity and American nationalism, and, most paradoxically,
political liberalism and support for pro-slavery polygenists. It was an amazing
reach, possible only because of Agassiz's charismatic and genial manner and
his highly visible lack of identification with any particular region or creed. It
is not surprising that his museum project was promoted in 1859 as an antidote
to the "conflict of parties" then agitating the country, and the zoologist was
applauded specifically for his ability "to bring all varieties of opinion into
harmonious and generous action."[72]

In the early 1850s Gray expressed private discomfort with Agassiz's pursuit
of popularity and his support for polygenism. Yet he considered his behavior
acceptable for a public scientist, undertaken in service to Harvard and American natural history. Later in the decade, however, faith in reconciliation and
support for the encompassing style were losing credibility among previously
moderate Northerners. Contrary to arguments about the "natural" climatic
limits of slavery, the implementation of popular sovereignty in the Kansas-
Nebraska Act had resulted in bloody struggles between Southern and Northern settlers rushing there to assert primacy. In the midst of the debate over the
Kansas bill, Charles Sumner, who had become Massachusetts' senator when
Gray's father-in-law declined to run, was brutally beaten by a South Carolina

congressman on the floor of the Senate. What had previously seemed mel-
lifluous, if vague, generalities were now recognized as "humbug" that merely
obscured an increasingly evil and dangerous situation. Northerners who established the Republican Party believed that the time had come to take hold of
the hard questions regarding the nature and the future of the Union and to
act resolutely, hoping that right would win out.[73]

Darwin and the Union's Struggle for Existence

For Asa Gray, the critical period extended from November 1858 to the following May. During that time he became directly aware of the impending crisis
of the Union, and he began the activity that would create the crisis in his
union with Agassiz. Gray's initiative was centered, appropriately, on the biogeography of North America—the character and significance of the continent's
pattern of natural diversity. Gray's interest in this issue was so great that it
induced him to make a leap into an intellectual unknown that he knew would
have disturbing consequences for his most cherished Presbyterian values. He
decided to become a public advocate of Charles Darwin's new theory of evolution by means of natural selection.

In the fall of 1858, President James Buchanan and his southern allies were
struggling to maintain their leadership in both the Democratic Party and in
Congress. Patronage was one of the few tools they had left in their desperate
effort to keep a multisectional governing coalition together. In New York, party
leaders broke precedent to dun federal employees for campaign contributions.
At the Assay Office, the high-minded John Torrey refused this demand, and
his assistants emulated him. The assistants were immediately fired, and Torrey's job was put on the line. On hearing about Torrey's plight, Gray expressed
political passion for the first time in their long correspondence. He was
shocked that Buchanan would do "so outrageous a thing." He felt "personally
degraded and disgraced by . . . the spirit which animates the administration,
in matters both small and great." He counseled patience, arguing that "there
are *some signs*,—tho faint and distant," that "better times and a more honor-
able state of things" would come.[74]

In the midst of delivering this message on the effect that the degradation
of national politics was having on their lives, Gray urged Torrey forcefully to
"come on" to Boston immediately to hear him read a paper that would "knock
out the underpinning of Agassiz' theories about species & their origin."[75] In
December Gray had outlined his position at the small Cambridge Scientific
Club, and he now planned to go on record at a meeting of the American

Academy of Arts and Sciences, held in the supportive setting of his father-in-law's house. He pressed his case further at later meetings of each of these groups, and in April published a substantial paper that detailed his position.[76]

The subject Gray chose for undermining Agassiz was American biogeography. For the preceding five years he had been intermittently investigating the floras of the Northeast and the West and comparing them to those of Japan and Europe. In contrast to the 1840s, he now admitted that the plants of the eastern and western United States differed in important respects. What he rejected was Agassiz's claim that the division of the continent into natural "provinces" was either primitive or permanent. Instead he argued that differences between regions were contingent consequences of the "long and eventful history" of America. Between the Miocene Epoch and the present, Gray claimed, the northern hemisphere had cooled, warmed, and then cooled again. During warm periods, temperate plants in both eastern Asia and eastern North America migrated north and "interchanged." (Plants west of the Rockies were comparatively isolated because of cooler temperatures near the coast). He confirmed this claim by counting identical and similar species in Japan, the American West, and the Northeast, showing that Japanese plants overlapped more completely with the Northeast than with either the West or with Europe.

The paradoxical fact that two floras, separated by an ocean and 140 degrees of longitude, were so similar, was inexplicable, Gray argued, from Agassiz's theory that present-day species had originated essentially where they lived now. The distribution of species was in large part a consequence of migrations and invasions, which depended on the vagaries of climate and the peculiarities of topography. The present arrangement of living things on the earth was not permanent, and it held no deep philosophical significance.[77]

On its face, this was a specific argument in a limited field. Yet in both its origins and its implications, Gray's picture of the distribution of American life involved profound issues. Gray had initially been drawn to the question of the distribution of American plants in 1854 by a letter from Charles Darwin, who was searching for data supporting his ideas. In 1857 Darwin outlined his theory to Gray—his first exposition of his ideas outside his immediate intellectual circle. As a result of these interactions, Gray looked at both identical and similar species and realized that the distribution of closely related species of plants fit Darwin's model. Gray was well aware of the danger that evolutionism posed for the belief that God had ordered life intelligently and beneficently. In the 1840s, he had savaged evolutionary speculations both in print and on the lecture platform on these grounds.[78] Yet Darwin's theory seemed much more fully grounded in evidence (prominently, Gray's own), and it provided a clear and realistic picture of the operations of nature—one that was certainly preferable to the intellectual fog he associated with Agassiz.

Gray came out publicly as a Darwinian in April 1859, stating in a footnote that he was "disposed . . . to admit that what are termed closely related species may in many cases be lineal descendants from a pristine stock."[79] A month later he explained and defended Darwin to his Cambridge colleagues, and the following winter he prepared a long review of Darwin's book for the *American Journal of Science*. He framed his discussion as a comparison between Darwin's "legitimate attempt to extend the domain of natural or physical science" and Agassiz's a priori assumption that "the scientifically unexplained [was] inexplicable." In letters to English friends he questioned Agassiz's intelligence and ridiculed his laziness; but, above all, he criticized his colleague for being "a sort of demagogue [who] always talks to the rabble."[80]

The American crisis over Darwin extended through 1860 with discussions at the American Academy, a critique of *Origin* by Agassiz in the *American Journal of Science*, and three essays by Gray in the *Atlantic Monthly*. It disappeared abruptly at the end of the year as the more prominent American crisis finally came to a head. Yet the coming of the Civil War made explicit the degree to which evolutionary and national issues were linked in the mind of Asa Gray.

Gray expressed his political perspective clearly: "God save the Union, and confusion to all traitors." He signed up to drill and to guard the Cambridge arsenal—in spite of the fact that he was over fifty—and scraped together money for war bonds. He anticipated that the war would lead to the end of slavery, and he expected that most "ill-usage of negroes" would soon end.[81] Gray told Joseph Hooker that the war "has made me intensely, and even proudly *national*," but his nationalism was in no way soft or romantic. It was in fact explicitly Darwinian, and he expressed this most clearly and forcefully in correspondence with Darwin himself. Gray's premise was that "natural selection crushes out weak nations." Were the United States to break up, its parts could expect a future of rivalries and wars, along with domination by European powers—most notably England. Americans would only be "secure and respected" if they were strong, and they would only be strong if they stayed together—if necessary, by force. In the long run, Gray believed, the South would have to be "renovated and Yankeefied." He summed up his position with a blasphemous Darwinian beatitude: "Blessed are the *strong*, for they shall inherit the earth."[82]

Louis Agassiz had a quite different reaction to the crisis of the Union and to the war. Sometime in 1859 he quietly dropped "natural history," with its implication of a utilitarian national purpose, from his proposed Museum of Natural History and Comparative Zoology. He opened an institution dedicated to studying the timeless interplay between difference and similarity throughout the natural world. For Agassiz, the war was primarily an opportunity to advance his vision of science. He told his student assistants that the best way for them to help in the war effort would be to stay at their desks, as examples of continuing American devotion to learning.[83] He toured the North, lecturing on methods for studying zoology. Finally, he used the hope that scientists could provide advice to the military on inventions to arrange the creation of the National Academy of Sciences, whose purpose was in fact almost completely honorific: its charter designated fifty men, chosen largely by Agassiz and Coast Survey head Alexander D. Bache, as the nation's scientific leaders.[84]

It was Agassiz's involvement with the National Academy that led Gray finally to break with him completely. Gray was unenthusiastic about a national scientific organization that could so easily become dominated by the government. But he was more unhappy that Agassiz was using the academy primarily to make invidious distinctions among American scientists. He expressed particular concern that Agassiz had kept Spencer Baird, Agassiz's primary rival in both museum development and zoological leadership, from membership. From Gray's perspective, it was hypocritical for the politically disengaged Agassiz to use an explicitly *national* academy during wartime to further his personal ambitions, and it was anti-American for a foreigner to discredit the individual who was most involved in advancing a national zoological enterprise analogous to the one Gray envisioned for botany.

In 1864 Gray attended the National Academy's meeting, held in New Haven, and forced the election of Baird over Agassiz's strong opposition. Returning together by train to Cambridge, the two Harvard men argued violently. At the dispute's climax Agassiz told Gray that he was "no gentleman"; Gray, considering himself "foolishly and grossly" insulted, broke off all personal contact.[85] Gray had, in fact, not behaved like a gentleman. A traditional element in mercantile etiquette was that loosely associated partners did not interfere in each other's independent schemes. From Agassiz's perspective, what he did to Baird, his fellow zoologist, was none of the botanist Gray's business. Gray, however, viewed the matter differently. As a naturalist, he had begun to take a perspective that extended beyond plants to encompass all of living nature. And as a nationalist, he was ready to fight anyone whom he thought was undercutting efforts to use science to expand and strengthen the United States. By comparison with these causes, Harvard collegiality seemed a paltry concern, and he gave it up.

In 1865, American naturalists could look back with considerable pride on their achievements since the beginning of the century. Yet they could also recognize that they stood at a point of transition. Asa Gray had succeeded in organizing American botanists into a single national network of collectors and taxonomists. Zoologists were moving, more haltingly, in the same direction. But it was not clear whether this proprietary system could be maintained indefinitely. The problem of financing such an enterprise for animals was chronic and severe. The larger difficulties were those inherent in nineteenth-century proprietorships: maintaining control over a growing establishment, dealing with new competitors, and, most intractably, transferring power to a second generation of leaders.

A similar mixture of achievement and uncertainty can be seen from a nationalistic perspective. On the practical level, Torrey and Gray had outmaneuvered British imperial botanists and were well advanced in their project to provide an overview of North American plants. On the ideological plane, they had reinforced continentalism and had participated in the successful effort to establish that temperate North America was both a natural and a political unit. The very success of these endeavors created a new situation, however. After 1865 the nation's natural unity was a given. Interest shifted from distribution and geography, and from similarities and differences, to what, in a variety of senses, was "inside" America—to the meaning of what existed, and its future.

In the course of pursuing their organizational and nationalistic goals, naturalists—most notably Gray—embraced the wonderful new intellectual resource of evolutionism. Charles Darwin's particular formulation of the "development hypothesis," and his ability to make it scientifically respectable, had served important purposes in the secession crisis. The next generation of naturalists would work to apply this protean concept to the new problems before them.

Culturing Fish, Culturing People:
Federal Naturalists in the Gilded Age, 1865–1893

The Innocents Abroad, Mark Twain's account of his participation in a package tour of the Mediterranean in 1867, mixed shrewd satire and bumptious ignorance. The latter element predominated in an early set piece in which Twain expressed amazement and pride that his ship would include an official "COMMISSIONER OF THE UNITED STATES OF AMERICA TO EUROPE, ASIA, AND AFRICA," but then learned that this supposedly magnificent government "potentate" was in fact "a common mortal, and that his mission had nothing more overpowering about it than the collecting of seeds, and uncommon yams and extraordinary cabbages and peculiar bullfrogs for that poor, useless, innocent, mildewed old fossil, the Smithsonian Institute."[1] Not only did Twain get the name of the Smithsonian Institution wrong and attribute antiquity to an organization that had been functioning for only twenty years, but he was also misinformed about his fellow passenger's goals. Although the commissioner might have seemed innocuous, his enterprise was not. It was, in fact, an integral part of the federal government's hectic involvement with national development—a subject Twain would satirize a few years later in *The Gilded Age.*[2]

In the two decades after the Civil War the center of American activities in natural history shifted from Massachusetts to Washington, D.C. Federal scientists appropriated Gray's and Agassiz's project to identify North American organisms and thereby to assist in their exploitation. In a number of important respects, however, the move south transformed work on the nation's natural history and future. The most obvious and fundamental shift was from a private endeavor that utilized federal resources while keeping state power at arm's length, to an enterprise of the American national government pursued by bureaucrats officially responsible to the people and their representatives. The second change was in scale. Federal science involved staffs and budgets at least an order of magnitude greater than those seen in Massachusetts. The class identity of the enterprise also altered. Leadership among naturalists in the antebellum period had been located in individuals closely affiliated with Boston's elite. Federal science after the Civil War, by contrast, was pursued by a more geographically and socially mixed lot.

These differences in structure entailed equally significant differences in behavior. In place of Gray's and Agassiz's centralized proprietorships, government science operated in a loose, cooperative fashion. As bureaucrats, federal naturalists justified their work more explicitly as a contribution to national political and economic development. Finally, these naturalists articulated a comprehensive and bold philosophy that linked biology and evolution to democracy and public service. At least some were quite self-conscious about the ways their locale, activities, and ideas were integrally related.

The activities of the federal naturalists can be illuminated through a focus on "culture." Although this term held, by the 1880s, both the anthropological and literary-philosophical meanings that have been familiar for the last century, the primary usage of the term in Gilded Age Washington involved actions rather than conditions. Naturalists, like most Americans, privileged a "biotechnological" meaning of culture—as the ways that organisms were grown, altered, and, with varying degrees of intent, artificialized. In what follows I emulate both this multiplicity and hierarchy of meanings. I describe the local culture of Washington naturalists and indicate their participation in culture, but emphasize their efforts *to* culture—to improve American organisms, ranging from fish to people.

THE STRUGGLES OF SPENCER BAIRD

Prior to the Civil War, relatively little work in natural history was done in Washington, largely because not much of anything happened there. The nation's capital was an aggregation of government buildings and boardinghouses, with a population that rose and fell according to the congressional calendar. The "general government," as it was called, had few activities, and pork-barreling congressmen moved as many of these as possible out of Washington to their home districts.[3]

The only significant scientific enterprises in antebellum Washington were the Coast Survey and the Smithsonian Institution. Their leaders—Alexander D. Bache and Joseph Henry—were both physical scientists with comparatively little interest in living things. Moreover, both men accepted the view that they should disperse their resources around the country as much as possible. Such a policy, they believed, would put research money in the hands of the most capable individuals, build up science nationally, and maintain the tenuous political support their organizations needed. Thus only a fraction of federal science money went to natural history, and nearly all that small amount was provided to non-Washingtonians such as Agassiz, Gray, and Torrey.[4]

The young zoologist Spencer F. Baird learned about these constraints on the pursuit of natural history in Washington soon after he arrived at the Smithsonian in 1850. Henry had hired him primarily to run the institution's publication program. He authorized Baird to pursue his interest in building up a zoological collection, but he did not want such work to interfere with his own mission to foster discovery by others. Henry regarded Baird patronizingly as "a young dog" who could be controlled with "a strong arm and a few hard knocks." He kept him in line during a bitter controversy in the early 1850s with librarian Charles Jewett over the Smithsonian's mission. He pressured Baird throughout the decade to accommodate Agassiz's demands for extended loans of large numbers of specimens. As noted in the previous chapter, all involved understood that these loans could facilitate Agassiz's plans to build a great "national" museum in Cambridge, an institution that could easily overwhelm any efforts undertaken in the isolated political capital.[5]

Under these circumstances, Baird did the best he could. He sought volunteer collectors throughout the country, as well as in Canada and Mexico, and he guided the inexperienced with a pamphlet on techniques for collecting and preserving zoological specimens.[6] As the son-in-law of the army's inspector general, he was able to recruit officers posted to remote areas, and he induced the army quartermaster to provide free transportation of specimens back to Washington. He used his status as a government official to place collectors on official surveys, most notably the expeditions of the mid-1850s to reconnoiter routes for a Pacific railroad. Baird's focus was resolutely empirical. He believed that the best way to learn what was interesting was to learn everything; his most important methodological innovation was to insist that his collectors gather both unusual and common animals and that they label each specimen with precise information on the location where it was found. Even more than other American naturalists, Baird was obsessed with details and cared little for philosophical generalizations.

By the late 1850s Baird had established some presence for natural history in Washington. The lavish funding of the surveys enabled him to print unprecedentedly detailed catalogs of the animals of the American West. In 1858 Congress transferred the government's accumulated specimens from the Patent Office to the Smithsonian and appropriated $4,000 annually for this "National Museum." Baird was able to support a handful of young assistants, in part by setting up a dormitory in the unused tower of the Smithsonian castle. Yet both Baird's objects and his assistants were there on sufferance from Henry, who lived down the hall; neither their futures, nor his own, were secure.[7]

A Golden Age in the Gilded Age

The Civil War transformed the nation's capital. Hundreds of thousands of soldiers, businessmen, and freelancers streamed through Washington in connection with the work of saving the Union. The resident population more than doubled, to over 150,000. A functioning national bureaucracy appeared for the first time, and the Republicans' interest in giving the federal government a more active role in developing the country—epitomized in the creation of the Department of Agriculture in the midst of the war—signaled that this new organizational structure would be permanent. The number of federal employees working in Washington nearly tripled between 1861 and 1871 (from 2,200 to 6,200) and then more than doubled during the next decade.[8]

This growth in the size of the federal government opened opportunities for entrepreneurs who understood how to operate the new mechanisms of power. From the perspective of outsiders such as Mark Twain, Washington after the war epitomized a "gilded age" of vulgarity, hustling, and corruption. Naturalists had a different perspective. Looking back from the 1920s, ichthyologist David Starr Jordan recalled a genuine "golden age of American government science."[9] The expansive federal apparatus offered naturalists opportunities to advance themselves and science while developing the country. The central figure in this golden age, Jordan emphasized, was Spencer Baird. By Gilded Age standards, he was a modest and guileless public servant, executing responsibilities efficiently and providing sound advice. Yet he advanced such a multitude of mutually reinforcing projects that by 1885 he oversaw a virtual empire of natural science within the federal government.[10]

The war gave the National Museum visibility. The free exhibitions in the Smithsonian castle provided one of the few uplifting diversions that citizens and soldiers passing through Washington could enjoy. The increased flow of visitors made the museum's claim to "national" status much more plausible. Baird used the museum's popularity, and his friendships with politicians and with generals such as George McClellan and William Emory (both of whom had collected for him in the 1850s), to increase the museum's appropriation from $4,000 in 1860 to $32,000 in 1875; after 1870 he gained substantial independence from Henry in guiding museum operations. He began to hire a permanent staff, including ornithologist Robert Ridgway, invertebrate zoologist Richard Rathbun, and ichthyologist George Brown Goode. Baird's men coordinated the federal presence at the politically important Centennial Exhibition in Philadelphia in 1876; in return, he was authorized to bring back many of

Figure 2–1

Spencer F. Baird, second from right, in the doorway of the unfinished National Museum building, 1880. To his left are General William T. Sherman and Peter Parker, Smithsonian regents; to his right, architect Adolf Cluss. Unidentified laborer in rear. Photograph courtesy of the Smithsonian Institution Archives, RU 95, Negative #78–10099.

its exhibits to Washington for permanent display and was promised (at some future date) a building in which to house them—something that would finally establish, in bricks and mortar, the permanency of the National Museum.[11]

During these years Baird also created the United States Commission on Fish and Fisheries. In 1869 he learned that New England fishermen were concerned about declining catches, and two years later he gained an appropriation of $5,000 to investigate the extent and causes of the problem. He rapidly transformed this limited inquiry into a permanent operation with himself as chief. The commission's explicitly economic mission (and Baird's friendships with influential New England senators George Edmunds and George Hoar) enabled its budget to grow to nearly $200,000 within a decade.[12]

With Joseph Henry's death in 1878, Baird finally took control of all his affairs. Up to the end, Henry had sought to separate the National Museum from the Smithsonian Institution to preserve the latter's private status and disinterested mission.[13] Yet Baird's experience and connections were such that he was the inevitable choice to succeed Henry as Smithsonian secretary. In place of Henry's policy to fund any individual doing good work on any scientific subject, he concentrated the institution's resources on Washington-based natural history. He immediately induced Congress to provide $250,000 for the long-discussed museum building, and within three years had erected the largest structure he could get.[14] He shuffled staff among the Smithsonian, the National Museum, and the Fish Commission to increase the number and scale of projects. Although the budgets of Baird's three organizations each remained modest, when aggregated, they made up a large and continuously growing locus of expenditures (see figure 2–2). (Baird also continued to tap other federal funding sources, most notably the military and the census bureau.) Baird's empire extended across the continent and beyond: he displayed his political and geographic reach with his ability, in 1882, to build and launch the *Albatross* (figure 2–3), a two-hundred-foot-long steamer costing nearly $200,000, whose mission was to explore the Gulf Stream for the Fish Commission and the National Museum.[15]

Baird and his assistants were not alone in their efforts. Ferdinand Hayden, a midwestern physician who had collected for Baird in the 1850s, came to

Figure 2–2

Federal naturalists' budgets, 1870–90.

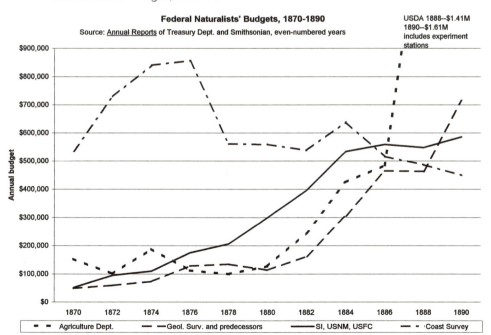

Figure 2–3

U.S.S. *Albatross*, 1890. Photograph courtesy of the Woods Hole Laboratory, Northeast Fisheries Science Center, National Oceanic and Atmospheric Administration, United States Department of Commerce.

Washington in 1867 following military service to lead a survey, first of Nebraska, and then of all the western territories. His discovery and description of the wonders of the Yellowstone region in 1871–72 enabled him to obtain annual appropriations of almost $100,000. John Wesley Powell, an Illinois naturalist who had lost an arm in the battle of Shiloh, used his friendship with Ulysses S. Grant to become head of a survey whose domain rapidly broadened from the geography of the Colorado River region to encompass both Indian ethnology and environmental planning throughout the West. Anglo-Missouri entomologist Charles Valentine Riley rode into Washington in 1877 to lead a commission to fight the great midwestern locust plague.[16] Army physician John Shaw Billings turned the surgeon general's library, located in Ford's Theater, into the national center for medical science information in the late 1870s when he began to use federal resources to publish the monumental *Index-Catalogue of the Library of the Surgeon-General's Office* and to develop the Army Medical Museum.[17]

Initially these men had to struggle to maintain their appropriations. In doing so, they sometimes competed with each other in ruinous fashion. Yet by the early 1880s their projects had matured into organizations similar to those of Baird. By 1881, Riley's Locust Commission had metamorphosed into the Department of Agriculture's Division of Entomology. In 1879 the four competing surveys of the West (those of Hayden, Powell, army officer George Wheeler, and geologist Clarence King) were combined to form a new United States Geological Survey; within two years, due in part to Baird's influence, Powell became the Survey's head. After 1880 the budgets of both the Department of Agriculture and the Geological Survey grew in parallel with those of Baird's enterprises.[18]

By the early 1880s more naturalists were working in and from Washington than any other place in the country.[19] They were, moreover, a cluster that was substantially independent of existing American scientific hierarchies. An essential element of the leadership exerted by Gray and Agassiz, and in Washington by Bache and Henry, had been their identification with the best American society and with European intellectual elites. A few of the younger Washington naturalists, most notably the Yale-educated geologist Clarence King, moved in these circles. But many more—including Hayden, Powell, Billings, and botanist Lester F. Ward—had come from nondescript midwestern backgrounds. Baird, in his search for assistants willing to work on the frontiers, had accepted a number of rough, independent men. These men had, to a significant extent, educated themselves before arriving in Washington. Creative but in certain respects naive, they continued to rely on reading and discussion for their intellectual development.

A SCIENTIFIC COMMUNITY

Naturalists lived good lives as participants in a rich and distinctive subculture within the nation's capital. Gaining a toehold in the bureaucracy could be difficult, and numerous individuals, such as the raffish ornithologist Elliott Coues, subsisted on part-time, temporary, or clerical appointments.[20] But those who established themselves were more secure, better compensated, more productive, and more engaged than nearly all other Americans involved in science.

Most federal scientists were insulated from political pressures. Salaries, which ranged from $1,000 annually for young assistants to $4,500 for a leader such as Baird, enabled many to buy substantial homes and to employ servants to maintain them.[21] At a time when American factory workers could labor sixty hours per week, federal employees were expected to be at their desks from 9 AM until 3 or 4 PM on weekdays, and a half-day on Saturdays, with time

2-4

Commission's Woods Hole facilities, viewed from the southwest, probably
e U.S.S. *Albatross* is moored at the government dock extending out into the
rbor. The laboratory is on the left; the power plant, with chimney disguised
, in the center; Baird's "residence" stands just to the left of the flagpole.
with figure II-2, p. 96. Photograph courtesy of the Woods Hole Laboratory,
Fisheries Science Center, National Oceanic and Atmospheric Administra-
:d States Department of Commerce.

ich.[22] Each bureau had its own amenities: the Geological Survey
d a permanent catered lunch called the Great Basin Mess, and the
it of Agriculture established an informal kaffeeklatsch around the
delivery.[23] Organizing and publishing data were straightforward.
te scientists agonized over clerical and printing costs, federal natu-
oth a surplus of assistants and easy access to the Government
fice. In the mid-1880s, Baird's enterprises generated more than
nd pages of material annually, with often sumptuous illustrations;
ent of Agriculture was not far behind.[24]
were able to finesse even the two great inconveniences of life
n: the oppressive summer heat and the tedium of bureaucratic
ojects allowed staffers to get out of the city during the summer
ime time enabling them to display their interest in local prob-
paid summer work included excursions through the Piedmont,
out West, and, periodically, journeys to Europe for consulta-
rences. Baird, characteristically, found an ideal solution, estab-
ommission summer laboratory in Woods Hole, Massachusetts,
where he and his family, and the powerful Massachusetts

senator George Hoar, vacationed. In 1882 Baird induced Congress to build a harbor there as a "port of refuge" for distressed fishermen; it would, coincidentally, be ideal for the new *Albatross* and other Fish Commission ships. On the village's waterfront the government built a laboratory, power house, and (since there were no inexpensive hotels) a residence for Baird's family and assistants. Museum man that he was, he made certain to include aquaria and educational displays open to the public (see figure 2–4).[25]

The problems of bureaucratic isolation and routine required more elaborate countermeasures. To some degree, the divisions among bureaus were counteracted by the combination of happenstance and Baird's planning that located most of the naturalists in a complex of buildings centered on the Smithsonian (see figure 2–5). Such proximity made real cooperation possible at a time when the telephone was still a curiosity. The more important element in creating community and fostering creativity, however, was the leisure provided by wives, servants, and the federal workday. Naturalists were able to construct a second evening world of clubs and societies. There they could communicate among themselves in circumstances that were independent of bureaucratic hierarchies and rivalries and enabled them to transcend the narrow intellectual horizons of official Washington.

The Megatherium Club and the Potomac Side Naturalists' Club (which sponsored hikes into the country) had brought together federal naturalists beginning in the late 1850s. These informal gatherings were overshadowed, however, by Joseph Henry's Philosophical Society of Washington. Its formal-dress Saturday evening meetings, dominated by physical scientists, affirmed Henry's intellectual and cultural elitism and his desire to distinguish the life of the mind from the world of work.[26] Within months of Henry's death, the naturalists established a set of institutions that gave the capital's scientific "Society" both a new focus and a different tone. The center for the new circle was the Cosmos Club, founded in late 1878 by Powell and his associates. Its name, taken from the German geographer Alexander von Humboldt's project to provide a comprehensive description and philosophical interpretation of the earth and life, expressed, only half-jokingly, the aspirations of federal naturalists. It was a very different gathering from the Philosophical Society. Located initially on Pennsylvania Avenue and, after 1883, on the east side of Lafayette Square, it was an easy stopoff near the intersection of the two horsecar lines that naturalists would use to get from the scientific complex on the Mall to the neighborhood around Thomas Circle where many of them lived. The club's tone was set by Powell's frontiering associates. In contrast to the mixed-sex government offices, where businesslike gentility prevailed, the Cosmos Club was limited to men and hence was more relaxed for those who were there. Over drinks and dinner, cards, chess, and cigars, men from different bureaus planned cooperative ventures, worked out their differences, assessed up-and-comers, and plotted strategy for dealing with legislative and executive officials. Moreover, it truly was a place for informal discussion of the "cosmos"—of intellectual developments and of the social ramifications of science.[27]

Naturalists established, in addition to the Cosmos Club, their own formal organizations—though ones less formal than the Philosophical Society. These included the Anthropological Society of Washington, the Entomological Society of Washington, and—most significant in the present context—the Biological Society of Washington. The Biological Society was organized in 1880 by a group composed of Riley, National Museum employees George Brown Goode and Richard Rathbun, Columbian College zoology professor and National Museum curator Theodore Gill, and Lester Ward, at that time a librarian in the Treasury Department. Baird was immediately named the society's sole honorary member. The group met every other Friday evening at the Smithsonian to hear papers and discuss general issues. The Smithsonian published the society's proceedings. At the Biological Society, Washington naturalists established an identity, not just as bureaucrats, but as members of a real scientific community.[28]

Baird extended the scope of this community through his efforts at Woods Hole. In a masterpiece of multiple intent, he induced leaders at Harvard, Johns Hopkins, and other academic institutions to purchase the land on the Woods Hole waterfront and then donate it to the Fish Commission for its

Figure 2–5

Map of naturalists in Washington, D.C., 1885. The Smithsonian castle stood on the south side of the Mall, which was broken up and landscaped in informal fashion, with curving drives, until the end of the nineteenth century. The National Museum, the Fish Commission (in the former Armory), and the Congressional Botanical Garden extended eastward toward the Capitol. The Department of Agriculture and the National Carp Ponds were on the west. The Geological Survey operated out of the National Museum. In 1887 the Army Medical Museum and Surgeon General's Library would join the complex, moving from the former Ford's Theater to a location (not numbered) midway between the National Museum and the Fish Commission. By contrast, the government's physical scientists were divided: the Coast Survey was located south of the Capitol, while the Weather Bureau occupied space west of the White House, and the Naval Observatory was in Foggy Bottom. Map redrawn from "Map of the City of Washington Prepared and Presented by Thos J Fisher and Co., 1884," and from material in Richard Longstreth, ed., *The Mall in Washington, 1791–1991* (Washington, DC: National Gallery of Art, 1991); residence addresses in *Proceedings of the Biological Society of Washington* 3 (1886): xi–xxiv.

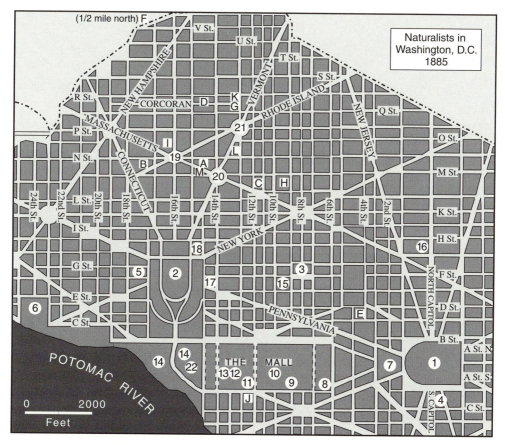

Naturalists in
Washington, D.C.
1885

POTOMAC RIVER

0   2000
Feet

BUILDINGS AND LANDMARKS
1 Capitol
2 White House
3 Patent Office
4 Coast Survey
5 Weather Bureau
6 National Observatory
7 Congressional Botanical Garden
8 Armory--Fish Commission
9 National Museum
10 Smithsonian Institution

12 Agriculture Department
13 Agriculture Department Greenhouses
14 National Carp Ponds
15 Surgeon General's Library
16 Government Printing Office
17 Cosmos Club 1878-1883
18 Cosmos Club 1883-
19 Scott Circle
20 Thomas Circle
21 Logan Circle
22 Washington Monument

RESIDENCES
A Spencer Baird
B Elliott Coues
C William H. Dall
D Grove K. Gilbert/WJ McGee
E Theodore Gill
F George Brown Goode/Tarleton Bean

G Otis Mason
H John Wesley Powell
I Richard Rathbun
J Robert Ridgeway
K Charles V. Riley
L Frederick W. True
M Lester Ward

Figure 2–6

John Wesley Powell, ca. 1880.
Photograph courtesy of the
Smithsonian Institution Archives, RU 95, Negative #
58254.

buildings. In return, Baird promised professors and their students working
space and materials for the indefinite future. This arrangement had the virtues
of enabling Baird to circumvent strict rules regulating the government purchase of land, giving scientific rivals an interest in the success of his plans,
getting professors and graduate students to work on problems in which the
commission was interested, and, most grandly, forming a working national
zoological community around a federal nucleus.[29]

GUIDING NATIONAL DEVELOPMENT

Appreciating the historical significance of the federal naturalists requires examining both their daytime work as bureaucrats and their evening engagement with the mysteries of the cosmos. During the day, they pursued fundamentally national goals. These ranged from educating Americans about the
continent's animal and plant resources, working to increase the variety and
range of useful species, and dealing with biological crises that threatened the
expansion of civilization, up to participation in the government's efforts to

increase the territory of the United States. Although some projects were immediate responses to political demands, leading naturalists had sufficient freedom to take, in many areas, a long-term outlook.

From their own perspective, the major task of federal naturalists was to learn
about North American organisms and to educate citizens about the range of
potential uses these resources embodied. They sought to compile a fund of
reliable knowledge that the government and individual citizens could use. A
principal example of this kind of work was *The Fisheries and Fishery Industries
of the United States*. Between 1884 and 1887, George Brown Goode, twenty
associate authors, and as many clerks published five eight-hundred-page
quarto volumes, plus two volumes of plates, describing, "for the use of the
reading public," everything known about this domain, including all the commercially valuable fish, the distribution of fishing villages, and modes of catch-
ing and processing. Such reports were designed to provide authoritative reference points for private initiatives and public discussions, and to form baselines
for future, more detailed studies. Powell included similar overviews in the
annual reports of the Geological Survey, and he envisioned the placement of
all the scientific agencies' reports in public libraries in every county in the
United States.[30]

By defining both their domains and the concept of utility broadly, the naturalists gave themselves a great deal of room. Baird argued, for example, that
in order to understand fluctuations in the supply of commercially valuable
fish, Fish Commission scientists needed to study oceanic food chains—including the life histories and conditions of existence of nearly all marine invertebrates. This reasoning justified the construction of both the *Albatross* and
the Woods Hole laboratory. The former enabled naturalists to reach the Gulf
Stream and sample the organisms living there. The latter was both a processing plant for material collected at sea and a center for more intense work
on selected aspects of marine organisms' life cycles.

In crucial areas, the naturalists went beyond the provision of information
to supply organisms themselves. The unique property of floral and faunal
"goods" was that a desirable variety or species could potentially be transformed, through reproduction, from an expensive rarity into a common object. The naturalists highlighted this promise of abundance in their budgetary
maneuverings. The Department of Agriculture operated a program to distrib-
ute "new and rare" vegetable and flower seeds to all who requested them, and
the Fish Commission established "fish culture" operations throughout the
country. These ranged from limited projects in which trout species endemic
to the eastern slopes of the Rockies were transplanted to Pacific-flowing
streams, and vice versa, to large-scale efforts, such as the "planting" of Euro-

pean carp in ponds throughout much of the country. Such projects were justified as efforts to expand the natural bounty available to Americans, especially those in modest circumstances, and, more grandly, as the "ultimate refutation" of the "gloomy doctrine of Malthus." It helped that voters were told to direct their requests for seed packets through their congressmen and that fish culture stations were themselves "reproducible," often into the districts of influential legislators.[31]

Federal naturalists' most dramatic work involved the collapses and explosions of populations that frequently resulted in the nineteenth century from the introduction of commercial agriculture into previously insulated ecosystems. Various fish, oysters, and bison crashed; pests, plagues, and blights, including San Jose scale, Texas cattle fever, bovine pleuropneumonia, and peach yellows, spread rapidly along national transportation pathways. Government scientists could do little about most of these problems, but instances of engagement, effective control, or at least accurate prediction were common enough to gain them considerable credit.

Baird, for example, played a crucial, if ambiguous, role in the effort to prevent the extinction of the bison. In 1886 he sent his taxidermist, William T. Hornaday, to Montana to shoot part of one of the last wild herds so that the National Museum would have specimens to mount and display. At the same time he supported Hornaday's publication of an extensive report on the "extermination" of the species and the importance of acting at once to preserve it; appointed "curator of living animals," Hornaday mapped plans to establish a bison herd in Washington's Rock Creek Park, as part of the proposed National Zoo. He ultimately became founder of the American Bison Society.[32]

The more paradigmatic, if less heroic, episode of federal activity with natural populations involved the midwestern locust plague. In 1877, besieged by constituents who had helplessly watched migrating clouds of grasshoppers destroy their crops, Congress established the expert Rocky Mountain Locust Commission. Federal entomologists, led by the dashing Charles Valentine Riley, traveled to the Plains region, where they investigated the problem with a great deal of publicity. The commission's reports summarized the biology of locusts, advised on the effectiveness of weapons against the insect (most notably the "hopperdozer," a horse-drawn mechanical locust trap), and informed farmers that the insects were quite palatable, boiled or fried. Riley's main achievement, however, was to predict, correctly, that the locust population would collapse as rapidly as it had risen. He was able to reassure midwesterners that the insect would not permanently bar them and their crop plants from the Plains.[33]

Figure 2–7

Entomologist Charles Valentine Riley, posed with the tools of his science, mid-1870s. Photograph courtesy of the National Agricultural Library, Beltsville, Maryland.

Finally, naturalists aided both the territorial and economic growth of the United States. Baird's involvement with expansionism was one of the crucial elements in his rise to prominence. At the end of the Civil War, American leaders renewed their effort to take over the continent. In 1867 Secretary of State William Seward, a lifelong advocate of empire, jumped at the Russian offer to sell Alaska. Seeking to persuade skeptical senators and opinion leaders that this subarctic region was not a frigid waste, he turned to Baird for support. Baird had sent collectors to Alaska only two years earlier as part of a Western Union telegraph survey, and he was able to provide firsthand expert information on climate, agricultural potential, and zoological resources. Seward utilized this testimony effectively in pushing the Alaska annexation treaty through Congress.[34]

A decade later, Baird helped the government assert Americans' claims to the valuable fauna of the North Atlantic. The consensus in the 1870s was that State Department negotiators had bungled a maritime treaty with Canada in 1871 because they had not understood migration patterns and fishing practices on the offshore banks. This failure had provided part of the argument for the creation of the Fish Commission. Thus when negotiations reopened in Halifax in 1877, Baird was there to advise American diplomats concerning the effects of different treaty changes on Americans' access to fish.[35] Animal behavior continued to be a major focus for American diplomacy into the twentieth century: the State Department repeatedly turned to naturalists to bolster the claim that Americans could safely "harvest" the valuable fur seals of the Pribiloff Islands, but that hunting by Canadians and Japanese on the open ocean would soon drive the species to extinction.[36]

EVOLUTIONARY CULTURE

Federal naturalists did not, however, define themselves solely as providers of technical services. They had constructed their evening world explicitly to counterbalance what Theodore Gill described as their "excessive (because exclusive) cultivation of special departments of botany and zoology." At both the Biological Society and the Cosmos Club, naturalists made self-conscious efforts to discern the "general laws and principles" that would make sense of what they were learning and that would give their activities larger significance. It was obvious to Gill and to his colleagues that those laws and principles were "how organisms have been evolved and how grown and developed." Evolutionary development, expressed in both the growth of the individual and "the lifetime of nature," provided the federal naturalists an intellectual touchstone. The science that encompassed general laws covering all organisms was not natural history, with its aura of empiricism and perceived affinity with the new fad of stamp collecting, but biology—a term, Gill explained, that truly expressed the common interests of the naturalists.[37]

Discussions of the meanings of evolution and biology were, for many federal naturalists, the highest form of culture. Like many American intellectuals, they were deeply interested in the writings of Herbert Spencer, T. H. Huxley, Charles Lyell, and other British evolutionists. Although "the night was stormy," more than seven hundred Washingtonians attended a memorial meeting a few weeks after Darwin's death; speakers included Powell, Riley, and Ward. Elliott Coues stood only slightly on the margins when he published *The Daemon of Darwin*, a long poem in the style of Goethe's *Faust* that featured an imagined verse dialogue between the shades of Darwin and Socrates. A few years later a group of naturalists and humanists in the capital joined to form the Society for Philosophical Inquiry.[38] They sought the foundations for a unified, science-conscious, modern high culture.

Still, the most self-aware naturalists believed that the ultimate value of these broad intellectual interchanges lay in clarifying and advancing the collective work in which they were engaged. They believed that, as bureaucrats, they were participating in an organizational structure that was realizing the deepest purposes of the huge and chaotic, yet now firmly united, American nation. They argued that conscious guidance of nature and humanity by an evolutionarily informed democratic government was the crucial event of their time. They placed themselves squarely within this evolutionary process. With some plausibility, they believed that they stood at a unique nodal point connecting description and prescription, knowledge and power, or, most broadly, nature and culture. Their efforts to "culture" America can be seen most clearly through examination of the activities of Lester F. Ward and George Brown Goode.[39] The former illustrates the intellectual breadth and radicalism that was possible among the federal naturalists; the latter displays how their thinking impacted, in multiple ways, on major streams of American life. Both functioned comfortably within the universe of possibilities open to federal employees.

Lester F. Ward and the Irrationality of Nature

Lester Ward is known today primarily as a sociologist with a strong commitment to social reform. His two-volume *Dynamic Sociology* (1883) was the first text on that subject written by an American. His employment as head of the Division of Fossil Plants at the Geological Survey beginning in 1881 is sometimes seen as a sinecure provided by Powell to enable him to advance his work in social theory. Powell was certainly sympathetic toward Ward's sociological writing, but he hired him as a botanist. We can see Ward as a wide-ranging, but not atypical, federal naturalist. In the 1880s his broader interests revolved around evolutionary theory; concern about the future of humanity, particularly in the United States, was one part of that interest.[40]

Ward stood at the egalitarian and freethinking edge of bureaucratic respectability. Born in 1841, the tenth child of a nomadic midwestern miller and farmer, he was the younger brother of Cyrenus Ward, a wagon-wheel maker who became an early leader of organized labor. Ward served in the Union army ranks for two years and was seriously wounded at Chancellorsville; as a disabled veteran he was able to find a job in the Treasury Department and began to attend Columbian College in the evenings. With his craggy face and full head of hair, Ward epitomized assertive American masculinity; yet he was notable as a scientific advocate for the equality and even superiority of women.[41]

Lester Ward's first significant government research project, conceived around 1870, displayed both his approach to knowledge and his central intellectual concern. On government time but on his own initiative he produced, in the spirit of positivist historians such as William Buckle, the first tabulation of historical statistics on the numbers and national origins of immigrants to

the United States. Published in the 1871 *Annual Report* of the Treasury Department, these charts provided an authoritative public reference point for
discussions of the changing ethnic composition of the human population
of the country.[42] During the 1870s, while employed as a librarian (that is,
documents and data manager) for the Treasury Department's Bureau of Statistics, Ward prepared the ponderous treatise whose title, *Dynamic Sociology*,
would indicate its aim to extend and critique Herbert Spencer's famous *Social
Statics*.

At the same time, however, Ward was pursuing a serious interest in botany.
This was the branch of natural history least developed in Washington around
1880: Henry and Baird had handed over the subject, along with their herbarium, to the new Department of Agriculture in 1869, but the practical men
there had little interest in common plants. Ward, collecting specimens with
the Potomac Side Naturalists' Club, made contact with Powell and Baird and
began to publish; in 1881 the National Museum distributed Ward's field guide
to plants of the midatlantic region. It was as a botanist that he became a
founding member of the Biological Society of Washington.[43]

In 1881, soon after taking over and reorganizing the Geological Survey,
John Wesley Powell sought someone to take charge of plant fossils. This was
a frontier science: exotic and exciting, but notoriously difficult. Some striking
species from particular strata could easily be identified, but most specimens
were mere impressions of leaves or fragments of stems, which were very difficult to characterize. What Powell needed was someone who could utilize
government resources to provide both daily expertise and basic intellectual
organization. Agassiz's former assistant, Leo Lesquereux, was the only true
American specialist in this area, but he was deaf and over seventy. Ward was
essentially finished with *Dynamic Sociology*; as an experienced practical botanist and bureaucrat, he was a plausible candidate for the new position.

Ward jumped at this chance to leave his desk in the Treasury Building and
join the naturalist community full-time. At the Geological Survey, located in
the new building of the National Museum, he turned the tools of the bureaucracy to the task of improving conditions for research in this domain and
orienting it in a direction that would make it intellectually more important.
As a Survey employee, he had easy access to specimens from around the
country and to a full supply of maps, books, and journals. The abundance of
clerical labor and the new technology of index card files enabled him to initiate the essential if workaday pretaxonomic project of compiling references to
all known specimens of plant fossils. His successors at the Geological Survey
maintained this research tool for a century.

Figure 2–8

The scientific bureaucracy at work in the 1880's. Lester Ward reviews text at his Geological Survey office in the National Museum; Annie C. Moorhead maintains the new information management technology of the era—an index card file, probably of fossil plant specimens. Photograph courtesy of the Smithsonian Institution Archives, RU 7321, Negative # 85-102757.

All these elements can be seen in a marvelous photograph (see figure 2–8), probably from the late 1880s. Ward, seated beneath a lithograph of a fern, at a drafting table near the window of his National Museum office, reviews manuscript text. A box of fossil specimens can be seen on the worktable immediately to his right. On the wall, above a map of the United States, is a bookcase filled with Geological Survey reports and books; the two identifiable volumes on the lower shelf are Charles Lyell's *Principles of Geology*. The books on the table are botanical reference manuals, possibly by Asa Gray. Sitting modestly to the side is Ward's young female assistant, Annie C. Moorhead; she is working, with a steel-nibbed pen, to update a file box filled with specimen reference cards. A field notebook, with doodles on its binding, is suspended on a metal rod.

Ward's most notable paleobotanical achievement was simple but enduring: to coin and popularize the term "paleobotany" as a way to give his subject an identity and to emphasize that it was fundamentally a biological, not a geological, science. His more substantial contribution was to distill what was known about plants and their evolutionary history from the masses of data he was accumulating. The Survey volumes in Ward's bookcase included two reports, totaling more than four hundred pages, summarizing the state of knowledge as he saw it in the 1880s.[44] He was also making respectable contributions, through paid fieldwork, to specialized discussions: he brought botanical evidence to bear, for example, on the major paleontological problem of characterizing the end of the Cretaceous.[45]

At the same time, mostly in nonofficial publications, Ward was reflecting on the meaning of evolution, with particular attention on the implications to be drawn from the history of plants. He presented an alternative to the narrative, common in zoology (as, for example, in O. C. Marsh and T. H. Huxley's discussion of the horse), that evolutionary change was a gradual and linear ascent. He emphasized that showy flowering plants had appeared quite suddenly in the Cretaceous, but had changed relatively little in the millions of years from that time to the present. Ward's representation of botanical history (figure 2–9) combined an emphasis on the transitions in predominance among groups with an amorphous depiction of fluctuations and continuities.

Ward located an important message in the midst of this flux. Flowering plants had arisen, not because of some internal process of improvement, but because of insects and their preferences. Ward argued that "psychic factors"—the mental activities, or desires, of animals—had functioned as an evolutionary wild card. They had sent plants, and animals, in innumerable novel directions, but they had produced evolutionary results that were often far from optimal, when viewed from a design perspective.[46] More generally, Ward emphasized how wasteful the Darwinian process of reproduction, variation, and natural selection was. He argued that it was "wholly at variance with anything that a rational being would ever conceive of, and that if a being supposed to be rational were to adopt it he would be looked upon as insane."[47]

For the author of *Dynamic Sociology*, these ideas had present-day implications. Ward argued that the arrangements existing in the natural world were not particularly valuable as models for human planning; on the contrary, the artificial was superior to the natural. More immediately, he emphasized that Herbert Spencer was mistaken in his belief that the present social order was anywhere near optimum. Human life could be improved through intelligence and communal action. In an 1886 lecture, jointly sponsored by Washington's Biological and Anthropological Societies, Ward noted that the key to agriculture had been "the creation of an artificial environment advantageous to man." He then emphasized that "the exact homologue of this culture of plants and animals, when extended to man, is education in its broadest sense."[48] Ward, as a Gilded Age federal employee, was quite realistic about the limitations of the American political system. Yet he was convinced that scientists functioning within a democracy were best positioned to provide society with rational guidance.

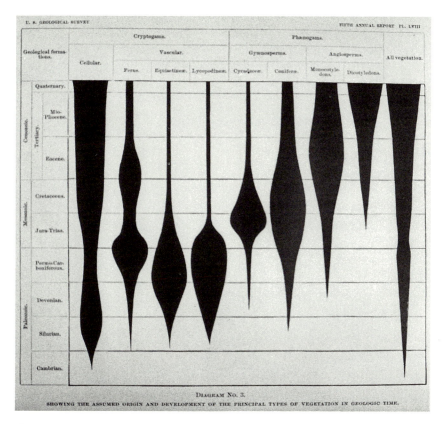

Figure 2–9

Lester Ward's diagram of botanical history, "showing the assumed origin and development of the principal types of vegetation in geologic time." From Ward, "Sketch of Paleobotany," plate 58.

George Brown Goode and the Power of Culture

G. Brown Goode, as he signed his letters, was a different kind of person from Lester Ward. Born in 1851 to a prosperous family that moved from southern Indiana to the New York–Connecticut border region when he was seven, Goode attended Wesleyan College and Harvard, and began to assist Baird in his ichthyological work in 1872. Goode was a meticulous gentleman who eschewed the masculine camaraderie that prevailed among the westerners and war veterans at the Geological Survey and the National Museum. A committed Christian throughout his life, he refused to publish one of Ward's essays in the Smithsonian *Annual Report* because he considered it too materialistic.[49]

Figure 2–10

George Brown Goode, about 1880. Photograph courtesy of the Smithsonian Institution Archives, RU 95, Negative #10739.

Yet Goode shared, and extended, much of Ward's vision: they did, after all, pass the same exhibits at the National Museum daily for more than a decade as they went to their offices to work.

Goode, like other Washington naturalists, was fascinated with the lushness and diversity of nature. As a major force at both the Fish Commission and the National Museum, he had unique access to marine specimens. Within this diversity, he chose to concentrate on the nightmarishly shaped fishes then being pulled from the ocean depths by the British exploring ship *Challenger* and by its American emulator, *Albatross*. Goode published two large and magnificently illustrated volumes describing the fishes that had evolved in the depths of the Atlantic Ocean. Equally important, he put these freakish animals on display at the National Museum. With their exaggerated profiles, needle-shaped teeth, and luminescent appendages, they were as exciting to the public as any of P. T. Barnum's old hoaxes, but they were indisputably real natural objects, visible only because of scientists' initiatives. Such spectacular objects were persuasive displays of federal naturalists' success in uncovering the unknown.[50]

At the same time that he was pursuing these evolutionary grotesques, Goode sought to articulate how nature was transformed by intelligent individual and communal human action, or culture. For Goode, culture was the application of technology to the organic. It was diverse in its settings and manifestations, but also open to improvement. Within this conception, Goode easily included the newer, conflicting meanings of culture, as both the variety of peoples' ways of life and as the distinguishing quality of advanced civilization.

The multiple yet interconnected meanings of culture can be seen in Goode's engagement with "fish culture." In a narrow sense, this term referred to the hatcheries that the Fish Commission was building around the country in the 1870s and 1880s. Goode, however, understood that these establishments were merely one element within a larger framework. This broad view of fish culture can be seen in the American fisheries project. Goode began this multivolume work with "the natural history of useful aquatic animals": the patterned diversity of marine organisms' identities, locations, and life histories. Next came an analogous account of the humans who interacted with the fish. On the basis of surveys, interviews, and observation, Goode and his colleagues outlined the distribution of fishing villages, the customs of fishermen, the equipment being used, the amount and kinds of catches, and relations that fishermen and other humans had established in the United States around the institution of the market.

Ultimately, Goode engaged with the most active elements of fish culture. The most focused of these was the new effort by scientists and culturists to improve human life by "planting" fish in new areas and by increasing the

Figure 2–11

Mancalias shufeldtii Gill, one of the deepwater species dredged in 1880 by Fish Commission scientists. G. B. Goode and T. H. Bean, *Oceanic Ichthyology*, figure 401

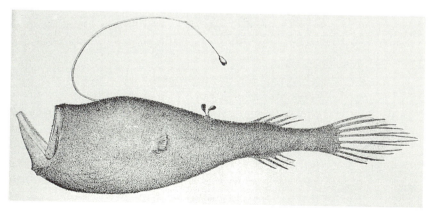

numbers of fry through artificial propagation in protected hatchery environ-ments. But Goode recognized that this scientific and technological labor by bureaucrats needed to be paired with organized restraint on the part of the American people in dealing with marine organisms. The most difficult, but necessary, task was legislative regulation of fishing.[51]

Although Goode devoted great energy to the production of texts, he did not think that writing was the most effective way to culture American people. Neither he nor his mentor, Baird, expected, for example, that merely publish-ing circulars would induce large numbers of Americans to cultivate carp. To demonstrate that the fish was easy to grow, and to give carp culture both status and a patriotic aura, the Fish Commission created the National Carp Ponds. The commission captured a strategically located, but previously neglected, piece of real estate—the sewage-filled swamp just north of the Washington Monument. Both the monument and new municipal sewers were finally being completed in the years around 1880. Baird and Goode reconfigured the swamp into a series of ponds and gardens decorated with fountains and shrubbery. Visitors to the obelisk would inevitably pass by the Fish Commis-sion's installation, where they could see both European and Japanese carp, learn how to build and maintain a pond on boggy farmland, and obtain fry— and become motivated to make fish culture part of their lives.[52]

The dimensions of Goode's concept of culture—and the importance of exhibitions in culturing the public—can be seen more fully in his work at the National Museum. Baird's primary collecting interest had been in zoology, but he was sufficiently omnivorous and opportunistic that he accepted anthro-pological artifacts, contemporary art works, patriotic memorabilia, and, as part of the arrangement that led Congress to fund the museum's building, the examples of modern industry displayed at the 1876 Philadelphia world's fair. When he became Smithsonian secretary in 1878, he appointed Goode assis-tant director of the museum. The younger man took on the substantial tasks of organizing the museum's collections, managing its assortment of regular and volunteer curators (essentially all the major Washington naturalists), and mounting a coherent and politically acceptable permanent exhibition in the new building that opened in 1881.

For Goode, "the chief requisite to success in the development of a great museum [was] a perfect plan of organization and a philosophical system of classification."[53] It is not surprising that he started from the basic premise articulated by Ward, shared among the Washington naturalists, that the great drama of the cosmos was humanity's evolving ability to construct an artificial, civilized life out of the materials provided by nature. He thus divided what he called a "museum of man" into eight parts. He began with anthropology—

the general characteristics and diversity of mankind. He then considered "the earth as man's abode," and the natural resources—mineral, vegetable, and animal—that it included. Next came "exploitative technology," including, on the one hand, mining, lumbering, hunting, and fishing, and, on the other, the "culture" of plants and animals. These extracted materials were then "elaborated" through manufacturing and utilized as food, clothing, shelter, medicine, and transportation. These industrial activities provided the base for "social relations," including communication, trade, government, and war. At the top stood the "intellectual occupations": art, literature, science, philosophy, and "education and reform."[54]

The most striking aspects of Goode's classification were its naturalistic populism and its evolutionary self-consciousness. He noted that "one of the standard instructions" long given to Smithsonian zoological collectors had been to focus on "those things which are most common." Similarly, he argued that the "evolution of civilization" would be found in changes in the "every-day" costumes and tools of "the masses of people," not in artistic rarities produced for aristocrats.[55] Progress had been the result of innumerable individual and communal efforts over centuries. But continued improvement now depended on education and reform—that is, on making the public aware of the structure of the cosmos and their place in it and on getting their political representatives to act intelligently. Thus Goode placed museums together with schools and libraries as the institutions that would lead humanity to the final stage in his classification: "the climaxes of human achievement."[56]

G. Brown Goode exerted a significant influence on American culture during the last two decades of the nineteenth century and beyond. More than four million visitors viewed the National Museum's displays of natural history, art, and industry between 1881 and 1900; these included casual tourists, congressmen and their families, and American and foreign cultural leaders. In addition, through Smithsonian publications and through direct outreach, Goode proselytized his vision of educational exhibitions among the groups of leaders organizing provincial museums and fairs.

The culmination of this effort was the planning of the 1893 World's Columbian Exposition in Chicago. At the request of the fair's president, Goode prepared a draft "system of classification," based largely on that of the National Museum, for the main exhibits. Goode recognized that compromise would be necessary, both with commercial interests and with the largely autonomous efforts being made by governments (including, prominently, the United States, on whose exhibit board Goode served). He expressed only pleasure, however, with what he saw as the exposition's primary aim: "to expound . . . the steps of the progress of civilization and its arts in successive centuries, and in all lands up to the present time and their present condition; to be, in fact, *an illustrated encyclopedia of civilization.*" These had been, he intimated modestly but accurately, his guiding principles during the preceding twenty years; the fair was furthering his and his colleagues' ongoing project to culture the American people.[57]

During the last generation, scholars have interpreted the World's Columbian Exposition, and Gilded Age fairs and museums more broadly, in a wide variety of ways. Nearly all, however, have taken for granted that the fairs were "glorifications of conventional wisdom . . . manned by the orthodox," and that part of this conservatism was a pervasive "evolutionary ideology."[58] Such an impression, more than a century later, is a mark of the success of Goode and his colleagues. It took real work for these museum people to insert any thematic coherence whatsoever into the jumble of objects and commercial products that crowded museums and expositions. Moreover, they took an outlook that was largely materialistic and historicist, associated with such controversial European intellectuals as August Comte and Herbert Spencer, and made it both comprehensible and palatable to millions of Americans.

To be sure, not everyone was persuaded. In *A Connecticut Yankee in King Arthur's Court* (1889), Mark Twain highlighted, not the continuity between early and modern industry, but the gap separating the traditional aristocratic order from the outlook of a modern gun factory superintendent. He drew deeply pessimistic conclusions about the future of what he would call, a few years later, "the damned human race." Few Americans, however, took Twain seriously.[59]

Conflicting Visions of American
Ecological Independence

THE BEAUTY AND MENACE OF THE JAPANESE CHERRY TREES

In August 1909, Tokyo officials presented two thousand ornamental cherry trees to their "sister Capital City" of Washington, D.C. Incorporating exemplars of oriental horticulture into the landscape of official Washington (in place of a recently planted grove of American elms) would offer symbolic compensation to Japan for the American demand two years earlier that Japanese immigration cease. The trees arrived in early January 1910, but within a few days, Charles L. Marlatt, acting chief of the Bureau of Entomology of the United States Department of Agriculture (USDA), reported that they were infested with crown gall, root gall, two kinds of scale, a potentially new species of borer, and "six other dangerous insects." He urged that "the entire shipment should be destroyed by burning as soon as possible."[1] Marlatt's recommendation went all the way to President William Howard Taft. Taft acceded to his expert's views, and on January 28, federal employees took the dormant trees from storage sheds on the Washington Monument grounds, heaped them into piles, and reduced them to ashes. At the time, as well as more recently, Marlatt's report was interpreted as a minor technical interruption into the intricacies of cultural diplomacy.[2] My aim is to examine this incident from a broader perspective.

Chapter 2 described the development of the community of federal naturalists during the Gilded Age and sketched both their technocratic labors and their designs to culture the United States. The grander schemes of these men faded after 1890, and they became less prominent in American cultural life. The circumstances of that relative decline will be discussed in chapter 4. Their technocratic activity and influence on the composition of the nation continued into the twentieth century, however. In this chapter I describe one major area in which federal naturalists continued to influence the future of life in the United States, in ways that were both immediately consequential for plants and animals and linked with the makeup of the human population. In 1909, Charles Marlatt was undercutting presidential policy in order to

Figure 3–1

Japanese cherry trees burning on Washington Monument grounds, January 1910. Photograph courtesy of the National Agricultural Library, Beltsville, Maryland.

further his bureau's campaign for plant quarantine legislation. This campaign was one element of a larger struggle within the USDA over the relative importance of introducing valuable "plant immigrants" and excluding undesirable weeds, blights, and "alien insect pests." Viewed from a still broader perspective, the cherry tree incident lay at the center of a reorientation in the relations that the human population of temperate North America had with species and varieties (including human varieties) originating in other parts of the world.

My framework is taken from Alfred Crosby's *Ecological Imperialism*.[3] Crosby has interpreted the European conquest of the Americas, Australia, and New Zealand as the replacement, between 1492 and the early 1800s, of plants, animals, and humans resident in these areas by types that had evolved in the more competitive environments surrounding the Mediterranean Sea. The earliest European settlers were only dimly aware that their easy success had a biological dimension. Crosby's central metaphor is so striking just because it conveys the distance between the oblivious self-regard of European colonists and the profound ecological forces that made their hegemony in the temperate "neo-Europes" possible.

How did Europeans become aware of the bases and implications of ecological imperialism? In the late nineteenth century, small groups of specialists in the neo-European ecological "colonies" began to appreciate the dimensions of this phenomenon and began to explore what might be done to gain more control over the flow of species and varieties into their territories. It is thus possible to extend Crosby's metaphor: in certain places, ecological imperialism prompted the descendants of human colonials to initiate movements for what can be called "ecological independence."

Taken as a whole, the United States led in these efforts. In contrast to the British dominions or the South American countries, it combined political independence, a developed scientific community, and a functioning national bureaucracy. Its resident organisms endured the continuing effects of ecological imperialism. Most important, however, at the end of the nineteenth century its leaders were rapidly establishing new kinds of relations with the rest of the world. Increased global trade, rapid steamship transport, and the creation of an American overseas empire combined to generate potentially far-reaching biological consequences. American agricultural explorers began to import large stocks of new germplasm, thereby increasing the quality and diversity of crops in both North America and in the United States' new tropical dependencies. At the same time, there was fear that the new international contacts would lead to an increase in foreign pests and parasites entering the country. The consequence of empire could be the destruction of the American agricultural economy, and the native biota, from within.

The leaders of the American ecological independence movement were located in Washington at the USDA. At the turn of the century, the Bureaus of Plant Industry, Animal Industry, Forestry, Entomology, and Biological Survey made up a massive biological science enterprise. Their employees were deeply engaged in shaping both the political and natural economies of the nation. In particular, leaders of USDA biological science worked systematically to manage species and varieties whose presence or absence could affect the well-being of American citizens.[4]

Investigation of Washington's agricultural scientists is particularly interesting in that it reveals a deep division among them over the biotic future of the country. Their differing perspectives were a function of both specialty and personality, and they were heightened by the interbureau rivalries that permeated the department at this time. On one side were botanists in the Bureau of Plant Industry who were cooperative, sensitive, and sometimes diffident about their careers as bureaucrats. They sought to increase the number and variety of useful kinds of vegetation in the United States and were serene about the country's ability to prosper in a global biotic system. They can, appropriately, be described as ecological "cosmopolitans." On the other side were zoologists in the Bureaus of Entomology and Biological Survey, who tended more toward competitiveness, aggressive masculinity, and careerism. They worked to protect Americans from foreign organisms considered pests and were concerned to preserve the distinctive biotic elements they saw in the continental United States. Theirs was, in a word, a "nativist" perspective.[5]

The issues at stake were technical, though not abstruse. At the same time, they were bound, through language and imagery, to ethnic sensibilities cultivated over centuries of political conflicts and ecological displacements. European-Americans identified with familiar and useful plants, whether the "native" spreading chestnut tree or "introduced" amber waves of grain. Conversely, attitudes toward foreign pests merged with ethnic prejudices: the hessian fly and the oriental chestnut blight traded meanings with their presumed human compatriots. The cherry tree incident—with its contrast between strong, symmetrical, but dull American elms and effete, twisted, and spectacular Asian exotics, its subtexts of racial inequality and (horti)cultural sisterhood, and its implications of a Yellow Peril hidden within beautiful packaging—was saturated with such meanings. USDA leaders' bureaucratic survival depended on sensitivity to the political power of connotations.

The direction that the American ecological independence movement took was highly contingent. Predominance, in fact, shifted from nativism in the 1890s, to cosmopolitanism in the following decade, and then back to the nativists in the 1910s. Ultimately, the nativists prevailed. For more than eighty years, both government policy and informed opinion have granted the protection of indigenous species and varieties priority over the introduction of exotics. An introducer bears a significant burden of proof and, if problems arise, shoulders a large amount of blame. Moreover, past movements of organisms are now stigmatized as "ecological imperialism," a morally suspect and politically retrograde category.

The focus of this chapter is on bugs, plants, and birds. As the cherry tree controversy implies, however, debates in this area were closely linked to attitudes about human groups. I will conclude by sketching how an appreciation of disputes over flora and fauna can lead to a reinterpretation of the efforts by "native" European-Americans in the 1910s and 1920s to limit human immigration.

AMERICA'S ECOLOGICAL OPEN DOOR

The effects of ecological imperialism on eastern North America are well known. From the time of the first contacts, European explorers and colonists spread pathogenic bacteria and viruses among indigenous peoples. They introduced Old World plants and livestock, and inadvertently imported rats, weeds, and destructive insects and fungi. By the 1650s the Indians were disappearing from New England, and the makeup of the fauna and flora in settled areas had altered dramatically.[6]

A second wave of plant and animal introductions took place during the nineteenth century, as the rising standard of living, a new interest in horticultural novelties among gentleman farmers, and improved transportation combined to facilitate the movement of organisms. This phase of ecological imperialism was less dramatic than the first, and its initiators were groups long resident in North America. But, as was the case in the 1600s, these introductions resulted from the haphazard activities of individuals with particular interests, and they had ecological consequences well beyond those conceived by their initiators.

Horticultural and agricultural improvers provided the main impetus behind the increasing flow of organisms into temperate North America during the nineteenth century. Nurserymen, tourists, and immigrants directed fruits, seeds, seedlings, and bulbs from Europe and Asia toward the United States.[7] They substantially increased the diversity of crop varieties, and, even more so, of fruits and ornamentals. However, some introduced plants (such as honeysuckle and water hyacinth) soon became weedy in their new environments. Imported plant material also harbored weed seeds (notably the tumbleweed, or russian thistle), and they provided transportation for undesirable insects (including the codling moth, cabbage worm, and various scales and aphids) and a significant number of blights and rusts.[8]

More isolated, but, in retrospect, more remarkable, were the campaigns by enthusiasts to bring Eurasian animals, notably birds, to the United States. Between 1850 and 1870, local "acclimatization societies" imported thousands of english sparrows. By the 1880s, bird lovers had attempted to introduce at least twenty Eurasian species. They soon succeeded in establishing both the ring-necked pheasant and the starling.[9] The most notorious animal introduction of the nineteenth century was the gypsy moth. Etienne L. Trouvelot, an artist employed by Louis Agassiz at Harvard's Museum of Comparative Zoology, worked in his spare time breeding hybrid silk worms that could survive Massachusetts winters. In 1869 he imported gypsy moth eggs from France to his home in the Boston suburb of Medford. Some of the fuzzy egg masses blew out an open window, and the moths established themselves in brushland behind his house. After two decades of acclimatization, the Medford moth population suddenly exploded, producing the spectacular fouling and deforestation that northeasterners have experienced periodically ever since.[10]

The federal government participated in plant and animal introductions in essentially the same ways as private individuals. By turns, the State Department, the Patent Office, and the Department of Agriculture collected and distributed foreign seeds and plants.[11] Chapter 2 noted that the Fish Commission distributed carp at the same time that it redistributed such American fishes as the rainbow trout and striped bass from one coast to the other.[12] Although the scale of government work was at times substantial, it remained a series of discrete actions rather than a systematic program of introductions. More important, federal officials made no effort (with the notable but quite limited exception of carriers of epidemic disease) to regulate the flow of organisms across the borders of the country.[13]

While Americans enthusiastically incorporated improved crop varieties and novel ornamentals into domesticated landscapes, they excoriated both introduced pests and those seemingly responsible for them. The hessian fly, russian thistle, and english sparrow all generated hatred that easily encompassed those deemed responsible for bringing them—whether the acclimatization societies, the scientific community, or members of the eponymous ethnic group. Yet neither scientists nor mainstream political leaders argued for greater national control over the movement of organisms. The invisible hands of nature and the market were equally obscure and wonderful in their effects. Asa Gray reported dispassionately in 1879 on the continuing introduction of European weeds into the United States. Cornell University horticulturist Liberty Hyde Bailey articulated the dominant perspective forcefully in 1894. He argued that pests were the inevitable by-products of the disturbance of nature by agriculturists. The only solutions were vigilance and time. Western farmers were plagued by the russian thistle because they had been too rushed—and lazy—to plant and plow properly. Government's only role was educational; the law, Bailey concluded, "cannot correct a vacancy in nature."[14]

THE BEGINNINGS OF A FEDERAL RESPONSE TO PESTS

Two years after Bailey's speech, USDA botanist Lyster Dewey prepared a bulletin promoting "legislation against weeds."[15] Federal scientists were coming to believe that a system driven by individual whims and by natural and market forces could be improved and that they, as experts employed by the "general government," should hold the central responsibility in this area. They began to consider how problems previously viewed in isolation—not only the introduction of promising exotic organisms into the United States and the exclusion of noxious alien species, but also the preservation of native species and varieties, the cultivation of desirable forms in new regions of the country, the improvement of wild types, and the control of pests—were in fact interrelated. But they held changing and conflicting views on where to place the emphasis: on weeding out the bad or cultivating the good.

Initially, USDA scientists were minor players within the Washington science bureaucracy. They gained local and national visibility only around 1888, when the department rose to cabinet status, the Hatch Act created a national network of agricultural experiment stations, and appropriations jumped (see figure 2-2). With most local and routine tasks delegated to the provinces, Agriculture Secretary Jeremiah Rusk was able to envision the scientific staff in Washington as the department's "central station," the site for its most complex and abstract research and for work on problems of national scope.[16]

A cohort of young agricultural college graduates staffed the department's ramifying divisions, which by 1890 included entomology, vegetable pathology, economic ornithology and mammalogy, botany, forestry, and pomology, in addition to the more autonomous Bureau of Animal Industry.[17] Housed together in a few buildings on the south side of the Mall (see figure 2-5), staffers formed a nascent technocratic culture organized around the production of reports. Yet they knew that divisions were in competition for funds and that each group's success in the long run depended on producing work that would interest potentially influential outside "friends." Some of the young agricultural scientists also rapidly learned, at the Cosmos Club and the local scientific societies, about the broader evolutionary and nationalist frames within which their work could be situated.[18]

The initial structure of USDA central station work was modeled largely on that of the Smithsonian and the Geological Survey, discussed in chapter 2. Basic research, in the form of data collection leading toward national surveys, was mixed with opportunistic problem solving that would display the immediate value of a division's activities to influential interests. C. Hart Merriam's Division of Economic Ornithology and Mammalogy, for example, explored the distribution of North American mammals and birds, hoping ultimately to produce a continental map of "life zones"—both a contribution to pure science and a set of parameters that farmers could use when thinking about new crops. At the same time, division staffers assessed whether species such as blue jays or woodpeckers were truly "noxious" to agriculture, and they reviewed the efficacy of bounties for reducing the number of pests and of closed hunting seasons for protecting game species. Entomologists and vegetable pathologists pursued similar dual programs: taxonomy and mapping on the one hand, and tests of insecticides and fungicides on the other.[19]

Beginning in 1894, however, USDA scientists shifted their attention dramatically toward the problem of introduced pests. The possibility of active intervention in this area had first arisen in California, a novel environment where American settlers, European and Asian vegetation, and oriental insects had converged within a few years. In 1881, California fruit growers lobbied successfully for a state plant quarantine law; eight years later, the successful control of the cottony-cushion scale by a USDA–introduced predator made clear the value and credit that could come from government pest control successes.[20]

These issues rose to national significance, however, only in the early 1890s. Due to increased trade and happenstance, tumbleweeds and wheat rust overwhelmed farmers in the northern Plains region, gypsy moths devastated the Boston suburbs, the "Mexican" boll weevil crossed the Rio Grande, and the (Asian) San Jose scale spread from California to the East on infested nursery stock. Panic about pests was greater than at any time since the Midwest's locust plagues of the 1870s. These problems in the natural economy were heightened because of the devastation to the nation's political economy in the wake of the Panic of 1893. The collapse of farm prices led rural leaders to demand immediate and comprehensive government action to save their livelihoods.[21]

Within this context, basic research on, for example, the natural history of peach yellows, seemed trivial. Work to promote quarantine laws, by contrast, made department scientists visible while it moved responsibility for dealing with pests to the legislative arena. It affirmed the value of cooperation and identified specific causes of the malaise of the period. Finally, agitation for quarantine work promised immediate results but was much cheaper than laboratory research—an important consideration, since the USDA budget had been cut one-quarter by the retrenching Cleveland administration.

These were the bases for Dewey's guidelines on weed control legislation. Entomology chief Leland Howard joined with a bulletin on insect control laws, and he went a step further with a stark warning on the enormous number of Mexican and Japanese insect species then poised to invade the United States. Entomologists proposed to stop the spread of the boll weevil by closing a fifty-mile-wide strip of Texas to cotton culture, and they argued for measures to keep the San Jose scale from spreading throughout the East. Howard and plant pathologist Beverly Galloway capped these efforts by organizing the National Convention for the Suppression of Insect Pests and Plant Diseases by Legislation; this group of experiment station scientists and horticultural leaders met in Washington under USDA sponsorship the day after McKinley's inauguration, proposing legislation to establish a system for inspecting both importations and interstate shipments of nursery stock for injurious insects and plant diseases.[22]

The quarantine movement entered a new phase in 1898, when the United States suddenly acquired a global empire. For reasons to which I will return, this change undercut the restrictionist activities of USDA entomologists and botanists. The department's zoologists, by contrast, were energized by their new imperial outlook. Theodore Palmer began to put North America's ecological position into a historical perspective, and he used that analysis to establish, for the first time, the principle that the federal government should control the nation's biotic borders.

As an adolescent naturalist in California, Palmer was deeply interested in questions of the geographic distribution of animals and, hence, in whether organisms were in their proper places. After receiving his undergraduate degree at Berkeley in 1888, he joined C. Hart Merriam in the "great adventure" of a biological survey of Death Valley. Under Merriam's aegis, he rapidly advanced to become assistant chief of the Division of Biological Survey in 1896, with responsibility for its "economic" programs. A clubman and collector, he established close and enduring ties with leading bird and sportsmen's organizations.[23]

In 1898 Palmer became concerned about the ecological backlash that might result from American imperialism. On the one hand, he argued, the federal government was now responsible for protecting "our island dependencies" of Hawaii and Puerto Rico from new noxious species; on the other, he worried that the "increase in the means of communication" between these islands and North America, consequent on the establishment of imperial ties, could soon result in the "calamity" of the introduction of animals such as the mongoose into the United States.[24]

To convey the depth and scope of these threats, Palmer explained to the half-million recipients of the USDA *Yearbook* the animal aspects of what would later be called ecological imperialism. He described how, since Columbus, Europeans had repeatedly devastated biotas in America, Australia, and the Pacific Islands by introducing pigs, goats, rabbits, rats, and cats. The recent importation of the mongoose into Jamaica had had similar effects. Brought there in 1872 to control rats, the animals soon expanded their diets to include chickens, fruits, and a number of indigenous animals. The decimation of these species had led to a rapid increase in ticks and injurious insects. An erroneous 1892 report that said the USDA was planning to import mongooses to control gophers had led individuals "ignorant of the animal's past record and anxious to try some new method" to obtain specimens privately; these importations had been prevented, but only "by the most strenuous efforts."[25]

Against this backdrop, Palmer described the "ill-directed" efforts of acclimatization societies to import European songbirds. He deplored the introduction of the english sparrow and the starling, noting that the recently introduced great titmouse "is said to attack small and weakly birds, splitting open their skulls with its beak to get at the brains, and doing more or less damage to fruit, particularly pears."[26] He drew two lessons. The particular vulnerability of island organisms to predatory exotics was such that "unusual care" needed to be taken when considering introductions to Puerto Rico and Hawaii.[27] More generally, he argued, "some restriction" should be placed on the ability of private individuals and groups to introduce species that "may become injurious" in the continental United States. "Since," he pointed out, "it has been found necessary to restrict immigration and to have laws preventing the introduction of diseases dangerous to man or domesticated animals, is it not also important to prevent the introduction of any species which may cause incalculable harm?"[28] Power over exclusion would be part of the duties of the Department of Agriculture.

With the passage of the Lacey Act in early 1901, Palmer gained this power. The law was a striking example of the ability of scientific bureaucrats to reshape congressional initiatives. Conservationist congressman John F. Lacey had naively proposed that the Fish Commission expand its hatchery program to include game birds. Lacey's bill included an acclimatization provision: that the government "aid in the introduction of . . . foreign birds." Secretary of Agriculture James Wilson, however, succeeded in completely reversing the bill's thrust. He suggested that it be "broadened" to include not only hatcheries but all kinds of preservation, and (pointing to Palmer's account of the problems caused by the sparrow and the mongoose) he persuaded Lacey that introductions of foreign species needed to be regulated, not aided.[29]

The main purpose of the Lacey Act was stated clearly in its heading: "to enlarge the powers of the Department of Agriculture." The law curtailed the acclimatization societies, asserted political control over biotic borders, and gave the federal government—more specifically, the scientists working for the Biological Survey—the power to establish a comprehensive policy on faunal introductions. It ended four hundred years of biotic laissez-faire.

ECOLOGICAL COSMOPOLITANISM IN THE BUREAU OF PLANT INDUSTRY

The Lacey Act was nonetheless an isolated initiative. During the long tenure of James (Tama Jim) Wilson as secretary of agriculture (1897–1913), the USDA moved in a quite different direction. Wilson believed that American farmers had suffered so much from recent natural and economic disasters because they grew only a few varieties of a few crops. They needed new varieties of staples that were more resistant to disease and better adapted to the wide range of American climates. In addition, cultivation of new crops such as soybeans and sorghum would relieve price pressure on wheat, corn, and cotton. Lastly, Wilson hoped to expand the agricultural potential of the United States by enabling its climatically exceptional regions—Florida, the desert Southwest, and the tropical possessions—to produce citrus, dates, and other products previously imported from abroad.[30]

This program to restructure the country's domesticated flora reflected an ecologically cosmopolitan perspective. For Wilson, the world beyond the continental United States was a source of potential treasure, not danger. He thus initiated a vigorous search for unfamiliar foreign plants, a program whose realization depended on the new activist foreign policy of the McKinley and Roosevelt administrations. It also benefited from the expanding global presence of American capitalists. Wilson sought, in a literal sense, to bring home the fruits of empire.

Wilson's chief implementer was the gifted scientific entrepreneur Beverly Galloway. Thin and myopic, prone to depression, and a fancier of violets, Galloway was prominent among the "hermaphrodites" whom the manly bachelor chemist Harvey Wiley complained were in control of the USDA. Galloway came to Washington in 1887 from the University of Missouri as an assistant in the Section of Mycology. He gradually gathered a young staff (figure 3-2) that combined initiative and camaraderie, and he was continually pushing to broaden his group's responsibilities, budget, and personnel roster.[31] In the mid-1890s, Galloway coordinated his aims with those of other USDA divisions, most notably in his work with Leland Howard on the 1897 convention to control pests and plant diseases. Yet soon after Wilson became secretary, Galloway began to develop a broad program to promote new crops, breed new varieties, and extend the parameters of plant culture.[32] In late 1897 he asked Wilson to appoint David Fairchild, his former assistant, to organize a new office devoted to "systematic plant introduction."[33] It was a perfect, but fortuitous, convergence of organization and individual.

Fairchild, son of the president of Kansas State College of Agriculture, had resigned from the USDA four years earlier to pursue mycological graduate work in Italy and Germany. While crossing the Atlantic, however, he met Barbour Lathrop. Brother of the president of the Chicago Board of Trade (the most powerful force in the nation's agricultural economy), Lathrop, a forty-six-year-old bachelor, had diffuse passions for both intelligent young men and crop improvement. He gave Fairchild $1,000 for a trip to Java to study termite fungus gardens. He then followed the younger man there, pressured him to abandon his research, and invited him on a tour of the Far East, with the vague goal of finding new fruits to grow in California or Hawaii.[34]

Figure 3–2

Staff of the Section of Vegetable Pathology, U.S. Department of Agriculture, 1893. Front: David Fairchild, Beverly Galloway, Walter Swingle. Rear: Joseph James, Theodore Holm, Merton Waite, P. Howard Dorsett. Photograph courtesy of Special Collections, Fairchild Tropical Gardens, Miami, Florida.

Fairchild returned from this romantic odyssey in 1897 with polished manners, valuable social connections, a cosmopolitan outlook, and a love for the tropics. He took the position of head of the Section for Foreign Seed and Plant Introduction (SPI) enthusiastically. His office was immediately involved in coordinating the activities of official plant explorers: Neils Hansen and Mark Carleton traveled to Russia to collect wheat varieties and fruits adaptable to the northern plains, Walter Swingle went to the Middle East for figs and dates, and others fanned out to Japan, the tropics, and Europe. Fairchild also refined the delicate task of appropriating foreign plants. He gave his "special agents" the less sinister title of "agricultural explorers," and he provided them with elaborate official commissions, drawn by hand on heavy parchment with ribbons and gold seals, known vulgarly in the department as "Dago Dazzlers."[35]

From Fairchild's perspective, "systematic" plant introduction was quite different from previous casual attempts, private or governmental, to broadcast unfamiliar varieties. SPI staffers sought to discriminate among the myriad of varieties they confronted and to select only those with particular promising characteristics. They established acclimatization gardens in different regions of the country to nurture unfamiliar types and built up an extensive network of academic and private cooperators willing to propagate and evaluate new specimens. In some cases they also worked—most notably with Carleton's durum wheat—to alter commercial processing and marketing techniques. Finally, they struggled to make Americans' tastes more cosmopolitan. Fairchild was continually pushing prominent people to consume artichokes, avocados, mangoes, and other tropical products regularly. Durum wheat succeeded so well because its primary end product, pasta, was soon popularized by the new Americans born in Italy.[36]

Plant introductions gained greater significance because they were enmeshed in Galloway's broader program for plant improvement. Division staffers Herbert J. Webber and Ernst Bessey emphasized that otherwise unimpressive foreign plants could be valuable raw material in the production of useful hybrids; small Russian apples, for example, could be bred with large American varieties to improve the latter's hardiness. As Barbara Kimmelman has shown, USDA scientists were at the center of American hybridization research in the late 1890s; with access to unprecedented stocks of germplasm, and with great enthusiasm for Mendelism after 1900, they transformed the art of breeding into the science of genetics.[37] SPI provided the material on which the breeders were able to work.

Initially Galloway and his followers were hampered by the USDA's fragmentation. Wilson's interest in bureaucratic consistency had led him to locate plant exploration within the Division of Botany; but that organization's chief, Frederick Coville, was notably unenthusiastic about the work. Staffers continually worried that he was undermining their efforts.[38] Moreover, all the department's research divisions were hampered because the Washington-area hothouses and fields were controlled by the separate Division of Gardens and Grounds, run by the septuagenarian landscaper and Granger, William Saunders. When Saunders died in September 1900, Galloway immediately advised Wilson to merge the USDA's five plant science divisions into a new Bureau of Plant Industry. Consolidation, he argued, would lead to "more sympathetic union" among researchers and would enable them to tackle larger problems. Wilson agreed, and Galloway became head of a botanical establishment with more than two hundred employees.[39]

The scope and direction of Galloway's vision in the early 1900s can be seen in presidential addresses he gave to the American Association for the Advancement of Science (AAAS) and the Botanical Society of America while at the height of his national scientific influence.[40] He argued that the creation of the Bureau of Plant Industry marked a new era for botanists. It showed that governments, faced with the problems of population growth, urban concentration, and the international commercial system, were beginning to see in botany the potential key to the reorganization of the nation's food production system. Scientists could respond to this interest by turning their attention to such "applied" subjects as exploration, breeding, plant diseases, and the co-adaptation between plants and environments. In rhetoric laden with the discipline's longstanding gender role anxieties, he chided his academic colleagues for wanting to keep science "pure and undefiled," and for being "handicapped on every side by a sort of immaculateness." The "fully progressive" botanist, by contrast, was a "man of affairs" able to persuade "keen, analytic, practical men" that his work was worthwhile. These new kinds of botanists, Galloway hoped, would work toward "endless harmonious expansion," in both science and society.[41]

The Return of the Nativists

As Galloway's vision of botany became grander, his willingness to cooperate with the Division of Entomology declined. In 1895, Galloway and entomologist Howard each controlled budgets of about $30,000. Within ten years Galloway gained, through expansion and aggregation, a budget of $740,000, nearly ten times that of the entomologists. Galloway's protégé Fairchild reflected this change in interbureau relations. In 1898 he had expressed conspicuous concern about the problem of "objectionable introductions," pleading that "carefully conducted Government importations" would not bring in pests. That worry soon faded. By the mid-1900s, Frank Meyer, his most intrepid explorer, was sending back a wide range of novelties from China and Siberia. Fairchild began to advertise acquisitions widely in the aptly titled *Plant Immigrant Bulletin*, and he distributed large amounts of exotic nursery stock to a network of private collaborators throughout the country. His office became the center of an international plant exchange network.[42]

During these same years, however, federal entomologists became increasingly anxious about what they came to call "the insect menace." The economic and social devastation wrought by the boll weevil became evident around the turn of the century. Insects were being implicated in the spread of malaria, yellow fever, typhus, and other epidemic diseases.[43] The individual most concerned about defending the country's borders against pests was Charles L. Marlatt, the Bureau of Entomology's assistant chief. Recognizing that imported plant material was the major conduit for new insects, he di-

Figure 3-3

Charles L. Marlatt, about
about 1920. Photograph courtesy
of the National Archives.

rected his campaign against nurserymen and, ultimately, against the chief
advocate for plant immigrants, David Fairchild.

It was a highly personal conflict. Marlatt and Fairchild had grown up together in Manhattan, Kansas, and had come to Washington in the same year;
Marlatt was best man in Fairchild's wedding. Both experienced the Orient.
But while Fairchild traveled widely and easily with Lathrop as a tourist-collec-
tor, Marlatt used his honeymoon in 1901–2 to inspect Japan and eastern
China for scale insect species that might endanger American fruit trees. He
saw the devastation that had resulted from Chinese attacks on Westerners in
the Boxer Uprising, and on his return to America he witnessed the slow death
of his young wife, Florence Brown Marlatt, from an infectious illness she
contracted on their trip. He was thus deeply aware of the dangers foreign
organisms posed to Americans, and he approached his official duties with
great seriousness.[44]

Marlatt took charge of the USDA campaign to control plant imports in
1909. Leland Howard, although a master of persuasion, had been unsuccessful after a decade of efforts to build a consensus in support of a plant quarantine law. Marlatt used different tactics. First he tried to shepherd a quarantine

bill through Congress quietly. When that did not succeed (the nurserymen learned about it after House passage and sidetracked it in the Senate), he shifted to the opposite approach. He decided to raise the consciousness of both influential elements of the public and forces in the government about the danger posed by infested nursery stock. The Japanese cherry trees were heaven-sent for this purpose.[45]

The originator of the cherry tree project was David Fairchild. An enthusiast for *japonisme*, he had privately imported some of the unfamiliar trees (and a Japanese gardener) to adorn his Chevy Chase estate. In 1908 he began to promote the idea of planting a "field of cherries" in the newly constructed West Potomac Park.[46] When the Japanese government appropriated the idea and expanded it as part of their 1909 diplomatic offensive, Fairchild became the liaison, accepting the tree shipment from the Japanese in Seattle, arranging transport across the country in refrigerated railroad cars, and providing planting advice to federal landscapers. Marlatt undercut Fairchild's efforts completely. The Bureau of Entomology held authority to inspect the USDA's own plant introductions.[47] Marlatt used this power to make official Washington aware, once and for all, that exclusion of pests was the highest priority in horticultural commerce.

Japanese leaders recognized the long-term benefits of the tree-planting project and declined to take public offense at the American action. They proposed a second shipment, and this time they countered experts with experts. Representatives of the Imperial Quarantine Service, the Imperial Horticultural Station, and the Imperial University supervised cultivation and assured the USDA that all specimens had been selected from an area free from scale insects, raised in ground free from nematodes, sprayed with fungicides and insecticides, and fumigated twice before packing. This time Department of Agriculture officials approved every tree, and they were planted in 1912 around the Tidal Basin, along the Potomac, and on the White House grounds, "as a living symbol of friendship between the Japanese and American peoples."[48]

From one perspective, Marlatt was still not satisfied. A few years later he complained that this second shipment had introduced "the oriental fruit worm, which is now widespread in the eastern half of the United States, and is occasioning losses estimated well into the millions."[49] He raised no objections at the time, however. His focus was on the larger goal of gaining permanent control over commercial activity, and in 1912 he succeeded in engineering passage of the Plant Quarantine Act. This law gave the power to regulate plant imports to a new Federal Horticultural Board; for nearly two decades, that board was controlled by Marlatt.[50] Between 1912 and 1920, the

Federal Horticultural Board changed American policy regarding the continent's biotic future. The acquisitiveness and cosmopolitanism fostered by Galloway and Fairchild gave way to ideals of autonomy and isolation.

Initially, plant quarantine was a focused program, limited to specific infestations from particular countries. Marlatt, however, did not consider that approach sufficient. On the one hand, organisms innocuous in their natural settings could become rampant in North America; on the other, scientists were still unfamiliar with many dangerous parasites. Chestnut blight epitomized these dangers: American scientists realized only in the 1900s that a valuable native hardwood, and, after Longfellow, a central symbol of the bountiful American countryside, was being exterminated by a strangling fungus. After identifying the pathogen for the first time, they traced it to China and concluded that enthusiasts for chinese chestnut trees had unknowingly imported it a generation earlier. From Marlatt's perspective, "[I]t is the unknown things that you cannot find that we have to protect this country from." He believed that total exclusion was the only way to protect American plants from such a multitude of invisible enemies.[51]

Marlatt's influence was limited while Galloway controlled USDA plant science. In 1914, however, after rising to the position of assistant secretary of agriculture, Galloway left the government for the less pressured setting of Cornell (where he had a nervous breakdown the next year). His successor, William A. Taylor, was more interested in the systematic improvements that geneticists could induce with the material they had than in the hit-or-miss search for new germplasm that he perceived in exploration.[52] Marlatt was soon orchestrating a broad exclusion policy. At a convention of the sympathetic American Forestry Association in January 1917, he called for new security measures. He reviewed the now standard list of foreign "plant enemies," adding the japanese beetle, which had just appeared in New Jersey. He recalled that the "virgin lands of the New World" had originally been free from such pests. Protecting the "standard products of our soil," he asserted, was more important than accumulating the "novelties and curiosities of the plant world for our gardens, lawns, and parks." A few months later he proposed Quarantine #37, a rule designed to end nearly all private importations of plants and bulbs.[53]

David Fairchild, as the chief government advocate for "novelties and curiosities," did what he could to prevent this change. Addressing the same elite forestry audience, he articulated the basic principles of ecological cosmopolitanism. He argued, first, that although any given foreign variety might seem to have minimal value, the total economic worth of plant imports was substantial — and would only be known in the future as a result of continued importation. His second appeal was to American principles of justice and charity: "[I]t would be eminently unfair to assume that because we do not know that these little apple seedlings from the old world or from Japan are as clean and free from disease as any which we can produce in America, they represent undesirable immigrants and should be excluded from the country." He pointed to the inextricable commingling of evil and good: chestnut blight had led Americans to discover new varieties of chinese chestnuts and oriental pears resistant to blight. Finally, he argued—with rhetorical contrast between the ancient and the modern—that no one could hold back the future:

> We can say to ourselves, "let us be independent of foreign plant production. Let us protect our own by building a wall of quarantine regulations and keep out all the diseases which our agricultural crops are heir to and have the great advantage over the rest of the world." But the whole trend of the world is toward greater intercourse, more frequent exchange of commodities, less isolation, and a greater mixture of the plants and plant products over the face of the globe.[54]

Loss of some cherished species was unavoidable; the best that could be done was to foster research, ingenuity, and attention to individual diseases.

Fairchild's opposition was futile. In November 1918 the secretary of agriculture promulgated Quarantine #37. Ostensibly this prohibition on private imports had no impact on the Department of Agriculture's own plant introduction program. In fact, however, the new regime subverted Fairchild's activities. It put the Office of Foreign Seed and Plant Introduction under suspicion as a potential source of disease, disrupted the delicate exchange networks that Fairchild had built up, and forced him to focus on the processing of commercially promising introductions. The *Plant Immigrant Bulletin* was terminated for lack of funds, but construction began on a new "Plant Detention Station" outside Washington, where "plant immigrants will be received and carefully grown, watched and propagated, to be sure that all alien enemies are excluded." In 1924, at age fifty-five, Fairchild gave up. He retired to Florida and joined a new private patron—meat industry heir Allison Armour—on a yachting expedition in search of new plants. Galloway, who had quietly returned to Washington as Fairchild's assistant in 1916, remained behind to maintain a policy of "rational plant exclusion."[55]

Marlatt, on the other hand, ended his career in bureaucratic triumph. He succeeded Leland Howard as chief of the Bureau of Entomology, and in 1928 also became head of the new Plant Quarantine and Control Administration. When the mediterranean fruit fly appeared in Florida the next year, Marlatt obtained emergency funding of nearly $6 million, and near totalitarian power, to fight it. He banned interstate shipment of all Florida produce, prohibited planting of vegetables in infested areas, required picking of all fruit prior to ripening, and organized a massive spraying campaign. Within a year the insect had been wiped out.[56]

To be sure, Quarantine #37 did not end ecological imperialism in North America. Biotic inequality and the potential for movement continued, resulting in introductions ranging from dutch elm disease to walking catfish and asian longhorned beetles. Yet the convergence of scientific thinking and state power that occurred between 1895 and 1920 did mark the end of the United States' status as an ecological colony. Since that time ecologically imperial forces have been hedged by at least some of the claims of political sovereignty.[57]

ECOLOGICAL INDEPENDENCE AND IMMIGRATION RESTRICTION

Thus far this account has been limited to the nonhuman. Yet it should be clear that attitudes about foreign and native organisms were intimately linked, through both everyday experience and analogies of policy, to views on "alien" and "native" humans.

Commonplace symbolic connections between geographically identified organisms and humans were omnipresent and powerful. Americans perceived the english sparrow as an avian Cockney pushing aside larger but better-mannered American birds. The gypsy moth's devastation of respectable neighborhoods confirmed casual prejudices about southeastern Europeans. The introduction of pulpy tropical fruits, carrying the aromas of the Far East, really did increase sensuality among Americans raised on tart apples and bland pears. These specific associations could operate at high cultural levels, and their implications could be molded consciously: the Japanese cherry trees were, by turns, tokens of equality, infectious interlopers, and sanitized ambassadors. Their successful introduction did, to some degree, neutralize the assertion of American racial superiority implicit in the Gentlemen's Agreement.

The density and pervasiveness of these thickets of meaning, however, make their overall import difficult to assess. As in the case of the nonhuman alone, we can gain a clearer conception of the efficacy of these modes of thought by focusing on the transformation of policy. The crucial arena in which ecological ideals intersected with demographic designs was, of course, immigration. The tendency in general histories has been to emphasize the opposition between those who welcomed newcomers and those who wanted to keep them out, and to view the Immigration Act of 1924 as the culmination of a decades-long narrow-minded campaign for the exclusion of those who were different.

By looking at American immigration policy through the lens of the debates over ecological independence, we find a more complex, but more meaningful and relevant, narrative.

For a century after the founding of the United States, Americans had the same mixture of interest and distrust, and the same laissez-faire policy, toward foreign humans as they did toward foreign plants and birds. At the end of the nineteenth century, however, federal passivity about immigration was challenged by the same two competing approaches discussed above. Like Palmer and Marlatt, nativists (in the original, political, meaning of the term) had little interest in the possible benefits that foreigners might bring to the United States. On the other hand, they were deeply concerned about the damage that aliens might wreak on American life. Like the ornithologists and entomologists, they sought to deal with these problems by excluding entire groups from the United States. The paradigm of the nativist approach was the Chinese Exclusion Act, passed at the insistence of "native" California workingmen in 1882, a year after the state's plant quarantine law. It kept out all but a trickle of Chinese laborers, and by excluding women, it was designed to prevent the Chinese "race" from establishing itself permanently in North America.

Moderates on immigration were more cosmopolitan. They were comfortable with the entrance of some foreigners from most countries, and they recognized the value of immigrant merchants and laborers to the American economy. In place of blanket exclusions, they envisioned a policy of individual selection, similar in its fundamentals to that pursued by Galloway and Fairchild with plant immigrants. Botanists exercised discrimination constantly in their collecting choices, and they used plant introduction gardens to cull less desirable species and varieties. Bureaucrats dealing with humans had relatively little influence over the composition of the immigrant stream. As a consequence they concentrated on sorting those who had arrived. They proposed to assess each entrant's inherent quality, and then to welcome superior specimens, treat those with minor defects, and send inferior beings back to their homelands. A policy of selection was never instituted for Asians. But it provided the rationale for the operation of Ellis Island, the United States' major immigration facility, during its most important years.

Ellis Island was built in the early 1890s to improve the government's ability to collect taxes from immigrants. Its mission was revolutionized, however, by Theodore Roosevelt. As a nationalist and a naturalist, Roosevelt was deeply concerned about the biotic future of the country—including the preservation of large herbivores (on which he closely collaborated with Merriam and Palmer of the Biological Survey), the fecundity of European-Americans, and the quality of immigration.[58]

On becoming president in 1901, Roosevelt sought to transform Ellis Island into a site for immigrant selection. His image of this artificially augmented mudflat became clear in his choice of immigration commissioner for the Port of New York. William Williams was in most respects a typical patrician Wall Street lawyer. From Roosevelt's standpoint, however, he had important relevant experience: a decade earlier, as a State Department employee working in close consultation with the USDA Division of Economic Ornithology and Mammalogy, he had defended American management of the Pribiloff Islands fur seal herd before an international arbitration tribunal. He was thus familiar with problems of migration, understood the management of large semi-confined populations, appreciated the principles of selection, and knew how to use experts to implement and bolster policy.[59]

During his two terms as commissioner (1902–5, 1909–14), Williams developed procedures to cull the Ellis Island herd. He believed that at least 25 percent of entrants were "of no benefit to the country" and would "lower our standards."[60] The key to finding that bottom quartile lay in the system of medical inspection. Williams eliminated the informal "corrupt" arrangements whereby immigrants had previously bypassed the doctors. He doubled the medical staff, formalized the details of the inspection procedure, specified actionable defects, and introduced new tests—most notably the eversion of every entrant's eyelid with a buttonhook in the search for the common eye infection, trachoma. In 1909, when Williams returned for his second tour, his staff introduced jigsaw puzzles as a means to pick out feebleminded entrants. The American Genetic Association, whose president was David Fairchild, strongly supported these measures to improve the quality of immigration.[61]

The disruptions of World War I ended American efforts to manage European immigration through individual selection. The interruption of transatlantic commerce made the absence of newcomers seem the norm for the first time in decades, and wartime "100% Americanism" heightened their alienness. When immigration suddenly recommenced in 1919, on a scale magnified by the war's devastation, a political consensus rapidly formed around the need for exclusionary measures similar to those directed against Asians. In 1921 Congress established a cap on all European immigrant groups, limiting entry to 3 percent of those already resident in the United States. This was refined and made permanent in the Immigration Act of 1924.[62]

Discussions about the biological aspects of immigration restriction have focused on the extent to which the details of the system established in the 1920s were racist.[63] They clearly were, but such racism was a historically secondary phenomenon. The more fundamental change in the biology of immigration was the introduction of a stringent limit on the total number of European entrants. In 1913, nearly 1.2 million foreigners came to the United States; twenty years later, only 23,000 arrived. Federal policy made the United States, for the first time, a nation made up almost entirely of "natives." This new demographic policy had numerous causes. But the scientific bureaucrats already involved in making the United States ecologically "independent" provided both a framework for conceiving this transition and some of the impetus for realizing it. For better or worse, they worked to shape American life.[64]

PART II

SPECIALIZATION AND ORGANIZATION

WHITMAN'S AMERICAN BIOLOGY

In July 1890, Charles O. Whitman stood on the shore of Vineyard Sound in Woods Hole, Massachusetts, where he could watch the ferry taking Methodists to their annual camp meeting and vacationers to their summer colonies on Martha's Vineyard. Whitman was happy where he was. The year before, after two decades of wandering and uncertainty, he had become professor of biology at the newly created Clark University in Worcester, a few hours from Woods Hole by train. More important was the fact that his own camp meeting—or summer colony—for scientists, the Marine Biological Laboratory (MBL), was becoming a success. In the two years since its creation in 1888, it had grown from fifteen participants, most of them women, to a group of forty-seven that included Edmund B. Wilson, Thomas Hunt Morgan, and faculty from Harvard, Princeton, and MIT. To further broaden the influence of his undertaking, Whitman determined to publish a number of the general talks being given each Friday evening the laboratory was open. The first collection of *Biological Lectures Delivered at the Marine Biological Laboratory of Wood's Holl* appeared in early 1891.[1]

Whitman began this volume with his own manifesto. "Specialization and Organization, Companion Principles of All Progress—The Most Important Need of American Biology" expounded the scripture of Herbert Spencer.[2] It began with the evolutionary philosopher's well-known premise that "a society is an organism" and "an organism is a society." Most of the essay laid out the case, by 1890 familiar to any zoologist, for the second of these propositions. It described how organic evolution had been a gradual progression from free but lowly "nomadic" protozoa, to cells that "herded together" to form colonies, thence to the "Hydra community of cells," and finally to the complex division of labor within the higher organisms. "Progress in the organic world," Whitman argued, "is always from the less to the more heterogeneous."[3]

This natural, progressive process of evolution provided Whitman the template on which he built his real argument. He asserted that the evolution of science obeyed the same laws as the evolution of organisms. "Cosmogonists" who had "engaged single-handed with all the mysteries of the universe" had given way to the present, seemingly rapid breakup of scientific activity into ever smaller specialties. Whitman celebrated this process because "concentration of energy" had intensified the productivity of scientific labor. He

Figure II–1
Charles O. Whitman, about
1882. Photograph courtesy of the
University of Chicago Archives.

knew, however, that academic disciplines were not as well defined and con-
tinuous as the cells of a hydra. If more and more scientists were pursuing increasingly specialized inquiries, how would the "union" of American biology be maintained?

Whitman denied that biological science was "flying into disconnected
atoms." But he believed that it was important to counteract the "danger of
narrowness that may lurk in the tendency to specialize." The solution would
be found in "that kind of organic association which permits each unit to
work for itself while making it the servant of all the rest." Journals, which
linked geographically dispersed writers and readers and circulated internationally, were essential, but "the unity of action in so extended a body cannot be complete." What was necessary was a way to bring together the leading American naturalists and place them in "intimate helpful relations." A
"national marine biological station"—the MBL—would do this; as such it
was "unquestionably *the* great desideratum of American biology." The
result, Whitman concluded, would be to accelerate "that '*moving equilib-
rium*' of our specialized forces which constitutes progressive scientific life."[4]

Figure II–2

Woods Hole scientific institutions, early 1890s, from the southeast. The plain twin structures (actually one building) to the right of the water tower housed the MBL. The Fish Commission occupied the Queen Anne revival facilities near the flagpole. In this photograph, the institutions appear equivalent in size and prominence. In fact, as figure 2–4 shows, the Fish Commission facilities were significantly larger. In addition, the MBL was separated from the harbor by a street and strip of land, while the Fish Commission dominated the waterfront. Photograph by Baldwin Coolidge, Courtesy of the Society for the Preservation of New England Antiquities, Boston, Massachusetts.

This essay would certainly have irritated Spencer Baird. Whitman was presenting the idea of a seaside national center for American biologists in Woods Hole as new and describing the MBL as a self-sufficient, cooperative venture. He ignored the prior existence of the Fish Commission's laboratory, which Baird had established in that village only a few years earlier, in part as an effort to bring zoologists from around the country together. In 1890, in fact, faculty and graduate students from Harvard and Johns Hopkins, among other leading academic institutions, were working there, not at the MBL. Moreover, the MBL depended, in fundamental ways, on support from the government: the Fish Commission supplied Whitman's laboratory with both organisms and pumped-in sea water. The relationship between the two institutions was evident in their architectural juxtaposition, which can be seen by examining figures II-2 and 2–4 together.

An equally important, if less obvious, problem, from Baird's perspective, would have been the vacuousness of Whitman's characterization of "American biology." For Whitman, biology was "American" only in the location of its practitioners and in the hope that this imagined scientific community would produce some significant intellectual advance. It was not American in the more robust ways that mattered to the federal naturalists—either through attention to the organisms of the North American continent or through devotion to national purposes.

I have framed Baird's objections hypothetically because he was not on the scene, able to make them. He had died, in Woods Hole, in August 1887. In the summer of 1890, his successors were unable to focus on their laboratory: they were back in Washington, answering questions from a Senate committee investigating waste, fraud, and abuse at the Fish Commission, with particular attention paid to the facility at Woods Hole.[5] The laboratory had become a directionless white elephant. Whitman, through his manifesto, was providing a new basis for cooperative work at Woods Hole and, he hoped, for American biology.

This shift was a particularly dramatic instance of a much broader process occurring in the decade around 1890; namely, the transition in the leadership of science in the United States from the federal government to the rapidly expanding universities. Within this new setting, Whitman's Spencerian formulation was apropos. Individualism and specialization were ascendant, and the possibility, bases, and meanings of scientific "union," or organization, were problematic.

Part 2 of this book examines the situation that Whitman described in his manifesto. The present chapter deals, first, with the decline of federal leadership in the life sciences in the wake of Baird's death, and then with the burgeoning of specialization—the proliferation of scientific entrepreneurship. Its keynotes are individualism and enterprise. It describes the variety of places for work, kinds of activities, and relations between scientists and their supporters. It emphasizes the diversity of communities arising in biological science domains during this period; that is, it conveys the extent to which those sciences were in fact, as Whitman expressed it, a congeries of "disconnected atoms." The next chapter considers the conventional efforts of American life scientists to coordinate their activities. These attempts to develop societies, journals, and university programs were organized, in large part, around the keyword "biology." Success was mixed. The final chapter in this part looks in some depth at Whitman's ingenious and unique solution to the problem of specialization: his ability to create, in Woods Hole, a seasonal scientists' community, which combined the characteristics of the old camp meeting and the new summer resort. Woods Hole made "American biology," of the sort Whitman envisioned, possible, and made it more meaningful than, for example, "American chemistry" or "American psychology." Biologists gained a great deal, both socially and intellectually, from the Woods Hole experience.

Yet these gains were not without costs. Whitman's American biology was different from, and less meaningful than, the science imagined by the leaders of the Biological Society of Washington. The insulation of biologists from the larger problems of American life had negative consequences for both life and biologists. It is more difficult, retrospectively, to identify missed opportunities than to point to achievements. An arresting exemplar of a chance for national influence passed by, however, can be found in Whitman's migrations to Woods Hole.

Life Science Initiatives in the Late Nineteenth Century

THE ECLIPSE OF THE FEDERAL NATURALISTS

In chapter 3 I described how government scientists such as Beverly Galloway and Charles Marlatt exerted considerable influence over the development of the United States in the early twentieth century. But these people no longer made up a coherent scientific community, nor did they provide national cultural leadership comparable to their Gilded Age predecessors. In part, the federal naturalists were eclipsed by the expansion of other centers for science, most notably the new research universities. Some of their problems, however, were specific to themselves. A comprehensive analysis of the changes in federal science at the end of the nineteenth century is beyond the scope of this book, but it is possible to outline the events that led the bureaucrats to withdraw into their separate, often competing bureaus.[1] We can then begin to sketch the other American sites for life science.

The expansion of federal naturalists' activities was under challenge beginning in 1884. In that year, Congress established a joint committee (the so-called Allison Commission, named for William B. Allison, the Republican senator who chaired it) to investigate the organization and activities of a significant part of the government science establishment. On its face, the Allison Commission had relatively little to do with the naturalists. Its mandate, set by the structure of congressional committees, included the Coast Survey, the Weather Bureau, and the Navy's Hydrographic Office. The only major naturalists' organization included was the new Geological Survey; the commission took no official notice of any of Baird's enterprises, the Department of Agriculture, or the Army Medical Museum.[2]

Yet the commission was more of a challenge than it appeared. One of its six members was the wealthy Massachusetts congressman Theodore Lyman. In addition to being Louis Agassiz's student and then his son-in-law, and a member of the National Academy of Sciences, he had long led the Massachusetts Commission of Inland Fisheries and had opposed Baird's views on the decline of fish stocks. Immediately after his appointment to the Allison Commission he gave the presidential address to the annual meeting of the American Fish Cultural Association (held at the National Museum), where he pointedly questioned whether the Fish Commission's involvement in fisheries management was economically justifiable. Luckily for Baird, Lyman was defeated for reelection in November 1884 and then withdrew from public life due to illness.[3] But during the next two years, his brother-in-law Alexander Agassiz took up the opportunity for criticism the Allison Commission offered.

Agassiz privately advised committee minority member Hilary Herbert and publicly attacked John Wesley Powell in *The Nation*. In addition, he took the opportunity to complain directly to Baird about the "sophomoric spread eagle bosh" that Powell was presenting to the commission—in particular, Powell's argument that sending all the publications of the scientific bureaus to public libraries would educate the American people. Baird, who was then in the midst of producing, with Goode, their lavish volumes on fish and fisheries "for the use of the reading public," did not have to strain to see that he could come under the same strictures.[4]

Baird already had worries closer to home. The new Democratic administration of Grover Cleveland, devoted to economy and integrity, and antagonistic to bureaucrats who had grown fat through Republican patronage, saw in Baird an inviting target. In 1885, the new Treasury Department auditor, J. D. Chenowith, began questioning the expenses of the Fish Commission, in particular at Woods Hole. Baird's semi-official arrangements with private colleges and his decision to erect on the waterfront a summer "residence" for himself, his family, and his staff, even before building the government's laboratory, came in for particular scrutiny. Baird argued that the Fish Commission was a beneficiary of the private scientists' work, and he claimed that, given the expense and scarcity of summer lodging in Woods Hole, his construction of a "dormitory" had been an economy measure. At the same time, he mobilized the commission's friends to defend its value and his probity. Most effectively, he used his influence in the Senate to delay Chenowith's confirmation as auditor. Confronted by the Fish Commissioner's raw political power, the auditor ate humble pie: he traveled, hat in hand, from the Treasury Department across the Mall to the Smithsonian, affirmed that there were no problems in the Fish Commission accounts, and publicly begged Baird to withdraw the opposition to his appointment.[5]

Baird's victory was pyrrhic. His appropriations plateaued after 1884 (see figure 2–2), and his effectiveness as a leader declined as heart problems sapped his energy. In 1887 he traveled, for what he certainly suspected was the last time, to Woods Hole, and on 16 August, confined to a wheelchair, took a last tour of his domain. He spoke softly to scientists, collectors, mechanics, and sailors, all working there because of his initiatives. He watched the new steam-powered pumps sending seawater to the laboratory tanks, hatchery operations, and public aquaria. The boards creaked as he was wheeled down the government dock, past his ships with their tackle and catches. He could look with pride across the Great Harbor to the Pacific Guano Works, where menhaden were being rendered into fertilizer for southern cotton growers; the factory, smelly as it was, exemplified the new ways that he hoped fish could contribute to American prosperity. He could gaze in the other direction, out into the Atlantic toward the Gulf Stream, visited intermittently by the *Albatross*. After returning to the residence, he slipped into unconsciousness and died three days later.[6]

Within a few months Baird's empire broke up. The National Museum was least affected. With its own building, growing collections, and regular congressional appropriation, it possessed considerable bureaucratic security. G. Brown Goode had already been in day-to-day charge for some time, and he had received clear authority in early 1887 through his appointment to the new position of associate secretary for the museum. The rest of the Smithsonian was more problematical. The regents sought to go beyond the work of the naturalists and to reconnect with Joseph Henry's vision by appointing a physical scientist as secretary. Their choice, however—astonomer Samuel Langley—was poorly suited to build on the work of his predecessor. Personally shy and a Washington outsider, he made little effort to redirect the energies of the denizens of the Cosmos Club. Instead, he drew the Smithsonian away from its surroundings. One of his first acts, for example, was to rescind Baird's longstanding promise that the Geological Survey could, when it obtained an appropriation, build its headquarters on the Smithsonian's section of the Mall.[7]

The real change occurred at the Fish Commission, however. President Cleveland indicated his concerns by proposing that a Treasury Department lawyer succeed Baird; he apparently dropped this idea in favor of Goode's temporary appointment and a law making the commissioner a regular federal employee.[8] He then chose for the job, not a scientist, but the fish culturist and southern Democrat Marshall McDonald. McDonald was ordered to make the commission more immediately useful by emphasizing the search for fish and the operation of hatcheries and cutting back scientific work, whose relevance to the industry was at best indirect.

This change in direction had particular impact on Woods Hole. Baird had planned to use the laboratory to coordinate the activities of the nation's naturalists, but he had justified it to Congress on the narrow grounds that the *Albatross* needed a base for its expeditions to the Gulf Stream. When that ship was sent to the Pacific in late 1887 to search for new fishing grounds, the Massachusetts facility lost its primary government mission. It was merely a site for individual investigators, most of whom were professors. Between 1887 and

1890 the budget for scientific work at Woods Hole was cut nearly 50 percent. When the commission was investigated in 1890, critics attacked the laboratory's lack of practical purpose and its use as "an annex to the workshops of the National Museum." Senators approved the reduction in its appropriations.[9]

These institutional changes were ultimately accompanied by a shift in the Washington intellectual climate. The idea that wide-ranging explorations of the implications of evolutionism would advance public purposes faded after 1890. In early 1894 Goode and Langley removed essays by Lester Ward and anthropologist W J McGee from the popular supplement to the Smithsonian *Annual Report*, on the grounds that their discussions of the mind-body problem and human origins were speculative and materialistic. While Ward noted that most newspapers were on his side, this action gave federal naturalists a clear signal to refrain in the future from "metaphysics."[10]

What underlay these shifts? In chapter 2 I emphasized how creative the Washington naturalists had been in turning the Gilded Age into an golden age—culturing a lush growth of scientific activity on the rocky substrate of the federal bureaucracy. This growth had depended on the success of the naturalists in gradually increasing their appropriations and hence their numbers, and on their ability to maintain considerable autonomy in selection of people and problems. It had also required careful balancing between daytime and evening activities, between empirical investigations and speculative discussions, and between competitive bureaucratic entrepreneurship and scientific communitarianism. The vitality of these scientists depended on their ability to balance routine labor and intellectual free play, to interact across bureau boundaries, and to offer ambitious but unformed young men a reasonable hope that hard work would lead to advancement, both intellectually and bureaucratically.

When the growth of appropriations slowed in the mid-1880s, and both bureau and individual autonomy were increasingly put into question, the balance that the naturalists had maintained among their different activities and identities became untenable. Day jobs, data collection, and bureau identity came to predominate. The optimism of the naturalists faded—for some in the aftermath of Baird's death, for more with the intellectually and economically depressed circumstances of the second Cleveland administration (1893–97). Bohemians like Elliott Coues were shunted to the margins.[11] Originally ambitious technical assistants stagnated into middle age and beyond, like female employees, in dead-end jobs as bibliographers or curatorial assistants. Hornaday left the National Zoo after fighting with Langley in 1890. The entomologist Riley, who had married into money, quit the government in 1893 when his advancement at the Department of Agriculture was blocked. A year later, strained by repeated congressional attacks, Powell resigned from the Geological Survey. Goode's colleagues were convinced that his death at age forty-five in 1896 was the consequence of overwork.[12]

With the return of prosperity and the Republicans in 1897, federal science reemerged, but on a different foundation. The Department of Agriculture scientists discussed in the preceding chapter, as well as Geological Survey leader Charles D. Walcott, recognized that the road to continuously growing appropriations and hence bureaucratic success lay in linking research to well-defined problem areas of interest to influential interest groups. In the meantime, however, larger issues, and other people, were on their own.[13]

Concentrate on what is important and everything else is on its own.

From Agassiz to Burbank: A Cross-Country Tour

Even at their zenith, the federal naturalists were the visible guides of only a small part of the life sciences in the United States. Outside Washington, leadership anywhere was much less apparent than the diversity of initiatives that Charles O. Whitman hopefully characterized as "specialization." These included what were, to outsiders, a bewildering proliferation of neologistic identities, ranging from agrostology, bacteriology, cytology, dendrology, and ecology, to limnology, morphology, neurology, oology, and pomology, and so on. Enthusiasts for these subjects were scattered throughout the country (the Old South excepted). They had varied kinds and levels of education. Places of work ranged from university laboratories and hospitals to doctors' offices, farms, and nursery businesses. To be sure, American naturalists before the Civil War participated in a wide range of activities. As I argued in chapter 1, however, most of those endeavors were either local or tributary to a few national leaders, most notably Gray and Agassiz. The novelty in the last decades of the century was the ability of many young men, and a few women, to establish new kinds of enterprises that could acquire national, and even international, visibility. It is tempting to follow Whitman and assume that these enterprises were specializations *within* a developing discipline of biology. Such a belief ignores, however, the substantial amount of contingency involved.

To convey the diversity of American life science activities, I will describe these initiatives as they would have appeared to a hypothetical English scientist, with money, time, and broad interests in things organismal, on a tour of the United States in the year of Baird's death. The narrative is grounded loosely in the American tour that the naturalist Alfred Russel Wallace made in 1886–87; his actual itinerary, however, was shaped by family concerns, his modest income, and an inept lecture agent, and was thus more limited than what I present.[14] Like a tourist, I will expend the most energy and provide the greatest detail at the start; later destinations will be treated, not comprehensively, but to pick up what was missed or to provide comparison. To avoid repetition, I will mention only briefly the already discussed nation's capital and the "organizing" enterprises that will be examined in depth in the next chapter.

Massachusetts as Microcosm

Europeans had various points of entry to the United States in the late nineteenth century. Scientists not headed to Washington had only one: Boston. Since the 1840s it had been the most important locus for private interest in and work on organisms. It was not representative, but it was exemplary—expressing, in miniature, the national situation.

Historians who have studied Boston's Brahmins have argued that this elite—dominant in the nation's economy and politics in the first half of the nineteenth century—declined after the Civil War relative to more aggressive New York bankers and Irish-American politicians. This change is said to have resulted in, on the one hand, a shift in their attention to science and culture, and, on the other, an outlook that was increasingly conservative and limited.[15] Life science was an exception to this pattern. Chapter 1 showed how Brahmins' antebellum interest in botany and zoology had been part of their national economic and political leadership. These areas continued to develop after the war because they were supported by Bostonians who were deeply involved in national financial and industrial expansion.

Still, Boston dominance in some life sciences waned. In part this resulted from problems in managing the transition from the generation of Gray and Agassiz to the one that followed. An additional factor was the difficulty that any private individuals had in competing with the federal government in the years around 1880. It was not, however, the result of conservatism or defensiveness. Rather, the great problem was that Bostonians were profligate: they sought to multiply entities and opportunities to open a wide range of avenues of expression and influence. In contrast to the period before the war, life scientists in Boston suffered most from a lack of concentration—both financial and intellectual.

A prosperous scientific tourist would probably bypass Boston's crowded commercial center in order to find lodging at one of the numerous good hotels in the Back Bay. Perhaps the most appropriate would be the Agassiz Hotel on

Commonwealth Avenue. The name not only honored the great zoologist, but also recalled that, in the 1850s, Massachusetts leaders had justified the transformation of what was a public wetland into a residential enclave for the upper classes by using part of the proceeds from land sales to construct Agassiz's Museum of Comparative Zoology (MCZ), along with the Boston Society of Natural History and the Massachusetts Institute of Technology.[16]

The MCZ, across the Charles River in Cambridge, would be first on our tourist's list of sites, strung together here in one (impossibly) long day of visits. Louis Agassiz had used the Back Bay proceeds and private donations to construct a four-story building a few blocks from Harvard Yard in 1859. A renewed appeal to the state enabled him to double its size a decade later, but the building was unfinished and nearly all its objects were still in storage when he died in 1873. Louis's thirty-eight-year-old son Alexander took over management of the museum, as both a filial duty and as way to provide a respite from work as superintendent of the Calumet and Hecla copper mines. He used his ample resources, and those of his partner-relatives, to double the size of the building again and to support a curatorial staff that struggled to maintain the collections. He continued the series of expensive *Memoirs* and *Bulletins* begun by his father in 1863; by 1887 more than twenty-five of these volumes had appeared. Although he had severed most of the official ties between the MCZ and the state of Massachusetts, he opened a number of exhibition galleries within the museum building to the public. Mary P. Winsor has emphasized how little the collections of the museum were actually used for research, but a casual visitor could only have been impressed by the institution's growth and orderliness.[17]

On the fourth floor of the MCZ, our visitor would find a second, different group of workers. In 1886, Harvard president Charles Eliot appointed the mild-mannered zoologist Edward L. Mark professor of anatomy in the college. His primary responsibility was to teach undergraduates, especially those with medical ambitions. In addition he supported a small group of graduate-level research students. They studied, not systematics, but microscopic anatomy and embryology. This was a new area with great prestige in Germany, where Mark had received a Ph.D. It was also, by comparison with the museum, inexpensive. Mark's "Zoological Laboratories" and lecture hall occupied only four rooms; the laboratories contained a few expensive microscopes and a microtome, but not much more. Agassiz subsidized Mark and his students by giving them space and enabling them to publish in the MCZ's *Bulletin*.[18]

Searching out Harvard's botanists was a considerably more arduous task. Insider historian Samuel Eliot Morison later claimed that personal animosities had led cryptogamist William G. Farlow, physiologist George L. Goodale,

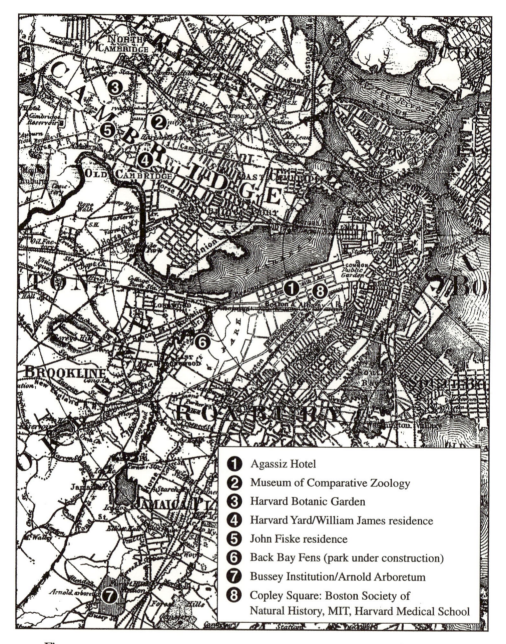

Figure 4–1

Boston and Cambridge, with sites mentioned in text. Adapted from "Map of Boston, for 1887. Published Expressly for the Boston Directory, by Sampson, Murdoch & Co. 155 Franklin St. Boston."

Figure 4–2

Museum of Comparative Zoology, mid-1880s, viewed from Oxford Street. Compare the size of this building with both the American Museum of Natural History (figure 4–4) and the Johns Hopkins University Biological Laboratory (figure 5–2) to gain a physical sense of Harvard's scientific preeminence. Photograph courtesy of the Ernst Mayr Library of the Museum of Comparative Zoology, Harvard University, © President and Fellows of Harvard College.

and dendrologist Charles S. Sargent to each "entrench himself in his special herbarium, laboratory, or arboretum," and he argued that this diversity prevented development of "a well-rounded program." A visitor in 1887, however, would see only a number of different enterprises.[19]

Goodale was most convenient. Though insignificant as a researcher, he was an urbane gentleman who lectured smoothly to large numbers of undergraduates and attracted support from wealthy female flower lovers. An ally of Agassiz, he was supervising construction of the Harvard Botanical Museum within the MCZ complex in 1887. He was negotiating with Brahmin patroness Elizabeth Ware, and Bohemian glass artists Leopold and Rudolf Blaschka, to acquire the spectacular display of glass flower models that, more than a century later, continue to draw visitors to the museum.[20]

Farlow had a separate operation a short walk away. He was a serious academic scientist who, after completing his medical degree in 1870, traveled to Scandinavia and Germany to learn the new techniques and problematics associated with the microscopic investigation of reproducing cells. During the 1870s he had worked on marine algae, and he cooperated with both the Fish Commission and the Department of Agriculture on immediately useful science. On becoming professor in 1879, however, he decided to emulate Gray and thus turned largely to taxonomy. He soon built his herbarium into the country's major repository for mosses, lichens, and fungi.[21]

The septuagenarian Gray was in fact still working. The Harvard Botanic Garden was a fifteen-minute walk west from the MCZ. While Goodale managed the garden, Gray continued to control the herbarium (attached to his home), from which he published and presided over the taxonomy of higher plants nationwide. In addition, he had gained a renewed reputation as a scientific philosopher with the publication in 1876 of *Darwiniana*, a compilation of his writings on evolutionary issues. Debates over the interpretation of evolution were heating up in Anglo-American circles in the latter 1880s, and Gray's unusual emphasis on both natural selection and on theism was, if not accepted, seen as a significant intervention.[22]

Attention to Gray's philosophical work would lead our tourist to realize, however, that he had not yet seen the most important evolutionary theorists in Cambridge. He thus went back to Harvard Yard to take a walk with William James. Twenty years earlier, James had accompanied Louis Agassiz to Brazil and soon after had been moved nearly to suicide by the bleak message he found in Darwin. After recovering, he had expounded and critiqued Spencerian evolutionism for Harvard undergraduates as part of the college's natural history program. By 1887, however, he had shifted to the philosophy department and was completing his massive *Principles of Psychology*; there, problems of evolution were merged with the larger issues of free will and the relations between mind and body.[23]

James's junior colleague, Josiah Royce, had taken over the grander aspects of evolutionary theory. Our visitor could ask to visit his course, "Monism and the Theory of Evolution in Their Relation to the Philosophy of Nature," an exposition that would reach a wider audience a few years later as *The Spirit of Modern Philosophy*. One of the students in the course that year was, in fact, the young Herbert Croly.[24] A still more popular evolutionary writer was Harvard professor of geology Nathaniel Shaler; his notorious gut course ("all the geology necessary to a gentleman") and his contributions to literary magazines explored the relations among biology, environment, and civilization, with regionalist and racist emphases.[25] Crossing Harvard Yard to the west, our tourist could then call at the home of philosopher-historian John Fiske. Fiske had recently become a notable proponent of evolution-based Christianity and was in the midst of arguing that the formation of the federal union in Philadelphia in 1787 was the culminating event in the development of human sociability.[26]

A visit to Fiske would put our visitor in a not-unanticipated quandary. Such broad conceptions of the meaning of evolution and the study of life could easily lead him into an indefinitely extended exploration of Cambridge investigations of religion, literature, history, and social change.[27] Deciding where life sciences ended and other subjects began seemed arbitrary but practically necessary. To make his itinerary manageable, our tourist would need to make an act of will to leave Cambridge for sites that were more explicitly organismal. In 1887, this would mean crossing the Charles River by carriage on the North Harvard Street Bridge, meandering south past the polluted mudflats that landscape designer Frederick Law Olmsted was beginning to construct as the natural environment of The Fens, and then through Brookline and the large estates around Jamaica Pond.[28]

There he would find a fossilized deposit of the ideals and activities of an earlier generation: the Bussey Institution. In 1842, Benjamin Bussey had bequeathed Harvard his country estate and the income from commercial properties in Boston to establish an agricultural school. Testamentary restrictions had kept the university from taking possession of the estate until 1869, and even then it had to appoint Bussey's grandson-in-law "instructor in farming" at a substantial salary. Searching for an appropriate mission, the Bussey's leaders dismissed the idea of training farmers' sons in their "trade" and emphasized instead the links between husbandry and science. They built an imposing Gothic Revival building, accepted donations from the Massachusetts Society for the Promotion of Agriculture, and initiated a research bulletin; but their optimism ended abruptly in 1872 when a fire destroyed much of Bussey's commercial real estate and thus cut their income significantly. By 1887 the Bussey Institution was an embarrassment, with one professor and a handful of mostly nonmatriculated students. In his annual report the following year, President Eliot noted sardonically that the institution had "settled down to a well-defined mode of life," "keep[ing] its buildings insured and in repair, and its farm from deterioration."[29]

Still, a trip to Jamaica Plain would be worthwhile because another institution was growing, around the moribund Bussey, which expressed the interests of contemporary wealthy Bostonians. In 1872, one-third of the Bussey land had been designated the Arnold Arboretum and had been put under the control of the stolid but determined Charles Sprague Sargent. Sargent had shown

Figure 4–3

Municipal laborers constructing Hemlock Hill Road, at the western end of the Arnold Arboretum, 1889. Courtesy of the Photographic Archives of the Arnold Arboretum. Photograph by the Boston Park Commission, 1889.

no interest in botany, or anything else, as a Harvard undergraduate, finishing eighty-eighth out of ninety in the class of 1862. But following three years of service in the Union army and three more touring Europe, he became deeply interested in estate horticulture. Managing his banker-father's Brookline plantings brought him into contact with the historian and amateur horticulturist Francis Parkman, as well as other influential suburban landowners. After being tapped to head the new arboretum, established by Harvard with a bequest (less restrictive than Bussey's) from New Bedford banker James Arnold, Sargent took private lessons on practical botany from Asa Gray. In return, he and his gardening mentor, Horatio Hunnewell, pledged the funds that enabled Gray to retire from teaching undergraduates.

Sargent developed the arboretum and enmeshed it in the world of upper-class Boston with the diligence and vision of the portfolio managers with whom he associated. He added 45 acres to the original 120 in 1882, and would grab 85 more, largely from the Bussey, thirteen years later. He worked with Olmsted to incorporate the arboretum into the Boston park system and to tap public funds, while retaining control by Harvard; the scenic Arborway would soon make the suburban locale accessible by carriage from the Back Bay. The arboretum itself was both a "museum of trees," displaying temperate North American sylva according to the natural system of classification, and a center for the promotion of naturalistic estate horticulture. Our visitor could ride along city-built drives and inspect the just-planted permanent collection of more than 70,000 trees and shrubs. He could also hear from Sargent his plans for his new weekly publication, *Garden and Forest*. This collaboration with Olmsted and the wealthy garden designer Mariana Griswold Van Rensselaer was designed to integrate botany, horticulture, and the design and management of large estates. Sargent also had a longstanding commitment, dating from the 1880 census, to provide the federal government with a report on trees and to advise on forest policy.[30]

At this point, our scientific tourist could give up his carriage and return by commuter train to the Back Bay. Not far from the Agassiz Hotel, on Copley Square, were three institutions that, in their combination of propinquity, difference, and independent development, neatly expressed the fragmented character of Boston's life science initiatives. The first, the Boston Society of Natural History, had been established in the 1830s as a vehicle for bringing together local enthusiasts for natural history and linking them with Harvard leaders. It had been the locus for some of the earliest debates over evolution and had been sufficiently influential in the late 1850s to claim part of the Back Bay public money. After the Civil War, however, it was unable to compete with the mushrooming MCZ as a research institution; instead, under the leadership of Agassiz student Alpheus Hyatt, it became Boston's popular science museum. In 1887, our visitor could accompany the parades of elementary school children to tour its exhibits, but could find little more in this substantial building.[31]

The Massachusetts Institute of Technology, on the other side of the square, stood a world away from Hyatt's genial displays of shells and skeletons. It had been established in 1861 to train young men in the expanding vocations of mining, railroad construction, and industrial engineering. While geology was important, life sciences were not. In 1883, however, the institute hired young William T. Sedgwick as professor of biology. He had broad interests, but he soon saw that his science, his city, and his institute converged at one point: the new subject of bacteriology. By 1887 Sedgwick was initiating a substantial program of experiments on sewage and would develop a major center for the new subject of sanitary engineering.[32]

The newcomer to Copley Square in the 1880s was the Harvard Medical School. Prior to 1883 it had been located on the south bank of the unsanitary Charles River, adjacent to the Massachusetts General Hospital's supply of poor patients. Beginning around 1870, however, Harvard president Eliot pressured the medical professors to shift their emphasis from the cultivation of clinical astuteness to the utilization of chemical and biological science. Eliot hoped to heighten the importance of laboratory research and to link the medical faculty with other life scientists. That new perspective justified the reloca-tion to the Back Bay, far from hospitals but more congenial for scientific investigators. The most important feature of the new building was its inclusion of seemingly ample space for hands-on instruction in chemistry, anatomy, and physiology.[33]

Professional pressures, however, pushed Harvard's medical scientists in a different direction in the 1880s. This can be seen in the careers of the two men most associated with the advancement of biological knowledge at the medical school during this period. In 1871 Leipzig-trained Henry Bowditch became professor of physiology, with the general expectation that he would devote significant energy to basic research. He set up a small laboratory but produced few results. As a full-time professor and a member of the Brahmin elite, he moved rapidly into administration, serving as dean for many years, coordinating construction of the Copley Square building, and, ultimately, leading a major capital campaign.[34]

The career trajectory of Brahmin embryologist Charles S. Minot is even more telling. After studying briefly with Louis Agassiz in 1872, Minot traveled to Leipzig to sample offerings in anatomy, physiology, and zoology. He then returned to Cambridge, where he obtained, not a medical degree, but a doctor of science in physiology. In the next decade he published widely on subjects ranging from vertebrate physiology and comparative embryology to the anat-omy and taxonomy of insects; he also taught histology and embryology, and agitated for introduction of a basic course in "biology" at the medical school. But in 1886 Minot learned that these activities, however scientifically merito-rious, were considered outside the scope of the Harvard Medical School.[35]

By the time our tourist arrived, Minot was conforming to his institutional surroundings—writing articles for a medical encyclopedia and beginning a program of research in human embryology. Five years later he would finally gain a professorship. Medical teaching and learning, not "biology," remained the touchstone for Harvard's medical school. Our visitor could conclude that although institutions devoted to nature education, the defense of urban civili-zation, and the cure of disease might exist in close proximity in the United States, the people inside them did not in fact have much in common.

The next stage of our hypothetical itinerary would be a quick side trip up the Massachusetts coast. While passing through the town of Lynn, our visitor could look out over the harbor to the Nahant peninsula, where Louis Agassiz had summered thirty years before. Salem, just to the northeast, housed the Peabody Academy of Sciences, developed enthusiastically by a number of Agassiz's estranged students in the mid-1860s. They had also established *Amer-ican Naturalist*, the first modern journal for enthusiasts in natural history. By

1887 the academy was moribund, and the journal, to be discussed in the next chapter, had moved elsewhere. But two significant residues of that effort at emulation remained. *Naturalist* cofounder and Salem resident Edward S. Morse had become a successful popular lecturer, presenting knowledgeable and interesting overviews of subjects ranging from the taxonomy of brachiopods and of Japanese pottery to the evolution of techniques for arrow release. An equally important Salem scientific entrepreneur was Samuel Cassino. In 1877, this self-educated local collector began to publish annually a list of the names, addresses, and interests of American natural history collectors. Cassino's *Naturalists' Directory*, which included adolescent hobbyists, commercial dealers, and such scientific eminences as Gray and Baird, constituted a national nexus for what was both an intellectual community and a commodity market.[36]

If particularly diligent, our tourist could continue up the coast to the fishing village, turned artists' colony, of Annisquam, hidden in a cove on the north shore of the Gloucester peninsula. There, for the preceding six years, Alpheus Hyatt, with support from the Boston-based Woman's Education Association (WEA), had operated an informal zoological summer school for both male and female students and teachers. In 1887, however, both Hyatt and the WEA were shifting their attention to Woods Hole. Annisquam was one of several impermanent efforts to combine zoological education with summer recreation.[37] Thus our tourist would likely turn south and west from Salem (passing by Woods Hole, which has been discussed already, and to which we will return in chapter 6), leave Massachusetts, and come to the grand "cottage" that Alexander Agassiz had dramatically situated in the mid-1870s on a bluff at the southern tip of Newport, Rhode Island. Like Annisquam, this was a private effort to combine marine zoology and summer leisure; it presented a sharp contrast, however, in its architectural solidity, limitation to Harvard men, emphasis on research, and continuation for more than a quarter-century. Agassiz half-heartedly offered to make his Newport laboratory a regular part of Harvard, but that possibility ended with his death in 1910; ultimately, it was meaningful only for the select individuals who had the good fortune to have been invited there.[38]

WHOA!!

Along the Eastern Seaboard

Our visitor's stop in New Haven, Connecticut, could be brief. Yale had the same confusing variety as Harvard, but fewer resources and less impact. The semi-autonomous Sheffield Scientific School provided a home for both zoologist Addison Verrill, a longtime Fish Commission collaborator, and fern taxonomist Daniel Eaton (grandson of Amos Eaton). The Yale Medical School was almost completely independent, operated by and for Connecticut medical practitioners. At Yale College, sociologist William Graham Sumner presented the classic "social darwinist" defense of individualism and free enterprise to an appreciative upper-class undergraduate audience. The university's really exceptional presence testified dramatically to the importance of idiosyncratic initiatives. The most famous life scientist at Yale was the United States' only "professor of paleontology," Othniel C. Marsh. Marsh received no significant support from the university. His uncle, London-based merchant banker George Peabody, supplied both his salary and the funds to build the Peabody Museum of Natural History. The federal government, in the persons of Marsh's longtime professional allies Baird and Powell, provided the bulk of operating funds and boxed specimens that made New Haven the chief center for study of American dinosaurs.[39]

New York City required not much more time. Although Manhattan was the commercial and financial capital of the United States, it was notorious for its lack of support for science. Columbia University, which will be discussed in the next chapter, did not establish a biology program until 1892; botanist Nathaniel Britton, who would organize the New York Botanical Garden in the 1890s, had only just been appointed an instructor at Columbia in 1887. The city's medical schools were numerous but unremarkable. A handful of charismatic physicians, most notably George M. Beard, the author of *American Nervousness*, occupied the broad territory where neurophysiology, psychopathology, and cultural anxieties overlapped.[40] The one notable institution was the American Museum of Natural History. It had been established in 1869, four years after the burning of P. T. Barnum's American Museum, by Albert Bickmore, one of Agassiz's students, and a group of Manhattan patricians, including William Dodge, J. P. Morgan, and Joseph Choate. Choate cooperated with Boss Tweed to obtain prime real estate on the west side of Central Park, along with a permanent city subsidy. But the museum was underfunded and undirected until the 1880s. In 1887 it had just begun to hire full-time curators, most notably mammalogist Joel Allen and ornithologist Frank Chapman. More important, its president, self-educated businessman Morris K. Jesup, was recognizing that large specimens, especially fossils, could provide the combination of spectacle and education that would give the museum visibility within New York society.[41]

In Philadelphia, the East's third major urban center, our tourist would find institutions similar to those in Boston, but with quite different priorities and prospects. The Academy of Natural Sciences had been, to a considerable extent, the model for the Boston Society of Natural History. In the 1860s it had also sought to engage the public with exhibitions. By the 1880s, however, the academy's leaders were turning in on themselves. Amateur curators cared for the collections, but communicated neither with other Philadelphia scientists nor, in any meaningful way, with the public.[42]

In contrast to this conservatism, University of Pennsylvania leaders were innovative to the point of capriciousness. In 1884 they created an undergraduate "school of biology." The program soon languished, however, as patronage went in different directions: first to the creation of a hygiene laboratory, important for the few years that it was headed by former federal medical activist John Shaw Billings; and, then, a few years later, to the endowment of the Wistar Institute of Anatomy and Biology, conceived as a museum of descriptive anatomy. None of these endeavors gained the resources or the leadership to have significant impact before the turn of the century.[43]

Edward Drinker Cope epitomized Philadelphia's blend of traditionalism, innovation, and flightiness. For nearly all his career Cope worked as an independent scholar from his home, utilizing a private fortune that was significant but not vast. He was an original evolutionary theorist and an individual who added significantly to the empirical foundations of vertebrate zoology and paleontology. He also took over *American Naturalist* in 1877 and made it a

Figure 4–4

The original building of the American Museum of Natural History, 1881, in the midst of the still undeveloped Upper West Side. Central Park West is on the right; Columbus Avenue, with elevated railway, on the left. Photograph © Collection of the New-York Historical Society, negative # 40836.

strong critical voice on the national intellectual scene. By 1887, however, Cope had lost most of his money through overspending on collections and through a series of failed mining ventures. Faced with financial ruin, he obtained a faculty appointment at the University of Pennsylvania in 1889, but he contributed little to the programs of that institution. His self-centeredness would lead him in 1890 to drag Marsh's name into the mud with public accusations of plagiarism; but Marsh gave as good as he got, and Cope came through no better than his opponent. Philadelphia gained little from Cope: his paleontological collection went, not to Penn or the Academy of Natural Sciences, but to the American Museum of Natural History; on his death in 1897, *American Naturalist* was taken over by Bostonians.[44]

For a comprehensive view of American biological institutions, our tourist would need to visit the Philadelphia suburb of Bryn Mawr and then go south, first to the eleven-year-old Johns Hopkins University in Baltimore and then on to Washington. The federal naturalists, however, have already been discussed, and Hopkins and Bryn Mawr will be examined in the next chapter. Therefore we can suppose that our visitor has had his fill of laboratories and museums in eastern cities and that his interest has turned to determining whether biological scientists working outside the older urban centers, in closer touch with the North American heartland, were significantly different. He takes a train back to the hustle of Manhattan and then heads north and west, out into the countryside.[45]

Our visitor's first stop, however, would lead him to realize that his distinction between urban and rural was problematic. On the west bank of the Hudson River, about ninety miles north of Manhattan, he could visit the farm of John Burroughs. For many years Burroughs had combined a sensibility drawn from his friendship with the romantic poet Walt Whitman with periods of close observation of birds, flowers, and country life. His seventh collection of essays, *Signs and Seasons,* had appeared the year before. Yet in conversation Burroughs would admit that he had gained his opportunity to observe rural nature and write about it because he had, for the previous thirteen years, been working for the federal government as a traveling bank auditor. He had just quit this position, at age forty-seven, to take up farming seriously. And this hard work was putting his literary-scientific productivity almost completely at an end. Burroughs had, in fact, just said good riddance to the scientific community with a broadside attack on the "human weasels" who robbed birds' nests in the name of science—in particular "that dessicated [*sic*] mortal called the 'closet naturalist,'. . .the most wearisome and profitless creature in existence." Burroughs affirmed his faith in science and nature, as well as his opposition to orthodox Christianity, but he spent the next fifteen years growing grapes for the metropolitan market.[46]

A less conflicted sense of the meaning of natural science in a rural setting could be found further west, in Ithaca. The initial input into the development of natural science at Cornell University—as at many other places—had come from Louis Agassiz, who visited for two months during the university's first year (1868) and placed one of his students, physician Burt G. Wilder, as professor of zoology. Yet Agassiz's and Wilder's precise, private, and urban scientific sensibilities never dominated Cornell. The university's other life scientists were agricultural in their orientation, and they presented a vision of field science that directly engaged the state and the public. Botany professor Albert Prentiss had studied at the pioneering State Agricultural College of Michigan; James Law, a Scottish veterinarian, established a program in animal medicine; and the enthusiastic undergraduate John H. Comstock soon became professor of entomology. By 1887, these men had already produced a significant group of naturalists and biological technocrats, including David Starr Jordan, Leland Howard, Daniel E. Salmon (first chief of the federal Bureau of Animal Industry), and entomologist-bacteriologist Theobald Smith.[47] Cornell's agricultural commitment was increasing dramatically at the moment our visitor arrived. In 1887 the Hatch Act provided each state $15,000 annually, plus mail and printing expenses, for agricultural experimentation. Cornell leaders gained control over New York's share and used a significant part of it to hire young Michigan horticulturist Liberty Hyde Bailey. Over the next quarter-century Bailey would develop and spread a vision of rural progress that combined Darwinism and uplift, utilized science in gardening, and linked rural cultivation with rural culture.[48]

Another kind of enterprise could be found a few hours north and west, in Rochester. Henry A. Ward had studied briefly with Agassiz in the 1850s but then moved back to western New York, where he developed natural history into a business. Ward's Natural Science Establishment marketed specimens—from meteorites and bottled frogs to stuffed elephants—to collectors and museums throughout the nation. In 1887, Ward and his staff were assembling a permanent exhibition for the Coronado Hotel, a resort then under construction in San Diego, California. They were also producing *Ward's Natural Science Bulletin*, a unique combination of scientific journal and advertising sheet. Ward's business was an important place of apprenticeship for young naturalists. A number of his assistants (including William T. Hornaday and Carl Akeley) went on to careers in zoos and museums, while others (most notably geologist Grove K. Gilbert and entomologist William M. Wheeler)

Figure 4-5

Agricultural science on a grand scale: the Cornell University Barn, probably early 1880s. Photograph courtesy of the Division of Rare and Manuscript Collections, Cornell University Library.

became major scientists. Ward's apprentices learned more about the practical aspects of science than was possible in any college.[49]

Our visitor, along with about one million others annually, continued on to Niagara Falls, where he could see the untamed force of nature (patriotically illuminated each evening with red, white, and blue lights) and could read Frederick Law Olmsted and Calvert Vaux's recent proposal to heighten the gorge's natural scenic value by rearranging the shoreline. Then, in Buffalo, he boarded a steamer for a leisurely cruise up the Great Lakes. As the boat entered Lake Michigan, he could see vast rafts of pine logs, stripped from the North Woods, waiting shipment to Chicago.[50]

The Great Valley, and Beyond

Our visitor would be uninterested in Chicago.[51] Although it had become the central site for gathering and processing the plant and animal productions of the Midwest, it contained little life science activity prior to the development of the University of Chicago in 1893 (to be discussed in the next chapter). Our visitor therefore would disembark instead in Milwaukee, where, in 1887, he could find Charles O. Whitman. Whitman was there because of one more of the many singular American biological initiatives during this decade. In 1884, thirty-two-year-old neurasthenic farm machinery heir E. P. Allis, Jr., decided to give up business and devote himself to comparative anatomy. He did so in Gilded Age millionaire fashion: he installed a laboratory in his aunt's lakefront mansion, and he hired forty-three-year-old Whitman, who had just written a manual on microscopical methods (published by *Naturalists' Directory* compiler Samuel Cassino), as both his private tutor and the laboratory's director.[52]

The amateur Allis imagined that a laboratory could produce results in the same way that a factory produced reapers, and he pushed Whitman to maximize output with a staff of helpers doing assigned tasks. Whitman, the professional scientist, persuaded his student that biological research was an art, and he hired assistants with the assurance that they would have the opportunity to do independent work along with their chores. A significant outcome of the Allis-Whitman relationship was the *Journal of Morphology*, a new vehicle for the production of the expensive illustrated research papers that zoologists expected. The open question concerning both the laboratory and the journal, as with other ventures, was that of permanence. In 1889, soon after Whitman went to Clark, Allis left for the French Riviera. He spent the rest of his long life in the company of the prince of Monaco and other wealthy devotees of marine science. The Lake Laboratory closed five years later, and the *Journal* suspended publication when Allis withdrew his subsidy in 1903.[53]

Heading south through Chicago and the corn belt, our visitor could find another major midwestern scientific publishing center, in Crawfordsville, Indiana. John Merle Coulter was president of Wabash College, a typical denominational (Presbyterian) school; every month, from this small town forty-five miles west of Indianapolis, he sent out the *Botanical Gazette*. This magazine combined taxonomic reports with newer kinds of anatomical and physiological studies, and, as its middle-American title indicated, it also offered a mix of community news, gossip, and advocacy. The *Gazette* was gradually pulling American botanists away from their decades-long deference to Harvard, establishing a more decentralized mode of interaction.[54]

The Hoosier center for zoology lay sixty miles south in Bloomington, at the state-run Indiana University. In the 1880s, Cornell graduate and longtime Fish Commission collaborator David Starr Jordan was its president. He collected data on the distribution of species such as trout so that commission employees could help evolution along by moving desirable types into unstocked streams; his popular essay, "The Story of a Salmon," was on its way to making that animal a tragic hero for American children.[55]

From Bloomington, there were easy train connections west to St. Louis. Its important, and unique, site was the Missouri Botanical Garden. Like the Lake Laboratory, this was the personal enthusiasm of a provincial businessman; but in contrast to Allis's brief, intense, and sequestered engagement with science in Milwaukee, this institution's founder had been at work with national leaders in a public setting for nearly forty years. In 1851, Anglo-Missourian hardware merchant Henry Shaw, on vacation in England, had conceived the idea of creating in St. Louis something analogous to the Royal Botanical Garden. A few years later he wrote directly to the garden's director, Sir William Jackson Hooker, for advice, and the international botanical network immediately swung into action. Hooker wrote to Gray, and Gray contacted his major midwestern collaborator, German-Missourian gynecologist George Engelmann. With Engelmann's ongoing counsel, Shaw imported gardeners and plants to ornament his suburban estate. He opened it to the public in 1859, and gradually added a library, museum, and herbarium, as well as a non-naturalistic, heavily decorated park. Following Engelmann's death in 1884, Shaw subsidized the publication of his collected works and enticed Farlow's student, William Trelease, with an endowed professorship at the tiny Washington University to come to St. Louis. In 1887, Shaw was an octogenarian in declining health, living in the house that had looked out onto his garden for decades; on his death, two years later, Trelease became the garden's director. With the help of a local board of trustees, and ambiguous relations with the municipality, Trelease continued to develop an institution that combined basic work in taxonomic botany with maintenance of a site for civilized leisure and civic pride. St. Louis was and remained the horticultural entrepôt for the Midwest.[56]

Traveling out onto the prairie, where Shaw's shovels had found their market, our visitor would soon be a world away from emulators of English estates. It was not necessary to go there to inspect American agricultural science—that existed at Cornell and could be found, twenty-five years after the passage of the Morrill Act, in some official fashion in every state. The interest of the trans-Mississippi institutions lay in the degree to which bioscientific activities were bound up with the construction of identities in new settings. The most unusual example was in Davenport, Iowa, where one of Asa Gray's former collectors, Charles Parry, had built up the local academy of sciences around investigation of the area's ancient mounds. The Iowans argued for the antiquity of midwestern civilization, and for the present residents' intellectual superiority over an eastern federal Establishment, by asserting that the mounds had been erected by a high-achieving non-Indian race and that the archaeologists working for Powell and Baird at the Smithsonian were arrogantly erroneous in crediting the mounds to the barbarians that the present Iowans had displaced.[57]

The more common situation, however, was more local in orientation. The federally subsidized agriculture schools had become markers of the major social cleavages on the Plains. Iowa and Kansas had each established two state universities. If Iowa City, Iowa, and Lawrence, Kansas, were to be sites for traditional colleges designed to cultivate the children of bankers and ministers, then farmer interests were adamant that Ames, Iowa, and Manhattan, Kansas, should house schools that would be of, for, and by the agriculturists—even if almost no farmers sent their sons to those schools to learn their vocation. At the Iowa and Kansas agricultural colleges, major wars erupted in the 1880s and continued into the 1890s between egalitarian, practical, and often cranky populists and more professional and elitist promoters of scientific research and traditional education.[58]

But there were no general patterns. Nebraska created only one state university, and in 1884, botanist Charles Bessey left the arguments of Ames for the relative serenity of Lincoln. He responded to popular desires by writing bulletins on plant diseases, but he was also able to carve out a place for serious science. Bessey exerted a multifarious and growing influence by 1887. His

Figure 4–6

Missouri Botanical Garden, St. Louis, probably 1880s. The Museum, housing the herbarium and library, is at left rear; Henry Shaw's villa, Tower Grove House, can be seen in the center rear behind the cupola of the Observatory. Photograph courtesy of the Missouri Botanical Garden Archives.

elementary textbook was shifting the focus in high school and college botany from classification to anatomy and physiology. He identified and trained a handful of local students—most notably Frederick Clements—into scientists who would develop ecology. And as botanical subeditor of *American Naturalist*, he became a nationally known commentator with sufficient influence to share with Coulter and Britton, after Gray's death in 1888, the coordination of efforts to create national taxonomic standards and to declare complete independence from English botanical leadership.[59]

Again back on the train, moving from Lincoln toward the Rockies and California, our tourist could take in the dramatic scenery and natural wonders of the American West. In South Dakota he could visit the site where Marsh's collector, John Bell Hatcher, was excavating brontotheres.[60] Further west, he could join Ward and other Geological Survey employees exploring Yellowstone Park. Alternatively, he might accompany William T. Hornaday atoning for his bison hunt by collecting living western mammals for the new National Zoo (using the official Fish Commission train).[61] But he in fact did none of these things. Though they were important opportunities to learn about American nature, they would tell him little about the nation's science; these naturalists were all transients connected, either directly or indirectly, with the eastern institutions with which he was familiar. As a consequence, he speedily continued to the terminus of his tour—the new parks and cities of northern California.

Like most tourists, our man would initially be amazed at this setting. Its Mediterranean climate, unlike that of any other part of the United States, raised idyllic associations for individuals with northern European or New England origins. Our visitor would take the by then standard trip to Yosemite Valley and then travel to the giant sequoia groves. In doing so, he would pass through the Sierra foothills, ripped up forty years before in the gold rush, and traverse the Great Central Valley where, through the concentration of engineering and labor, wheat was being grown with unprecedented intensity.[62] It was a commonplace that California contained the grandest elements of nature together with the most rapid exploitation and development. By the 1880s Anglo-American culture had established itself in the San Francisco Bay area. A few individuals were reflecting, in informed ways, on life in this setting, and others were planning new developments. Our now exhausted visitor could interview only a few of the most prominent figures.

The mainstream of California thinking was represented by the state's most visible biological scientist, University of California professor of geology and natural history, Joseph LeConte. In LeConte's optimistic vision, California was a new beginning, and all things were possible there for white men of good will. A Georgia slaveholder and liberal Christian, LeConte had studied with Louis Agassiz in the early 1850s and then returned south to teach and be close to his plantation. This comfortable world was destroyed in the war. In the winter of 1865 LeConte had wandered for weeks through the Georgia swamps escaping Sherman's soldiers; his books and papers were burned and his income evaporated. A professorship at the new university in Berkeley in 1868 offered the possibility of a new life. He sought to show that reconciliation and progress were inextricable: that evolutionism undergirded true religion; that preservation of nature was compatible with the development of natural resources; and that true democracy required leadership by an educated elite. For more than thirty years, LeConte was the most popular and influential teacher at Berkeley; his books and magazine articles circulated widely. His evolutionary philosophy of compromise and optimism made sense to a large, moderately liberal, and prosperous audience, both in California and more broadly.[63]

California was also open to innovators, from less prominent backgrounds. Some of these men were less interested than LeConte was in maintaining balance. The range of possibilities was symbolized by two charismatic figures who were active in California in 1887; they would become important nationally in the next two decades. After listening to LeConte, our well-informed visitor sought out John Muir and Luther Burbank.

Muir, a Scottish Wisconsinite, embodied the nature romanticism of Emerson transplanted to the rawer and grander scale of California. As an undergraduate at the University of Wisconsin in the early 1860s, Muir had aspired to emulate the romantic German naturalist Alexander von Humboldt. In 1867, inscribing his journal, "John Muir, Earth-Planet, Universe," he walked from Indiana to Florida in the hope of exploring South America, but then decided to go to California instead. He made his reputation as the rural workman who explored the glacial history of the Sierras. He guided intellectuals such as LeConte, Asa Gray, Joseph Hooker, and Emerson himself through the magnificent scenery, and he published evocative travel sketches in genteel magazines. In 1880, however, Muir married and took over management of his father-in-law's substantial orchards a few miles east of Berkeley. In 1887 he was comfortable but in crisis, unable to combine commercial horticulture with editing the lavish, multivolume *Picturesque California*. On a deeper level, he was torn between his agriculture-based prosperity and his convictions regarding the superiority of nature. Over the next five years Muir would transfer management of his ranch to relatives and return to the public stage as founder of the Sierra Club and as an uncompromising advocate for wilderness preservation.[64]

John Muir was a rebel; Luther Burbank also stood out, albeit in the opposite direction, as a self-conscious revolutionary. His experience of New England had been, not the meditations of the Transcendentalists, but the stony realities of Massachusetts farming. Burbank was substantially self-educated and read Darwin enthusiastically—not, however, *The Origin of Species*, but his *Variation of Animals and Plants under Domestication*. Burbank approached biological problems, not from a philosophical or romantic perspective, but from a deeply practical one. He wanted to find plants "willing to take a step forward" in evolution. In 1873 he selected the seeds that would grow into the Burbank potato, and two years later he abandoned Massachusetts for the garden setting of the Sonoma Valley, fifty miles north of San Francisco. In 1887 Burbank was a prosperous but still obscure nurseryman. Like Muir, he was trying to escape from the daily demands of commercial growing; but while Muir turned to picturesque nature, Burbank began to concentrate full time on experiments to transform plants. A horticultural Edison, he would mix the botanical resources of the world together to form "new creations." At the turn of the century he would become a national celebrity, turning down a professorship at Berkeley and then accepting a huge grant-in-aid from the just established Carnegie Institution of Washington. He exemplified the possibilities of biological invention.[65]

Snacking from a sack of Burbank's hybrid plums, our exhausted visitor would climb onto the train in Oakland to head back to the East and to Europe.[66] He found some structure in what he saw, but the overwhelming impression was of variety. The geographic expanse covered—even excluding the scientifically dormant South—was immense, and the number of sites for life science within that expanse was continually increasing. Individuals initiated innovations, and gained support, within frameworks that were primarily personal or local. There were a great variety of possible audiences and patrons for life science initiatives, including physicians, farmers, horticulturists, collectors, philosophers, romantics, and local boosters. Our visitor could certainly be pardoned if—contrary to Whitman's disclaimer a few years later—he did perceive a series of "disconnected atoms," each moving in directions that seemed independent of the others.

The situation in the life sciences in 1887 was a particularly vivid instance of a broader phenomenon, described a generation ago, in a classic work of cultural history. Harvard literary scholar Howard Mumford Jones depicted the late nineteenth century in the United States as "the age of energy."[67] Jones was a better historian than he was a physicist, and so for him energy meant both a quantity of work and what he termed "varieties of American experience"—the

Figure 4–7

Luther Burbank posed before a row of his "new creations" (in this case plums) at his nursery, Sebastopol, California, about 1892. Photograph courtesy of Stark Brothers Nurseries and Orchards Company, Louisiana, Missouri.

diversity of initiatives he located between the Civil War and World War I. He presented a picaresque survey of American endeavors, encompassing (among many odd couples) William Dean Howells and Wild Bill Hickock, Huck Finn and Ben-Hur, the *Atlantic Monthly* and the *Proceedings of the Society for Engineering Education*. The first and last individuals on our tour—the plutocratic, Swiss-Brahmin zoologist Alexander Agassiz and the hardscrabble, Yankee-California nurseryman Luther Burbank—make up another pairing of this sort. Although linkages could be made among many of these initiatives, no one term—not "natural history," not "life science," and certainly not "biology"— encompassed all our visitor was able to see.

Academic Biology: Searching for Order in Life

Most of the individuals mentioned in the previous chapter were content to cultivate their own gardens or to interact, at most, with people in either geographically or disciplinarily adjacent fields. In the 1880s and early 1890s, however, a small number of American life scientists wanted to get things organized. They sought to coordinate, at least potentially, all those working on organisms throughout the country; moreover, they sought to achieve this result in ways that were largely independent of the longstanding partnership between naturalists and the state to inventory and manage the biotic contents of North America.

Efforts at organization during these decades derived partly from convictions about the structure of the world and of knowledge. Life, for many scientists, was, in spite of its diverse manifestations, at root a single principle; they believed that their efforts to understand organisms ought to reflect this basic fact. In part, however, this search for order was also a desire to re-form scientific hierarchy. From the "organizers'" standpoint, the landscape of the life sciences in the United States was too flat and featureless, with resources dissipated and duplication common. The single peak in Washington was both too low and too constricted; standing on it, moreover, required constant and ultimately debilitating political labor. The organizers wanted a scientific terrain that was more pleasingly varied and that highlighted creativity on the part of the individual.

"Biology" lay at the center of these impulses. It was a singular scientific term. By comparison with "botany," "zoology," or "natural history," it was a recent coinage, invented around 1800 simultaneously by scholars in both France and Germany. Its meaning and significance, however, remained uncertain throughout the century. For some it was an umbrella term for all the life sciences. For others it was a particular specialty, but one that was uniquely fundamental, an intellectual core around which all other inquiries stood as more applied, limited, or derived. Within that latter perspective, there was uncertainty concerning the content of that core: was it chemical, cytological, historical, or ecological? A larger uncertainty was whether biology, by any definition, was an ordinary workaday scientific discipline, or whether it expressed an a priori philosophical faith in the meaning of "life." A final problem was whether it was a science that already existed or an ideal that would be realized in the future with the development of new knowledge.

This chapter and the next examine American efforts between 1870 and 1900 to organize the life sciences, with particular reference to biology. I move, as Whitman did in his manifesto, from the "extended" to the more "intimate." Journals and societies were the most attenuated links among American life science enthusiasts, bringing them together through texts or through the annual experience of a convention. University programs created a more intense form of community, linking particular groups of individuals on a daily basis. But although a few people—such as E. B. Wilson and T. H. Morgan at Columbia—cooperated within one university for decades, departments were always limited in terms of numbers, and they were usually structured to avoid overlapping expertise; moreover, most participants in them were birds of passage—spending a few years as graduate students and then maintaining a sense of belonging only through the tenuous bonds of mentorship and memory.

These were the kinds of organizational efforts common in science during the nineteenth century. Journals had been part of the lives of American men of science since the 1810s, and conventions had formed a regular feature of the national scientific experience since the creation of the AAAS in the 1840s. Departments were the standard structural unit within the universities that emerged during the Gilded Age. It would have been surprising had biologists not taken these initiatives. However, given the diversity of "biology," these efforts were not—by comparison with the activities of scientists in other disciplines—particularly successful. Journals, societies, and university programs were established, but they were also resisted, and sometimes successfully challenged, by those not satisfied with the disciplinary and intellectual hierarchy that advocates of biology sought to establish. Both these organizational efforts and the resistance to them are important to lay out in relation to the truly novel developments to be explored in chapter 6.

American Naturalists

The first continuing cooperative effort to create a national community of biologists (using the word loosely) was *American Naturalist*. In 1866, Frederick W. Putnam, a former student of Louis Agassiz, was a curator at the Essex Institute in Salem, Massachusetts. He invited Edward S. Morse, Alpheus Hyatt, and Alpheus Packard, who had all studied at the MCZ a few years earlier, to join him in producing, as they declared through their subtitle, *A Popular Illustrated Magazine of Natural History*. This emphasis on the popular came out of an accurate assessment of the state of their science at that time. *American Naturalist* was not a research journal: the expense of lithography would be too great, and the audience too small, for a publication that needed to pay for itself from subscriptions. In addition, federal and state governments, and the seemingly deep-pocketed Agassiz, were already engaged with such work. Putnam and his friends hoped to do more for science and to meet their expenses by reaching a larger audience than heretofore. The way to do that was to cover a broad subject area (including zoology, botany, geology, microscopy, and gardening), to appeal to Americans from all regions, and, most significant, to function as a bridge between the "teacher" and "student" of natural history, or alternatively, to act as a link among "the great brotherhood of enthusiasts," primarily collectors (Putnam had already gone some distance in identifying those groups when he compiled the first *Naturalists' Directory*).[1]

The editors' problem was that they were not able to generate text that would interest, and thereby constitute, their imagined national community of "naturalists." They sought to be as popular as possible without losing their sense of themselves as men involved in discovery. The result, however, was a mélange of articles that ranged from the homely "Botanical Excursion in My Office" to geologist John S. Newberry's AAAS presidential address attacking materialism in science.[2] *American Naturalist* lost an average of $3,000 annually from the time of its founding. In its early years, Salem's Peabody Academy made up these losses, but financial responsibility fell increasingly on the shoulders of the editors in the course of the 1870s.

With the possibility of profiting from popularization fading, lead editors Packard and Putnam decided to redirect their publication toward a more deeply interested, if demographically limited, audience. *Popular* disappeared from the subtitle in 1871, and specialized literature review sections were added a year later. In 1877 Packard explained the magazine's changed mission: rather than serving the brotherhood of enthusiasts, it would be "a medium between the investigator on the one hand and the teacher and student on the other."[3] Yet under both Packard and Edward D. Cope, who purchased an interest in the magazine in 1879 and coedited it (with Packard until 1887, and then until his death in 1897 with J. S. Kingsley), *American Naturalist* sought to maintain its initial breadth. The editors emphasized that it was "*the only magazine in the world to-day which keeps its readers en rapport with the work of Americans in the field of the natural sciences.*"[4] They recruited a group of associate editors (varying in number from four to eleven) to advise, channel material, and report on their areas of work. Initially federal naturalists predominated, but by the mid-1880s bureaucrats had been replaced by (mostly young) academics. Midwesterners such as Bessey and (when he was in Milwaukee)

Whitman were included to show that the journal was genuinely national in its participation. Associate editors represented a fluctuating roster of fields including, at different times, anthropology, botany, embryology, entomology, geography, invertebrates, microscopy, mineralogy, physiology, psychology, and zoology, among others. The journal emphasized general articles, reviews, news, and commentary; the one constant was an effort to avoid detailed, specialized research articles.

"Natural history" remained on the title page to describe the magazine's scope, but the editors used it less frequently as it became associated with the collectors' culture they were leaving behind. One replacement term — "natural sciences" — encompassed the disinterested study of the earth and its inhabitants (thereby excluding chemistry, physics, and clinical medicine). But that term was, on the one hand, too broad, and, on the other, already appropriated by commercial collector Henry Ward. The increasingly visible descriptor of the magazine's domain came to be "biology." In 1877, for example, Packard assessed the magazine's first decade of work with reference to "the progress of biology" during that period. Four years later he and Cope laid out "the scope and aim of the biological sciences" and used "naturalist" and "biologist" as equivalent terms. This emphasis was problematical insofar as the magazine continued to include geography and mineralogy, but within a few years they became minor elements.[5]

As an associate editor of *American Naturalist* for seven years at the time he presented his Woods Hole manifesto, Whitman was speaking from experience when he argued that a journal was too "extended" — too dispersed, too formal, and too slow — to provide the basis for a functioning biological community. The deeper problem, however, was that while *American Naturalist* could engender such a sensibility only insofar as it was a broadly communal endeavor, its fundamental character was that of a labor-intensive, financially draining proprietorship. Youthful idealism, narrow participation, and the Peabody Academy's subsidies obscured this tension in the magazine's early years, and the independently wealthy Cope initially displayed benevolence in his efforts to broaden the roles of collaborators. But Cope was an individual who held strong opinions on many subjects, and it is not surprising that he used the journal he subsidized — from a patrimony that largely disappeared in the 1880s — to advance his positions. These included not only his well-known battles with Othniel Marsh and John Wesley Powell over western fossils and his strongly held neo-Lamarckism, but also a range of incisive but impolitic views that alienated a large segment of the *Naturalist*'s natural community. In 1884, for example, he complained that Johns Hopkins University biology professor Henry Newell Martin (whose efforts will be discussed below) was

Figure 5–1

Edward D. Cope, posed with an anatomical token, 1876. Photograph courtesy of the Ewell Sale Stewart Library, the Academy of Natural Sciences of Philadelphia.

favoring medical physiology rather than "proper university work" in biology. Three years later he began an editorial with the statement that "in all our four hundred colleges and universities, with a dozen conspicuous exceptions, the instruction in the biological sciences is but little more than a farce."[6] Few associate editors remained on the masthead as long as Whitman.

Following Cope's death, scientists interested in *American Naturalist* were concerned above all to overcome the problems raised by his proprietorship. Financial responsibility was dispersed among a group of Boston scientists led by the patrician Charles Minot, and control was divided among nineteen associate editors. The editorship was given to the genteel and junior MIT professor Robert Paine Bigelow. The predictable consequence was that *American Naturalist* both drifted editorially and foundered financially. In 1907 its owners were glad to sell it to the psychologist and editor of *Science*, James McKeen Cattell. With no stake in biology, and some interest in eliminating a competing "general science" publication, he made the *Naturalist* a success by making it a typical research journal, focused on problems of evolution and on the new and rapidly developing specialty of genetics.[7]

A Scientific Confederacy

The second way biological scientists sought to organize a single national community in the late nineteenth century was through the creation of a disciplinary society.[8] Annual conventions were certainly more personal, informal, and intense frameworks for interaction than were journals. They also involved different kinds of commitments. On the one hand, the time and expense of convention travel was much greater than that of subscribing to and reading a magazine. On the other, the labor and responsibility for maintaining a society with an annual meeting were both more limited and more easily dispersed among many individuals than was the case for a monthly publication like *American Naturalist*. No one went deeply into debt over a society, and rotation of officers made sharing of power straightforward. Although societies could easily claim to be "national" in participation, the real possibilities for conventions of working scientists in an era of train travel were limited to those who lived along or near the Boston-Washington corridor. The dynamics of annual meetings thus had significant consequences for both the intellectual extent and the geographic scope of a disciplinary community.

Conventions of scientists had been important in the United States since the 1840s. The AAAS sought to include all sciences and to accommodate both a professional elite and broad membership; it also aimed to be truly national and promote science in different regions by mounting its annual August meeting in cities ranging in size and location (in the 1880s) from Philadelphia and New York to Buffalo and Minneapolis. The size of the AAAS—and the growing importance of specialization—was such that in 1882 it divided its section for "natural history" into four parts: geology and geography, biology, histology and microscopy, and anthropology.[9]

The AAAS was unsatisfying to many biological scientists. The late summer meeting time interrupted fieldwork and/or vacations. The convention included a large contingent of physical scientists, and it was dominated by bureaucrats and the miscellaneous collection of professors who associated with them. Thus there was a positive response when Alpheus Hyatt of the Boston Society of Natural History and Samuel Clarke of Williams College announced in 1883 that "a number of American workers in biology" were planning to create an "association of American naturalists." Their idea was to develop personal contacts among professors by scheduling gatherings over the collegiate Christmas break, restricting attendance to invited members, and limiting meeting locations to the Northeast. The organization in fact organized itself as the "Association of Naturalists of the Eastern United States," before adopting the grander "American Society of Naturalists" (ASN) around 1886. Its disciplinary scope, like the original name, was based on the *American Naturalist*.[10]

What were the professors to talk about? They limited their agenda to "the business of the naturalist"—to technical, professional, and educational issues—and excluded presentation of research reports. Leaders thought that methods were more likely to form a common ground for real discussion than the "multifarious" research interests of the anticipated members. The consequence of this nondisciplinary foundation, however, was that the ASN developed in a highly improvisatory fashion. After a few years the combination of shoptalk and yuletide sociability lost its novelty. Coteries with common disciplinary interests then began to see that they could share research results in some formal setting during the time they were already together. The politics of the medical profession induced the medically oriented members of the ASN to take the first step: they set up the American Physiological Society in 1887 in the aftermath of an ASN meeting. Over the next few years, other cliques created the American Morphological Society and the American Psychological Association. In 1889 the ASN officially invited these groups to meet in conjunction with its own gathering.[11]

By the early 1890s, therefore, the ASN was structured as a confederacy. The direction of the whole was largely the sum of the decisions of smaller disciplinary groups. This arrangement worked well for participants during that decade: meetings were flexible and congenial for most of those who attended. Scientists who felt excluded either left or formed a new organization under the ASN umbrella. While the geologists largely withdrew in the five years following the creation of the Geological Society of America in 1888, medical (physiology and anatomy) and biosocial (psychology and anthropology) scientists remained. All these specialists could reasonably think of themselves as participants in biology.[12]

For most members of the ASN and its affiliates, however, meetings were like journals in that they involved bounded commitments but provided only a limited field for asserting either identity or influence. If biology was merely an informal common referent in a society that sponsored symposia and smokers, it would have little operative significance. Scientists, like other academic professionals, were most deeply concerned with the places where they worked most of the year and, equally important, reproduced themselves, in the increasingly formalized activity of training the next generation. In the last quarter of the nineteenth century, that nexus of activities occurred in a new set of institutions: universities. In the universities, the term biology took on much more specific, and hence controversial, meanings.

MEDICAL REFORM, UNIVERSITIES, AND URBAN LIFE

The primary proponents of "organization" in the American bioscientific world of the late nineteenth century were not life scientists, but the organizers themselves—the presidents of the newly prominent urban universities. Harvard, Johns Hopkins, Clark, Columbia, and the University of Chicago (to name the most important) were organized or substantially reorganized between 1870 and 1900, becoming, collectively, a new kind of institution in the United States. University presidents believed—sometimes from philosophical conviction, but also by virtue of their offices—that knowledge was a coherent thing and that university departments could be structured in a way that would approach that ideal of coherence. Organization also served economizing aims: as specialists in fields ranging from physics to poetry sought places for their activities in universities, presidents hoped to create frameworks that would enable their schools to claim coverage of as much intellectual territory as possible with a minimum number of professors. Finally, presidents sought to structure their institutions in ways that would maximize the synergy that they believed existed between academic programs and professional schools. The argument that collegiate education could be linked productively to graduate, medical, legal, and engineering programs was one of the most prominent distinctions between the new universities and antebellum colleges; presidents were on the lookout for ways to realize this hopeful rhetoric.[13]

On all these grounds, "biology" was a word of power in American academic administrators' offices. August Comte, the great intellectual rationalizer of the early nineteenth century, had made it a major disciplinary category, equal with mathematics, chemistry, and sociology, in his influential structuring of all knowledge. In place of multiple programs competing with each other and dissipating resources—a situation that was most notorious at Harvard during this period—biology could make up a single department in which resources could be allocated flexibly. Finally, biology was an ideal unit for linking academic science and the medical schools that were being drawn increasingly into the university framework—by both providing undergraduates with a new "premedical" training and advancing research that would, in the long run, increase medical knowledge and improve care.

University presidents were thus quite responsive to arguments for biology during this period. The problem that faced both them and the professorial advocates of the science was to specify what the subject would be. How broad was it? What would it include? What were the relations between the amorphous bulk of elementary subject matter and the cutting edges of research?

CHAPTER FIVE

The first effort to build an American research university de novo was in Baltimore, Maryland. The bachelor merchant Johns Hopkins, who died in 1873, bequeathed his $7 million fortune to endow a school and a hospital. Hopkins's trustees moved first on the educational front and brought in forty-four-year-old geographer and administrator Daniel Gilman as president and organizer. Gilman determined to emphasize science, research, and graduate training much more than other American educational institutions had done. But with resources only a fraction as large as those of Harvard, he needed to exercise caution in creating programs. Since the Hopkins trustees were continuing to plan a hospital, Gilman imagined that a medical school could link the two bequests. Thus, when traveling abroad in 1875 in search of faculty and advice, he was quite receptive to the argument of T. H. Huxley that he should appoint a "professor of biology" whose major role would be to "provide instruction antecedent to the professional study of medicine." He was glad to hire Huxley's former assistant, twenty-eight-year-old Henry Newell Martin, sight unseen, to implement this plan.[14]

On arriving in Baltimore, Martin took the concrete step of creating an undergraduate course in "elementary biology"—an introduction to comparative anatomy and physiology and to techniques of dissection and microscopy, similar to the one he had developed with Huxley at the London School of Mines. But college education was low among the university's priorities, and Martin did not consider biology a sufficiently robust rubric for the more important parts of his program. These developed primarily from his experience as a physiology student at Cambridge University and from his mandate to promote medical science. Martin's research dealt with techniques for isolating the mammalian heart. The high point of his career was the construction of the university's Biological Laboratory; at the building's dedication in 1884, Martin explained forthrightly that because "this university will at no distant

134

Figure 5–2

Johns Hopkins University Biological Laboratory in the 1890s. Photograph courtesy of the Ferdinand Hamburger, Jr. Archives of the Johns Hopkins University.

day have a medical school connected with it," his laboratory had been "deliberately planned [so] that physiology in it shall be queen, and the rest her handmaids." In fact, Martin worked to make sure that this queen would have a very small court. He declined to hire a botanist, deflected Spencer Baird's efforts to get the university to establish a museum, and delegated comparative anatomy to the mild-mannered zoologist William K. Brooks, at a salary one-fourth his own. Cope was essentially correct when he wrote that Martin had abandoned a broad conception of biology for a program oriented around the needs of the medical profession.[15]

The key to understanding the early history of biology at Hopkins is to recognize that Martin was never able to establish his physiological monarchy. Financial problems repeatedly delayed the opening of the Johns Hopkins Medical School, ultimately until 1893. As a consequence, Martin had little contact with either clinicians or medical scientists. There was no pool of medical students from which he could draw assistants, nor could he award medical degrees. Within this vacuum, a new, unplanned, and more "democratic" disciplinary polity arose. Numbers, and the experiences of graduate students on the ground, made the difference. During the university's first decade, Martin supervised only fourteen graduate students. By contrast, the zoologist Brooks, in spite of his subordinate status and lack of reputation, attracted twenty-three, and he drew more than fifty people to his summertime seaside Chesapeake Zoological Laboratory. Physiology students took courses in "morphology" and spent summers with Brooks studying, not dogs and frogs, but fishes and crabs. Graduate students, whatever their interests, interacted continuously within the laboratory and developed their own ideas about both research priorities and the structure of science.[16]

By the mid-1880s, some of Martin's and Brooks's students began to take seriously the idea that they were studying "biology"—seen as an intermingling of animal (largely invertebrate) physiology and morphology.[17] As they moved into professorial positions, they sought to replicate their fortuitous experience, which they considered a natural state of affairs within the life sciences. In many cases they were hired by administrators with agendas similar to that of Gilman. They thus struggled with problems not unlike Martin's, but they arrived at new and different solutions. This can be seen at three of the most significant Hopkins offshoots in the 1880s and early 1890s: MIT, Bryn Mawr, and Columbia.

William T. Sedgwick, introduced briefly in the previous chapter, originally stood closest to Martin's blend of biology and medical science. For four years he assisted in the Hopkins undergraduate biology course; his 1880 dissertation, however, was a standard physiological exercise, testing the effects of quinine on spinal reflexes. In 1883, MIT president Francis A. Walker decided that the opening of the new Harvard Medical School complex next door provided an opportune time for him to upgrade his school's "natural history" offerings in relation to premedical education. He hired Sedgwick as assistant professor of "biology," with a mandate to develop a course like the one at Hopkins, in the hope that MIT could draw students planning to study medicine at Harvard. Martin was delighted that his protégé had penetrated "the Hub." But Sedgwick had the same problems as his mentor. On the one hand, the premedical biology program attracted few students; on the other, it gave Sedgwick no handle for research. His solution, noted earlier, was original but idiosyncratic: he connected basic work in life science with MIT's engineering emphasis by focusing on the newly prominent problem of bacteria in sewage. Sedgwick found his calling in 1887 when he began to collaborate with MIT's chemists and the state Board of Health on a sanitary survey of Massachusetts rivers; two years later he initiated the undergraduate major in "sanitary engineering" that would form the core of the MIT life science program for the next fifty years.[18]

A different direction was taken by Sedgwick's close friend Edmund B. Wilson. At Hopkins, Wilson studied with Brooks, traveled to the Naples Zoological Station, and wrote a highly regarded dissertation on the embryology of the coelenterate *Renilla*. After two years of temporary positions, he found a professorship in 1885 at the new women's college, Bryn Mawr. Its leader, Martha Carey Thomas, the daughter of an influential Baltimore physician and Hopkins trustee, hoped to provide a female counterpart to the male-only Baltimore institution. She thus initially sought a life scientist who, like Martin, would prepare students for medical school. Sedgwick advised her, however, that physiology was "well worked over" on the research side, and, more pointedly, that that science's emphasis on vivisection would make it inappropriate for an undergraduate women's college. Wilson obtained the job. Bryn Mawr, like Hopkins, advertised a program in "general biology," but its emphasis was different. In place of Martin's focus on the premedical, Wilson stressed the academic. His program included general biology, comparative morphology, physiology, and histology, with a culminating course on such basic problems as evolution.[19]

In the early 1890s, Columbia sought to create the integrated program that Gilman had imagined fifteen years before. The dearth of academic life science endeavors in New York City ended in 1890 when Seth Low, the wealthy young former mayor of Brooklyn, became president of Columbia. One of his first initiatives was to fund the university's absorption of the College of Physicians and Surgeons (P&S). A few months later he began to organize a biology department, which, he hoped, would be "the connecting link between the Medical School and the scientific and philosophical work of the other parts of the University." He hired former Hopkins graduate student (and nephew of J. P. Morgan) Henry Fairfield Osborn from Princeton to organize this program. Osborn immediately brought in Wilson as his partner, and Wilson obtained a position in the medical school (but with funding from the general university budget) for his Bryn Mawr junior colleague and fellow Hopkins graduate Frederic S. Lee, who was promoting "biological" (that is, invertebrate and cellular) physiology. Conversely, three P&S physicians (anatomist G. S. Huntington, physiologist J. G. Curtis, and pathologist T. M. Prudden) were made professors in the Columbia "Faculty of Pure Science." P&S (located near Columbus Circle, a mile away from Columbia College's buildings on Madison Avenue) provided the space in which the Department of Biology would set up its laboratories. Low hoped that this combination of strong personalities and multiple institutional linkages would enable Columbia scientists to develop the study of life, to draw upon the opportunities of the metropolis, and to apply their knowledge to pressing urban problems.[20]

Parallel developments occurred at the University of Pennsylvania. As I noted in the previous chapter, in 1884 senior anatomist and zoologist Joseph Leidy joined with the wealthy young physician and university dean, Horace Jayne, to organize an autonomous "school of biology" to provide premedical training, create a foothold for women students, and support research. While the faculty (including Jayne, Leidy, Harrison Allen, and botanist Joseph Rothrock) were initially drawn from the medical school, in 1886 J. A. Ryder, one of Baird's collaborators, became full-time professor of comparative embryology. By the end of the decade the faculty also included C. S. Dolley, Leidy's protégé and a former Hopkins student, as professor of general biology, as well as E. D. Cope. In 1891 the university began to construct a marine laboratory at Sea Isle City, New Jersey.[21]

The most striking development of that period, however, occurred in Worcester, Massachusetts. In 1889 Clark University began operations and gained, immediately and briefly, the status of academic innovator. Its founding president, G. Stanley Hall, former professor of psychology at Hopkins, believed that his benefactor, the local merchant Jonas Clark, had given him a mandate to create a cosmopolitan scholarly community and that he could hire people in whatever areas he wished. On the one hand, he took seriously the idea that biology was and should be a single, autonomous science; on the other, he chose the men (mostly former Hopkins colleagues) he thought interesting. Approximately one-fourth of Clark's original faculty were life scientists: anatomist Franklin P. Mall, neurologists Henry H. Donaldson and Clifton F. Hodge, and morphologists J. P. McMurrich and Shosiburo Watase all came with Hall from Hopkins; they were joined by physiologist Warren P. Lombard, paleontologist George Baur, and (as head of the department) Charles O. Whitman. The result of Hall's opportunistic selection of faculty was an erasing of the distinction in graduate work between physiologist and morphologist that was still strong at Hopkins. The only rubric under which all could work was biology. United in one building, the professors shared graduate students and interacted constantly in informal circumstances, practicing academic biology on a daily basis.[22]

Whitman and Chicago

At first glance it is surprising that Whitman was the individual who gained this leadership position. For the preceding fifteen years he had either been out of the country or working in highly personal patronage relationships, far from university leadership. His publications involved either issues of technique or highly specialized embryological problems. Moreover, he sought to operate within the existing disciplinary categories. In 1886, when working as Alexander Agassiz's research assistant at the MCZ, he insisted hopefully to his superior that the Harvard zoology program *"must* expand in order to keep pace with the rapid advancement of zoology."[23] The realization that he had no future at Harvard, however, altered his perspective. Exiled that year to Milwaukee and his glorified tutorship, he began to think about biology in a deeper fashion than the Hopkins coterie. He sought to make his arguments explicit and comprehensive, and to link the intellectual and entrepreneurial. The nature of his ideas developed rapidly with his experience, however, from promoting animal morphology within an alliance with medicine to belief in biology as a multifaceted independent academic field.

Whitman's plans for biology went through a number of stages. In 1887 he advertised his interest in a university position through a talk at the ASN, which he published in *American Naturalist* and circulated widely, titled "Biological Instruction in Universities." While straightforward in emphasizing the need for expanding graduate work, his proposals for organizing the various branches of instruction were vague and contradictory. He divided the field of zoology into professorships of comparative anatomy, histology, embryology, taxonomy, cytology, and physiology; but then he discussed zoology as merely one part of biology, equal in status to botany, anatomy, pathology, and—again—physiology. These last were the five disciplines recognized in the German universities; Whitman's belief that they would provide a unified organizational framework ignored the fact that in Germany they were not only independent of one another but divided between two different schools: medical and "philosophical." Whitman had not yet come to grips with the problem that his former Harvard colleague Charles S. Minot, for example, was confronting at that moment, mentioned in the preceding chapter— namely, the tension between the concept of biology and the aims of medical education.[24]

This article, however, along with his other publications and research entrepreneurship, were sufficient to gain Whitman his professorship at Clark. His experience soon led him to the conclusion that biology was a natural intellectual unit whose rapid growth depended on cooperation among different specialists in an environment free from distraction by outside concerns. In 1891 Whitman replaced his fivefold German division of biology with the Hopkins dichotomy between morphology and physiology; he then criticized physiology for being "limited too exclusively to the practical ends of medicine." It would be necessary to correct this imbalance by promoting the creation of "biological physiology," focused on the functions of invertebrates; when this subject received its proper emphasis, biology as a whole would regain its natural balance and develop as it should.[25]

Whitman gained the opportunity to implement his ideas and began to face both the temptation and the challenge of what would later be called biomedicine in 1891, when President William R. Harper of the new University of Chicago offered him and a number of his Clark colleagues places in the Illinois venture. Harper, like Gilman at Hopkins, planned a medical school as an integral part of his university from the outset. When his search for life scientists brought him into contact with Whitman, he was able to lay out a grand scheme in which a huge endowment, perhaps $4 million, would soon be available to support research in biological, biomedical, and clinical areas. The endowment would be so large that none of the conflicts Whitman worried about with regard to physiology could materialize.[26]

Whitman accepted Harper's offer to come to Chicago, but he tried to make sure that organization of the program and personnel would conform as much as possible to his own ideas. He pushed for an independent "school of biology," within which work in various fields would be coordinated. Though the current status of different specialties (and the availability of individuals) dictated an initial division of the school into departments of zoology, anatomy, neurology, physiology, botany, and paleontology, he hoped ultimately to divide the field, as he said, "according to the essential nature of the problems and methods, rather than according to the systematic or geographic relations to be studied"; this would mean elimination of "zoology" and "botany" and their replacement by finer functional divisions such as cytology, embryology, and taxonomy. When the availability of cash limited the number of initial appointments, Whitman's priorities became clear; he delayed the search for a botanist and in the crucial science of physiology abandoned the plan to bring in his Clark colleague, the vertebrate physiologist Warren Lombard; instead, he chose the one physiologist in America he had commended the preceding year for pursuing "biological physiology"—recent German-Jewish immigrant Jacques Loeb. Whitman hoped that Loeb would work to eliminate the distortion that medicine had produced in this field and allow it to grow in close association with its natural intellectual relatives.[27]

CHALLENGES TO ACADEMIC BIOLOGY

There were thus in the early 1890s a cohort of professors who believed in academic biology and who were having significant success in organizing a range of life science enterprises in urban universities around it. There were also efforts to promote it at smaller institutions and to link it, at least generally, to national purposes. For example, Hopkins graduate and University of Georgia professor J. P. Campbell prepared a survey for the federal Bureau of Education that promoted the value of the new biology within the ideology that the primary aim of college was to discipline the mind. More pointedly, Campbell argued that biology would provide training in the aspects of scientific method that students could use to deal "scientifically" with problems of business, society, and politics.[28] Yet the efforts to organize university life sciences around academic biology were soon challenged with considerable effectiveness by adjacent scientific groupings: on the one hand, by biologists' erstwhile allies in medical reform, and on the other, by those specialists—most notably the botanists—whose status was being diminished.

Medical Competitors

Medical scientists' challenge to biology was almost completely a matter of intrainstitutional competition and hence generated little public debate. The processes at work can be seen through examination of events at the three most important institutions discussed above: Hopkins, Columbia, and Chicago. (The similar events at the University of Pennsylvania were sketched in chapter 4.) The Hopkins situation was straightforward. In 1893, pathologist William Welch and clinician William Osler finally opened their long-planned medical school on the grounds of the new Johns Hopkins Hospital, two miles east of the main university buildings. Martin, however, was not part of this initiative—he was pressured to resign his professorship due to alcoholism and opiate addiction. William H. Howell became professor of physiology, working at the medical school, and the morphologist Brooks took control of the physiological royal palace Martin had so lovingly designed.[29]

At Columbia, the relations between biological and medical scientists were never stable. The P&S professors rapidly became accustomed to university surroundings and began to encroach upon areas claimed by the biologists. As early as 1893 Wilson and the physiologist Curtis were involved in official negotiations to demarcate their academic boundaries. The following year anatomist G. S. Huntington announced plans to build a museum of comparative anatomy at P&S. When Osborn protested to President Low that a medical school museum would merely duplicate the biology department's plans, the medical professors demanded that the name of the biology department be changed to the less comprehensive "zoology." Osborn retaliated by using his positions at the American Museum of Natural History and the new New York Zoo to restrict Huntington's supply of anatomical material. When Columbia College left Midtown for Morningside Heights in 1897, the zoologists turned their backs on their P&S colleagues and ensconced themselves in Schermerhorn Hall.[30] Only Frederic S. Lee was left stranded in the no-man's-land between zoology and physiology. He tried to resolve this in 1894 with an article claiming that "more and more are physiology's claims to admission to Pure Science and Philosophical faculties being recognized. It should be placed and will be placed by the side of chemistry, physics, and the morphological division of biology." But his own claim to admission to Columbia's Faculty of Pure Science, with its supply of graduate students, was rejected. General physiology remained a one-hour-per-week elective for medical students, and around 1900 Lee abandoned the subject to study problems of fatigue and industrial efficiency. On this basis he was promoted to a professorship in the medical school in 1904, and he became head of the department five years later.[31]

Biology was least affected by medical science competition at Chicago, but not for lack of trying by President Harper. He had promised Whitman and his colleagues that the university would provide them with substantial laboratory facilities. But when local philanthropist Helen Culver offered to donate $1 million in 1895 to endow biology, he tried to persuade her to direct the money toward medical work instead. He declared at the dedication of the biology laboratories the following year that he was "laying the foundation of a school of medicine," and in 1897 he engineered an affiliation between the university and Rush Medical College. He brought in Hopkins medical dean Welch to reassure the "cultivators of pure science" that there was no need to worry that a medical school "will bring any elements unsuited to the highest university ideals." Within the next few years, however, as he created new professorships in anatomy and pathology, he denied promotion to embryologist William M. Wheeler and cytologist Shosiburo Watase, whom he considered a "luxury."[32]

Ultimately, however, Harper's efforts to develop medical science were unavailing. When John D. Rockefeller refused to contribute, the plan for a University of Chicago medical school collapsed; Rush remained into the 1920s a semiautonomous institution.[33] Whitman's division, housed in the magnificent Hull Biological Laboratories, was left largely undistracted in its choice of faculty and research programs. Whitman and his colleagues were able to produce a generation of graduate students to staff colleges throughout the country who perceived the University of Chicago as the natural environment for academic biology. They did not realize that their program had been established largely because of interest in scientific medicine and that its survival had depended on the failure of that interest to be realized.

Recalcitrant Botanists

While advocates of academic biology and medical science elbowed each other at the major universities, competing for administrators' attentions and for endowment, the conflict between biologists and their colleagues in botany had an important public dimension. Botanists had remained unconcerned with the activities of the biology advocates through much of the 1880s, when biology remained a series of local endeavors clearly linked to medical reform. When, at the end of the decade, the biologists broke out on their own and began to expand—both into more universities and into colleges—botanists suddenly recognized the danger of what Harvard botanical leader William G. Farlow later called the "biological epidemic." They responded sharply and openly.[34]

J. P. Campbell's Bureau of Education survey on biological teaching was what stimulated the botanists to action. When the survey forms were distributed in 1890, John M. Coulter's *Botanical Gazette* published an anonymous polemic that attacked graduates of Johns Hopkins who were "totally devoid of any botanical training and totally pervaded with an uncontrollable yearning to label their zoölogical courses with the word 'biology.' " This renaming was taking place, the author claimed, because zoologists were ashamed of "past iniquity in methods of zoölogical instruction which makes the very word 'zoölogy' distasteful to the teacher of to-day."[35] A year later, the botanists broke up the biology section of the AAAS, which they had dominated from its creation in 1882, and set up their own, explicitly botanical, group. The handful of zoologists who had supported the AAAS (rather than the ASN) were immediately rudderless.[36] When Coulter learned of Whitman's plans at the University of Chicago, he complained editorially about "the old story of zoology masquerading in borrowed plumage as biology." Finally, when Campbell's report was published, Conway Macmillan, an aggressive young ecologist at the University of Minnesota, used the pages of *Science* for a slashing attack on Hopkins, Columbia, Chicago, and other institutions that, he argued, were promoting a "sham biology" that ignored "one-half the content of the science."[37]

These criticisms produced results. Nathaniel Britton, prime mover in the project to build the New York Botanical Garden, had been an instructor and adjunct professor at Columbia from 1887 on. In 1891 he became a regular member of the faculty, with a position independent of the biologists. At Johns

Hopkins, the refusal to hire a botanist finally ended in 1893. And at Chicago, after five years of intermittent battles, Harper hired *Botanical Gazette* editor Coulter as a "head professor" in 1896, with a mandate to create a botany program. When ground was broken for the university's Biological Laboratories, Coulter emphasized that the erection of a large building for botany (alongside others for zoology, anatomy, and physiology) demonstrated that it was a significant and developing branch of science.[38]

The consequence of the medical scientists' competitiveness and the botanists' attacks was that by the middle of the 1890s "biology" was a less viable framework for university science than it had been a decade earlier. In these rapidly growing urban academic settings, specialization was ascendant. Life scientists, like urbanites more generally, presented the appearance of being isolated, disconnected atoms, pursuing fragmentary and competing projects. Such a picture, though accurate, was not complete.

A Place of Their Own:
The Significance of Woods Hole

Until the end of the 1880s, life scientists had the same kinds of opportunities that people in other scientific areas had for organizing intellectually broad enterprises on a national scale. Given the diversity of initiatives, however, advocates of a single science of biology could gain only limited success through these avenues. Charles O. Whitman's great insight was to see that the particular objects and problems of biology could be linked to an unprecedented and unique kind of scientific organization. Biology could be grounded in the experiences of a community of scientists that would be both real and close to the ideal. Unlike other researchers, who experienced the "scientific community" as an abstract aggregation of the contacts they made in schools, journals, and professional meetings, biologists could structure their scientific lives around one particular place—Woods Hole, Massachusetts—that could blend the rhythms of everyday life with dreams of intellectual utopia. In chapter 2 and the prologue to this part I described aspects of biologists' activities in Woods Hole that were related to larger developments. This chapter focuses directly on that unique place.

The basis for the Woods Hole experience was a social development specific to the late nineteenth century; namely, the extended annual middle-class resort vacation. In the last decades of the nineteenth century, a plethora of "summer colonies" developed in the mountain and seashore regions of New England. Business and professional families traveled to them for a mixture of reasons: to get away from urban heat and congestion, to get closer to nature, and, above all, to socialize in a leisured setting that offered a larger circle of associations than could be found at home but was segregated by ethnicity, class, and interests. The academic calendar enabled professors to participate in this annual migration. Although limited in income, most college teachers had the freedom to arrange their summers as they wished.

Whitman envisioned Woods Hole as a summer colony for academic biologists. Three months each year, the Marine Biological Laboratory (MBL) would provide scientists an alternative to urban existence, would enable them to work on living organisms, and would enhance their ability to pursue research in an unconstrained environment. Whitman's larger aim, however, was to get isolated, individualistic scientists to create a community—a grouping that would be national in potential membership, enduring in existence, informal in organization, and rich in possibilities for intimate interactions. The plumbing supply magnate Charles R. Crane, the major patron of the Woods Hole scientific colony before World War I, spoke a language understood throughout the business and professional world when he explained to John D. Rockefeller, Jr., that the MBL was "more than a Laboratory and had many of the elements of a Biologist's Club."[1]

Whitman's hope that a unified science of biology could develop in and through a seaside summer colony, or club, was wonderfully successful. In the 1890s, Woods Hole became the major gathering place for the American scientists who identified with biology and would dominate the national landscape of the life sciences well into the twentieth century. Its summer residents included E. B. Wilson, T. H. Morgan, Jacques Loeb, and Frank Lillie, men who would make major discoveries and would make biology the first basic science in which Americans were internationally preeminent. The MBL both made possible the experience that defined biology and created and sustained an American biological elite.

Yet, as I have indicated, this achievement also entailed certain limits. Americans who structured a significant part of their lives around resorts could easily become insular and detached from the larger world around them. Biologists shared in these tendencies. In their pursuit of intellectually exciting questions, they could lose track of what was occurring to other American organisms.

This chapter details the development of the biological summer colony in Woods Hole in relation to four factors: the summer resort movement, the interest of life scientists in marine organisms, the idiosyncratic history of Woods Hole, and the leadership role of Whitman. It emphasizes the degree to which Woods Hole became the center for an American biology. It then concludes by indicating the degree of isolation of the Woods Hole biologists from both their scientific neighbors and some recognized national problems at the beginning of the twentieth century.

SUMMER COLONIES

Nineteenth-century American cities were particularly unpleasant places to live during the summer months. The heat and humidity were notoriously oppressive, especially in contrast to northern Europe. Garbage and manure produced smells and flies. Residents feared malaria and other epidemic diseases. A further problem during the summer, from the viewpoint of the well-to-do, was the increased visibility of the immigrant poor, who thronged streets and parks in efforts to find alternatives to poorly ventilated tenements.[2]

Most economically comfortable urban American families escaped the cities during July and August. In the early part of the century they would travel a few miles to live with nonurban relatives or would board informally with nearby farmers. After the Civil War, however, many of the more convenient destinations became urbanized, or, like Coney Island, became accessible to large numbers of lower-class short-term visitors. Wealthier families and those with opportunities for extended leisure sought new places, further away but otherwise more attractive, to which they could return each year. Most resorts developed in economically depressed areas, including such mountain regions as the Berkshires and Catskills, or at various points on the New Jersey, Long Island, and New England coasts accessible by either rail or steamship. Vacationers could rent rooms or houses, and in some places they bought up the property of local farmers and built summer homes.[3]

While summer resorts developed primarily as the result of climate and separation from cities, a number of other elements were also of considerable importance. The most obvious of these was closeness to nature. Resort hotels and homes opened out onto landscape and seascape and were surrounded by trees, flowers, and wildlife. The emphasis was on the picturesque and exotic. Summer residents who bought up subsistence farms gradually reestablished forested panoramas. They planted unusual flowers and shrubs and admired the presence of nondomestic animals—birds especially, but also small mammals in upland areas and marine organisms on the coast. Yet with the possible exception of the "camps" built in the Adirondacks, nature was never overwhelming at eastern resorts. In comparison to the scale, monotony, and vermin of the Western frontier, the experience of nature in these places was comfortable and easy, suitable for the elderly, women, and children—an enlarged, Americanized version of the ideal English garden. Though limited in grandeur, resort nature made possible an experience of the living world that was intimate and sometimes intense.[4]

A second aspect of resorts was their relaxed, playful atmosphere. Vacationers were away, for the most part, from work and school; they could experience the pleasures of living in unhurried, seemingly unhierarchical surroundings. Since most resorts were segregated by class, ethnicity, and lifestyle, summer residents at least anticipated that they would be free from the social tensions common in the cities. Moreover, social conventions were simpler in many such places set temporarily apart; women could walk freely without fearing derogatory comments from teamsters, and men could cultivate roses without losing status as leaders in business competition.

Resorts were also important as places where the urban well-to-do could interact within a communitarian setting different from that of cities. "Colonies" of prosperous urbanites usually formed around old, often quaint villages. These villages usually had two sharply defined social strata—the wealthy who worked elsewhere and a nonmobile lower class that combined service roles with some traditional employment, such as fishing. Yet because the lower-class residents were limited in numbers, dependent on their wealthier seasonal neighbors, and often old-stock Americans, the summer people usually perceived the community as socially and economically homogeneous in comparison to the cities, crowded with unassimilated immigrants. With physical closeness and time on their hands, summer residents were able to develop personal ties. Since vacationers were independent of the local economic hierarchies, they perceived community primarily as the result of voluntary activities. Furthermore, at resorts it was possible to establish close relations on a regional, if not national, scale, transcending the limitations of municipal social networks; wealthy residents of smaller cities in particular could go to resorts to establish social ties with metropolitan elites.[5]

Summer colonies were thus highly artificial and modern settings whose success, paradoxically, depended on mimicking what late nineteenth-century urbanites believed to be natural and traditional. In her history of New England tourism, Dona Brown emphasizes how hard it was for farmers to learn the "old-fashioned" Yankee cooking their boarders anticipated, and how Nantucketers transformed themselves between 1850 and 1900 from cosmopolitans into quaint islanders as they came to rely on tourism. Sophisticated vacationers understood the ambiguities of their summer settings; but over time, some could lose track of the amount of development that had taken place.

SUMMERING SCIENTISTS

The previous chapter emphasized that biology developed as an urban science. In the 1880s its geographic center was in Baltimore, at the Johns Hopkins University Biology Laboratory (see figure 5–2). Ethan Allen Andrews, a student and later professor, forcefully recalled that this laboratory seemed "even more in the midst of the city than the other buildings of the University." Designed as it was for physiology, it combined the features of a city schoolhouse with those of a slaughterhouse. Its unadorned main entrance faced heavily used North Eutaw Street, and local dogcatchers periodically brought their animals to a small basement door for sale to the staff. Chutes for sanitary disposal of sacrificed material ran from the workrooms to a basement incinerator (see chimney with circular deflector), and the building's rear overlooked an alley where, Andrews joked, "nothing white lived . . . except a freak sparrow with white feathers."[6]

Biologists experienced with greater intensity than other comfortable urbanites both the difficulties and the aspirations that came with the approach of summer. They were reminded how limited was the range of material—largely insects and domestic animals—available for laboratory work. The heat and vermin that made city life unpleasant during the summer months made work with organisms disagreeable and sometimes dangerous. By contrast, the countryside offered biologists the possibility of coming into contact with a great variety of living animals in natural surroundings. Seashores were particularly appealing, since organisms were abundant, fresh, and either cheap or free for the taking. Because marine invertebrates were both more durable and less personable than higher animals, they were easier to manipulate while alive; they were, figuratively and sometimes literally, transparent to the sufficiently careful observer. They possessed unusual anatomical and physiological qualities and could provide valuable insights into the evolutionarily crucial early stages of embryological development.

The desire of urban biologists to leave their universities for the seashore did not, however, entail either that they would focus on marine organisms permanently or that they would congregate, with their families and students, at the same place every year. Life scientists were opportunistic regarding their choices of organisms and were potentially peripatetic during an era when equipment needs were minimal. Louis Agassiz had traveled with the Coast Survey to Florida to study corals and then collected and studied the animals near his summer home northeast of Boston. He spent most of the 1860s working indoors in Cambridge. Then, with money from the New York real estate speculator John Anderson, he established a summer school for nature study on the bleak island of Penikese, in Buzzard's Bay, in 1873. Alexander Agassiz closed the school a year later, after his father's death, preferring the scientific and social advantages of Newport.

The most advanced American academic biologists were in fact moving in other directions in the 1880s. Hopkins zoologist William K. Brooks had established the Chesapeake Zoological Laboratory in 1878. This was a deceptively concrete name for a series of camps, involving simple equipment and a handful of professors and advanced students, set up at different sites on the Chesapeake, on North Carolina's Pamlico Sound, and, by 1886, in the Bahamas. This kind of strenuous travel, however, was limited largely to energetic young men. Brooks himself seldom joined in these trips after he reached the age of forty-five. Instead he built a home in the Baltimore suburbs on the newly constructed bucolic Lake Roland. Working with specimens obtained on short outings or through students, he focused his energies on transforming biological material into scientific text. In 1888 he declined the invitation to spend his summers in Woods Hole as director of the new MBL.[7]

The Development of Woods Hole

Prior to the Civil War, Woods Hole, Massachusetts, was a nondescript Cape Cod village whose two hundred residents subsisted through sheep farming, crystallizing salt from seawater, and providing support for a handful of whaling vessels. In 1850, Joseph Story Fay, a Boston-born, Savannah-based cotton merchant, decided to summer there. He bought significant tracts of property in and around the village and began to improve the scenery by planting trees— most notably European beeches—on the pastureland he owned.[8]

The direction of local development changed dramatically in 1862. That year a group of Boston and New York businessmen created the Pacific Guano Company, and they built a factory on Long Neck, a peninsula opposite the village. There Irish workers mixed fossilized bird droppings, sulfuric acid, and fish scraps into a powerful fertilizer. Ships began to carry guano, potash, sulfur, and decaying fish into Woods Hole's harbor on a regular schedule. In 1872 the Old Colony Railroad extended a spur to the village to service the factory. The arrival of the Fish Commission accelerated this process of industrial development. As I noted in chapter 2, Spencer Baird obtained a substantial congressional appropriation for harbor improvements and made Woods Hole the base for his fleet. The village began to anticipate a future as a significant federal and commercial port—a rest stop for sailors and a center for ship maintenance. Between 1850 and the 1880s the village's population nearly tripled.

In the 1880s, therefore, Woods Hole was not much of a vacation spot. Joseph Fay's daughter-in-law Elizabeth demurely recalled that when she moved there in 1876, "in certain winds the terrible odor from the guano works was most offensive." Street life was dominated by the dirty and rough fertilizer workers. Year-round industrial activity meant that seasonal housing was scarce. Baird solved his housing problem by having the government build quarters for his family and dormitory space for his male staff. Academic scientists unwilling to live in bachelor quarters, however, had to scrounge for accommodations. While Hopkins, Princeton, Harvard, and Williams College provided money for Baird's laboratory in return for workspace, few professors went to Woods Hole regularly. In the MBL's first year, all its participants were either unmarried or teachers anticipating a one-time student program.[9]

At the end of the decade, however, the ambience of Woods Hole changed dramatically. Marine activity declined after the *Albatross* left for the Pacific in 1888. The sudden bankruptcy of the Pacific Guano Company in early 1889 had deeper consequences. Laborers left, village rooms stood empty, and for the first time in a generation, the smells of phosphate, sulfuric acid, and putrid fish gave way to those of salt and seaweed. Discerning developers could begin to imagine a resort. A group of wealthy Bostonians demolished the factory on Long Neck, renamed the area Penzance Point (after the Cornish peninsula recently romanticized by Gilbert and Sullivan), and built large summer cot-

Figure 6–1

Pacific Guano Works, Woods Hole, Massachusetts, about 1875, seen from the future site of the Fish Commission. The same location about twenty years later, redeveloped for luxury cottages, can be seen in the left rear of figure II-2. Photograph © Mystic Seaport, Mystic, Connecticut.

tages on these now prime waterfront parcels (see figure II-2). A factory workers' dormitory was renovated to become the Breakwater Hotel.[10] At the same time, Charles O. Whitman was conceiving his idea that Woods Hole could become the locus for a cooperative summer colony for biologists.

WHITMAN'S DESIRES

I have discussed certain aspects of Whitman's career earlier. But to appreciate the reasons for his interest in the MBL and the nature of his development plans, we need to investigate his unique life more deeply. Alienated from his origins, and a longtime academic nomad, Whitman was searching in biology for a perfect community.

Whitman was born in 1842 in Woodstock, Maine, an isolated village in the foothills of the White Mountains. His young parents, prompted by evidence of his development, had married only four months earlier. They anticipated, however, that they would soon transcend their mundane imperfections, believing in the Adventist prophet William Miller's prediction that Jesus would return to earth for the Last Judgment in the fall of 1843 and that the faithful would then begin a new, eternal life. When the Second Coming did not occur, Whitman's father Joseph was among the most severely disappointed and was briefly incarcerated in an asylum. Described as an "Adventist of the hardest kind," he lived in isolation from his neighbors, working as a carpenter and anticipating the imminent return of Jesus.[11]

Whitman established his identity by resisting the miraculous decorporalization that his father had anticipated for him and by searching for a meaningful alternative perspective. As an adolescent, he deserted Adventism for the progressive rationalism of the Universalists, and he rejected biblical village craft work in favor of education and movement out into the wider world. After attending a nearby academy and then teaching for a number of years in local schools, he worked his way through Bowdoin College. He left Maine on graduation in 1868 to teach near the modern factory town of Lowell, Massachusetts, and three years later he obtained a position at the city-run Boston English High School. His interest in life science emerged in conjunction with a personal religious crisis in the early 1870s. Disturbed by the relativistic implications he saw in the writings of English positivists, he sought in the study of organisms a way to find meaning and purpose in the natural world. He attended Louis Agassiz's natural history summer school on Penikese in 1873, and two years later, at the late age of thirty-two, he decided to become a zoologist.[12]

Whitman's pursuit of learning and discovery involved more than a decade more of wandering and isolation. He was unusual among American zoology students in deciding to go to Europe for graduate work. Studying with parasitologist Rudolph Leuckart at the University of Leipzig, he took up the study of fertilization and early embryonic development, using as his material the fish-leech *Clepsine*. After a brief return to the United States he spent two years as a professor in the exotic setting of the University of Tokyo, and then visited the Naples Zoological Station. Following his return to America in 1882, he continued to work, in Cambridge and Milwaukee, in the highly technical areas of cytology and embryology.

During these years Whitman sought, quixotically, to create communities of autonomous investigators. He left Japan when university officials made clear that he could not treat his students as if they were colleagues.[13] In Milwaukee he demanded that his student-patron E. P. Allis organize the Lake Laboratory as a communal, not an industrial, endeavor. He jumped at the unpaid position of director of the new Marine Biological Laboratory (after both Brooks and Williams College professor Samuel Clarke turned it down) because he saw the possibilities the laboratory might offer for interaction among American life scientists, himself included. He was speaking both generally and confessionally when he wrote in 1890 that "isolation in work" was becoming "more and more unendurable."[14] Witnessing how Woods Hole's environment and economy changed with the closing of the guano works and the downsizing of the Fish Commission, and seeing the local importance of the MBL grow, he formulated the vision with which I began part 2 of this book. Intellectually, it was an expression of Herbert Spencer's philosophy. More immediately, it reconfigured the ideal group experiences Whitman had had in Adventist camp meetings as a child and that he knew were the norm among the Methodists who traveled through Woods Hole each summer on their way to their own annual gathering on Martha's Vineyard.

THE BIOLOGICAL COMMUNITY

From the perspective of his scientific contemporaries, Whitman did not seem very entrepreneurial at Woods Hole. The MBL had no research agenda, no formal ties between its classes and university degree programs, no outside institutional affiliations, and minimal leadership. All Whitman offered was a place for scientists to come each summer, with their families and students, where they could live as biologists. This seemingly minimal effort, however, was energized by Whitman's living faith: his organicist belief that biologists, like organisms, would interact to produce a whole larger than the sum of the parts and would thereby individually become enriched. Success, while natural, would not be automatic. It depended on a favorable environment, the right people, intimate associations, and continued autonomy.

The basic requirement for the successful development of a biological summer colony in Woods Hole was an environment that would facilitate, in Whitman's phrase, "rest and investigation." The MBL supported investigators by providing bench space in a series of hastily constructed, camp-style buildings and by stocking the laboratories with glassware and common chemicals. Unpolluted salt water and live organisms were supplied to all participants courtesy of the Fish Commission. A library, tended voluntarily by Mt. Holyoke professor Cornelia Clapp, grew rapidly. The laboratory acquired a few small boats and hired pilots who could take scientists beyond the village's piers and beaches in search of animals. Scientists had to bring their own microscopes, microtomes, and other instruments.[15]

Rest was a function of the larger local setting. The village's air and water were, of course, clear for the first time in a generation, and its streets were no longer crowded with fertilizer workers and sailors. Summer housing was more available than a decade earlier. The laboratory maintained primitive dormitory accommodations for students, and the scientists who returned annually soon established friendly relations with locals and thereby gained priority in claiming summer space. Permanent residents could joke among themselves about the "little fellows" who spent their time "bug hunting," but they grew dependent on this population that returned predictably each June and provided employment not only for domestic workers but also for sailors, fishermen, and property managers. Good scientists realized that Woods Hole was a place where they could feel at home.[16]

Attracting that select population was a challenge that Whitman understood how to meet. The Fish Commission's continuing arrangements with Hopkins and Harvard ensured that some well-connected academics would appear each year but limited the MBL's ability to attract men from those institutions. Whitman drew, first of all, from the urban universities created after Baird's death (Clark, Columbia, and Chicago), and then from women's colleges and smaller institutions. He attracted desirable individuals by offering free "guest" laboratory space. His hope was that professors who appreciated their experiences and wished to become full community members would be able to persuade their home institutions to make an annual contribution ($100 per investigator) to the laboratory and would also bring along students who would pay for instruction. Between 1888 and 1895 the number of investigators at the MBL grew from 7 to 63, and students increased from 8 to 163. In the

Figure 6–2

MBL main laboratory, about 1890, with blackboard, zoological charts, student benches, worktables with chemicals, and exposed pipes and framing. No artificial lighting was provided. Photograph courtesy of the Marine Biological Laboratory Archives, MBL/WHOI Library.

latter year, 30 "co-operative colleges and societies" were recognized in the annual report. Colleges and students supplied about half the laboratory's $14,000 operating budget. Biologists thus tapped the growing pool of higher education money, yet could reasonably view the MBL as a cooperative venture they supported themselves.[17]

If cooperation was the necessary basis for success, community was the consciously envisioned outcome. Whitman had emphasized that the effectiveness of "a social or an organic body" resulted from having its "points of union multiply," and Woods Hole in its early years was notable for the density of its community network.[18] Classes were informal in both organization and activity. The flimsiness of laboratory walls encouraged the sharing of results. Professors and students joined in collecting trips, sailing excursions, and picnics. Younger male scientists swam naked together off the wharf of the former guano works. Perhaps the most significant contribution to community life was the creation of a dining club, open to both MBL participants and the men from Harvard and Hopkins who continued to work at the Fish Commission.[19] On a more formal level, the Friday evening lectures provided a sequence of specific foci for common discussion. In 1894 the summer colonists came to-gether (as, in Whitman's words, "a national convention of students, teachers, and investigators in Biology") to establish the Biological Association. This cooperative organization, open to scientists and their friends and families, took on the twin missions of raising supplemental funds for the laboratory and lobbying for the interests of the summering scientists.[20]

The particular characteristics of Woods Hole life were evident in the great-est challenge that Whitman faced during his directorship. The MBL was a peculiar hybrid in its origins. Although Whitman had guided daily operations of the laboratory from its birth in July 1888, the institution was formally con-trolled by a group of scientists, educators, and feminists who had conceived it in Boston the year before. Boston Society of Natural History curator Alpheus Hyatt, Harvard botanist William Farlow, Brahmin medical embryologist Charles Minot, and MIT biologist William Sedgwick had envisioned the MBL as an annex to the seemingly dominant Fish Commission laboratory. They had planned a place where Massachusetts teachers could learn the ba-sics about marine life and where a few scientists could work on particular projects in a setting that benefited from proximity to the government facility.[21] Since their own scientific networks and summer plans were already estab-lished, they did not anticipate spending all their vacations in as poorly devel-oped an area as Woods Hole. The summer-camp laboratories that Whitman constructed did little to make the village more appealing to them, nor did the rapidly growing population of aggressive young New York and Chicago scientists. Moreover, some of the founders were not happy that the institution they created was becoming the center for what Farlow would call the "biologi-cal cult." Coming down from Boston on the "Dude Train" for a few days at a time, these trustees would feel marginal to the intense interactions among the summer colonists.

It is not surprising, then, that by the mid-1890s the Boston trustees were resisting Whitman's efforts to expand the MBL. They decided to stop his pell-mell erection of cheap workspace and instead proposed a capital campaign to build a single permanent facility that they believed would ensure the labo-ratory's true "national" preeminence. The conflict worked itself out in a man-ner that was typical of summer colony politics. The trustees had normally conducted the laboratory's important business in Boston during the winters, far from most of the colonists. A supposedly routine meeting held in Woods Hole on 6 August 1897 got out of hand when Farlow and Minot, who had come down for the weekend, had to leave in order to catch the day's last train back to the Hub. A rump assembly dominated by the summer residents then abruptly added a large number of individuals (mostly students) to the control-

Figures 6–3 and 6–4

Two views of the biological community, 1895. Above, students, staffers, and family members on a collecting trip. Compare the MBL's ship, *Vigilant*, in the rear, with *Albatross*, figure 2–3. Below, a group portrait at the laboratory. A University of Chicago booster (with letter sweater) situated himself prominently in both pictures. Photographs courtesy of the Marine Biological Laboratory Archives, MBL/WHOI Library.

ling MBL Corporation and called a special meeting ten days later to revise the bylaws. When that meeting opened, trustee leader Farlow tried to declare it illegal. After being outmaneuvered, he walked out; the colonists then seized control and reorganized the laboratory as a cooperative association run by the active summer participants. Whitman's prosperous brother-in-law, L. L. Nunn, guaranteed that the laboratory would run in the black. The biological community had seized control of its own development from a group that Whitman had, in a characteristic metaphor, dismissed as "extra-ovate."[22]

Woods Hole and American Biology

Scientists associated with the Woods Hole summer colony produced research of major significance. The cell lineage studies of the 1890s formed one of the first instances of an independent and conceptually fruitful tradition of American laboratory science. Interest in experimental embryology led by the end of the century to Jacques Loeb's spectacular invention of artificial parthenogenesis and, in the years prior to World War I, to Frank Lillie's creation of the foundations for modern understanding of sexual reproduction. MBL–affiliated cytologists such as E. B. Wilson and Nettie Stevens linked chromosomal mechanics to the phenomena of Mendelian heredity; T. H. Morgan's interactions with these colleagues formed the background for his Nobel Prize–winning work on *Drosophila* genetics.[23]

Woods Hole's importance, however, extended beyond the research performed there. Its greatest impact was as the center of a unified discipline of biology. Yet its role in defining that discipline was unique. Chemists, psychologists, and other scientists were isolated most of their lives in their university departments, government bureaus, or corporate laboratories. They interacted with colleagues through reading, correspondence, and the yearly meetings of their professional societies, where formal papers alternated with the momentarily intense interactions of conventioneers. Such rhythms pushed younger scientists in most disciplines to define their work carefully so that it would stand out, and they induced each field's leaders, in reviewing membership lists and preparing programs, to make explicit decisions about definitions and boundaries. Admission to societies and programmatic statements became major issues for these disciplined academics.[24]

Biology, centered on the annual gathering at Woods Hole, was different. Apart from their focus on academic research, biologists were catholic in their enthusiasms. They discussed a wide variety of subjects, all of which they considered parts of their science. These included heredity, evolution, cell cleavage, neuroanatomy, and behavior; in most cases, moreover, biologists ranged freely over organisms from protozoa to primates. As Jane Maienschein has emphasized, biologists also gladly accepted multiple causation in each of these areas; their only strong antagonism was toward explanations that seemed to reduce a wide range of phenomena to a single unifying factor. As a result of their eclecticism, American biologists made few efforts to provide a comprehensive definition of their science.[25] Biology was difficult, if not impossible, to define. Fortunately, however, definition was not necessary. Individuals who had established positions within the community centered on Woods Hole and who had demonstrated that they had good minds and did interesting research were, barring strong grounds to the contrary, biologists. Biology was the totality of subjects that Woods Hole scientists discussed. Whitman's inaugural lecture asserted the evolutionary conviction that organic social organization would eventually result in intellectual unification. Until that progressive development was completed, biology was defined in terms of its community.[26]

Woods Hole's importance was thus ultimately more social than intellectual. It formed the center for a cohesive academic elite that worked to make biology the core of the life sciences in America. Between 1892 and 1910 an average of 63 investigators worked each summer at the MBL, sometimes up to 30 more at the Fish Commission.[27] In addition, an average of 134 students attended the MBL's introductory course. Though investigators came from an average of more than twenty different institutions, the active local leadership—embodied in the MBL's board of trustees—was in the hands of professors from the University of Chicago and Columbia, the two leading metropolitan universities; by about 1905, Harvard, Hopkins, Princeton, and Pennsylvania were also playing major roles.

A number of measures indicate the extent to which the Woods Hole elite dominated the life sciences. The fifty highest-ranking zoologists in the first edition of *American Men of Science* (tabulated in 1903) included twenty-three who served as MBL trustees; if individuals closely associated with the Fish Commission and those who worked for a substantial number of years at the MBL (but left before being elected trustee) are added, that figure rises to thirty-three. The ASN had twenty-four presidents between 1891 and 1915; 71 percent of these were trustees of the MBL. All except one president of the American Morphological Society (1890–1902) were MBL trustees, as were six of the first seven presidents of the amalgamated American Society of Zoologists (1914–20). MBL leaders edited the *Biological Bulletin, Journal of Morphology, Journal of Experimental Zoology,* and the *Journal of General Physiology;* Columbia psychologist James McKeen Cattell, who edited *Science, Popular Science Monthly,* and *American Naturalist,* was a longtime summer resident.[28]

The wealthy businessman Crane understood the significance of describing the MBL as a "biologist's club." It was a place where a select group of people became friends, talked about their long-range plans, hammered out difficult disputes, and established a consensus to present forcibly to the outside world; even those who were not there each year could trust their old swimming companions to make decisions maturely and with due consideration of all the views that mattered. And because biology was centered on a club, formal organization was unnecessary. As Toby Appel has noted, the ASN became an anomalous professional society after the turn of the century, and the "American Biological Society" that the young physiologist Albert P. Matthews proposed in 1906 received no significant support. The leaders of biology worked out their field's problems informally during the summers.[29]

Neglecting American Life

Woods Hole, like other resorts, was an ideal world. Its scientific colonists were freed from workaday cares of elementary teaching and from outside distractions, and their days were rich in interactions with knowledgeable, like-minded colleagues and with students eager to learn. They were working with a comprehensible collection of manageable organisms and were able to address questions that were intellectually stimulating.

These qualities were, however, also the MBL's chief limitations. The problem with resorts was that people who frequented them could easily forget how artificial they were and could come to see life there as the norm. They would no longer willingly participate in a broader, more problem-filled world. The issue for biologists was similar. The exclusivity of commitment to Woods Hole varied from individual to individual, but insofar as scientists identified biology with life at Woods Hole, they became disconnected from other scientists, especially those in "applied" areas. They also became detached from immediate problems, especially those posed locally or nationally. Issues that were specifically American came to appear to be distractions from the deeper, more fundamental questions of academic biology.

This detachment was inherent in the original distinction that Whitman drew between the MBL and the Fish Commission and was reenforced in the 1890s as the commission's undistinguished leaders limited the reach of the government laboratory, even within the realm of "economic" problems. MBL leaders talked about the importance of building links with medical and agricultural scientists, but ultimately made only half-hearted efforts to bring in guests from those areas. Bacteriologists such as William T. Sedgwick and Edwin O. Jordan, who had been enthusiastic participants in Woods Hole activities in the first few years, drifted away from the MBL by the mid-1890s. Fisheries scientists in the late 1890s viewed their colleagues at the MBL as self-indulgent faddists.[30]

It is hard to argue conclusively about actions that MBL scientists could have taken but did not. The positive choices they made at any moment regarding subjects for research were so multiform that any claim regarding the roads they did not travel would be weak. The fraying of the thread connecting American academic biologists to the continent, its organisms, and their problems can be seen, however, in the poignant life of one of the more unusual summer migrants to Woods Hole. In the early 1910s, a passenger pigeon named Martha Washington became famous as Americans gradually realized that she was the last living representative of her species—unable, in her solitude, to reproduce. Following her death at the Cincinnati Zoo in September 1914, she was shipped to the Smithsonian Institution in a three-hundred-pound block of ice, mounted, and displayed as a symbol of the casual neglect of nineteenth-century Americans toward the continent's animals. The fact that she spent her best years in Woods Hole was forgotten.

At the end of the nineteenth century, a small but influential group of individuals at the nexus of zoology, government, and business became concerned about the extinction of distinctive American animals. The bison, a large mammal that had become identified with the North American landscape and the defeat of the Indians, drew the greatest attention. Naturalists, bureaucrats, and ranchers intermittently discussed the problems involved in protecting and breeding the remaining few hundred animals. These efforts culminated in 1905 with the creation of the American Bison Society under the leadership of Baird's old taxidermist William T. Hornaday—now head of the New York Zoo. Within ten years the bison had made it onto the front of the $10 bill and the back of the nickel, and the numbers of live animals approached five thousand.[31]

If the bison was the great success of turn-of-the-century wildlife preservation, the passenger pigeon was the tragic failure. At the beginning of the nineteenth century, Alexander Wilson had calculated literally billions in one flock in Kentucky. By the end of the century, well-situated Americans believed they had completely disappeared. In April 1900, for example, John F. Lacey, the Iowa congressman mentioned briefly in chapter 3 in connection with animal introductions, stood before the House of Representatives to lament that "the wild pigeon, formerly in this country in flocks of millions, has entirely disappeared from the face of the earth."[32]

He was not quite correct. As he spoke, a few pairs of pigeons were making preparations to breed. These were caged, a few blocks from the University of Chicago, in the backyard of Charles O. Whitman. In June these birds traveled by train with Whitman to their summer roosts in Woods Hole. Whitman had begun to study behavior and evolution in birds around 1896. In his yard he built cages for pigeons, doves, and other species, and he was able to obtain a handful of passenger pigeons from a Wisconsin fancier, whose own flock soon disappeared. He tried to breed his passenger pigeons and later claimed that he made substantial efforts to obtain more animals. His interest in these birds was part of his motivation in lobbying the University of Chicago to establish a "biological farm."[33]

But the preservation of the passenger pigeon was never near the top of Whitman's priorities. His major concern after 1895 was to complete and correct Darwin by demonstrating that evolution was a directed and progressive process. Unfortunately he published little on this subject prior to his death in 1910, and his thinking had little influence. His second interest was the preservation of his vision of Woods Hole. His presence was essential for maintaining the MBL's original values and autonomy; as a result, twice every summer he crated his birds for the three-day trip in a poorly ventilated baggage car between Illinois and Massachusetts. Though he tended the birds personally, losses in transit were inevitable.[34]

In 1904 Whitman reported that although his passenger pigeons had initially reproduced successfully, they "lately have failed to accomplish anything."[35] He attributed this failure to inbreeding. However, A. W. Schorger, who reviewed knowledge of the passenger pigeon exhaustively in 1955, noted that Whitman probably gave his birds inadequate protein; moreover, their conditions of existence were not the best. Whitman noted that when his birds began to nest, they repeatedly flew against the screens of their cages in their urge to find new locations, only settling down after a few days; moreover, their annual migration from Chicago to Woods Hole possibly coincided with one of the species' nesting periods.[36]

In contrast to the protectors of the bison, Whitman worked in almost complete isolation. He did little to learn what lore existed regarding their care, and he wrote nothing about his birds, or the impending demise of the species, in the ornithological press. Though he responded matter-of-factly to queries from pigeon enthusiasts around the country, he apparently never considered the possibility of making the preservation of the passenger pigeon an urgent

Photo by C. O. Whitman (University of Chicago)   October 16, 1906.

Mr. W. B. Mershon,
    Dear Sir:—I am much chagrined over my carelessness in overlooking your request for
a photo of a young Passenger Pigeon. I had best of intentions, but crowded work threw this out of
mind. I should have attended to it at first, had it been easy to get at the picture. I have been
away all summer and found things misplaced on my return. I fear it is now too late, but send the
picture to be used if you are still able to do so. I shall be very much interested to see your book.
I still have two female pigeons and two hybrids between a former male pigeon and the common
Ring-dove. The hybrids are unfortunately infertile males.   Very truly,
                                                              C. O. Whitman.

Figure 6–5

Photograph of one of Charles O. Whitman's passenger pigeons (possibly an offspring of Martha), reprinted with Whitman's comments, in W. B. Mershon, *The Passenger Pigeon* (1907), facing 198. Whitman's comments are as follows:

October 16, 1906.
Mr. W. B. Mershon,

    Dear Sir: I am much chagrined over my carelessness in overlooking your request for a photo of a young Passenger Pigeon. I had best of intentions, but crowded work threw this out of mind. I should have attended to it at first, had it been easy to get at the picture. I have been away all summer and found things misplaced on my return. I fear it is now too late, but sent the picture to be used if you are still able to do so. I shall be very much interested to see your book. I still have two female pigeons and two hybrids between a former male pigeon and the common Ring-dove. The hybrids are unfortunately infertile males.

Very truly,
C. O. Whitman

public cause (see text of the letter in figure 6–5). Efforts to search systematically for wild birds did not begin until 1909, when it was certainly too late.[37]

The bird who would become famous as Martha was among those who belonged to Whitman. She spent her prime summers, making some futile motions toward reproducing, in the midst of the annual gathering of the nation's most innovative biologists. In retrospect, the likelihood of preserving the species, even beginning in 1896, was not very great. But Whitman and his colleagues, although uniquely positioned—through possession, expertise, and status—to act, did little. Interested in basic scientific issues, they saw Martha as, at best, a tragic curiosity from an era they were leaving behind. In 1902 Whitman sold her to the Cincinnati Zoo. He retained a few birds, as he noted in his 1906 letter to the enthusiast W. B. Mershon, but they died a few years later.[38]

PART III

THE AGE OF BIOLOGY

A View from the Heights

In early 1892, when Columbia University was in the process of establishing its new Department of Biology, former U.S. Geological Survey leader and freelance intellectual Clarence King published an essay in *The Forum*, a New York literary magazine, titled "The Education of the Future." "What," he asked, "will the Columbia student do at the end of the next century?" Sounding themes his close friend Henry Adams would later memorably elaborate in his autobiography, King contrasted the rearward educational perspective cultivated by classicists since the Renaissance with the "ever-accelerating" rush into "new intellectual conditions" so evident in the America of the late nineteenth century. Also like Adams, King emphasized the twin nineteenth-century discoveries of conservation of energy and organic evolution. He noted that the first had already been translated into a myriad of important inventions. By contrast, scientists had as yet done "almost nothing to make [the] astonishing revelations" of evolution "conduce to the physical and moral welfare of the human race." In the future that imbalance would change. "This," King proclaimed, "is the age of energy; next will be the age of biology."[1]

Such a lapidary comparison would certainly have impressed such forward-thinking readers as the organizers of the new Columbia University Department of Biology. Today, attuned as we are to biotechnological achievements, it sounds wonderfully prescient, if a bit premature. The devil, however, would be in the details. What exactly would characterize the coming age of biology? And how would scientists and other Americans make the transition into this new era?

King himself was poorly equipped to deal with those questions. As a physical scientist, he interpreted the age of energy broadly, to include all of contemporary power and communications technology, from the dynamo and the phonograph to soon-to-arrive air transport. By contrast, his vision of the age of biology was limited. The only implication he saw in evolutionary science was what would soon be called eugenics—and that of a stern and narrow sort. From King's perspective, society was afflicted by an "army of incompetence, of insanity and disease." Scientists already possessed "a new table of biological commandments." Making the next generation "strong and fair" required only "the rigorous application of scientific biological restraints." It was merely a matter of a "quickened sense of moral responsibility."[2]

166

Figure III–1

Clarence King alone at the helm, sometime in the 1890s. This elegiac image ended the *Clarence King Memoirs* (New York: G. P. Putnam, 1904), compiled by King's friends. Photograph courtesy of the Huntington Library, San Marino, California.

King himself was not in a position, professionally or personally, to realize that goal. During the preceding decade he had moved away from the scientific world to pursue mining wealth and to participate in international high society. At the same time he had begun an enduring intimate relationship with Ada Todd, a young African American whom he married (without telling her his real name) in 1888. When he was urging educated white Americans to accept "biological restraints" and to honor "organic purity," his secret household in Brooklyn contained two (ultimately five) mixed-race children.[3]

The strains induced by this complicated existence were substantial. A year after preparing his prophetic essay, King committed himself to the Bloomingdale Insane Asylum, on Morningside Heights in northern Manhattan, for two months of treatment. He did little further work as an intellectual until his death in 1901, but found his greatest moments of happiness when living with his decidedly non-"fair" wife and children. The relations between knowledge and action in human biology were neither simple nor direct.

Five years after King's prophecy, Edmund B. Wilson and his fellow academic biologists, along with other Columbia faculty, took over the Manhat-

Figure III–2

Columbia University, around 1900, looking northwest across Amsterdam Avenue and 115th Street. Schermerhorn Hall stands to the right and in front of the large chimney. The domed building in the center is the Low Library. The large house to the right of the library steps is the Bloomingdale Asylum's Macy Villa, where King lived during his incapacitation in 1892. Photograph courtesy of the Library of Congress.

tan ridge top, with its views south toward the city's population centers and west across the Hudson River, until then occupied by the Bloomingdale Asylum.[4] Would they fare better than their self-appointed Moses in envisioning and realizing the age of biology in America? I examine that question in the remainder of this book. In the first four decades of this century, some leading biologists, and a larger number of their students and followers, left the insulated academic environments they had fashioned to explore the import of their new biology for American life. What did they imagine? How did they communicate with the public? What, ultimately, did they change?

These are large and unwieldy questions. I articulate manageable answers by focusing on three developments. The first, detailed in chapter 7, is quite specific. In the early twentieth century, a number of young men trained in biology descended, figuratively and literally, from Morningside Heights down into the great American metropolis a few miles south to spread biological knowledge among the striving masses. By the 1920s they had made biology, in the form of a year of high school study, part of the lives of urban middle-class—and ultimately nearly all—Americans. I emphasize how close high school biology was, in its origins and conception, to the thinking of leading academics and indicate how it was designed to advance, in a more focused way, the agenda of Lester Ward and G. Brown Goode; namely, to culture modern Americans.

Chapter 8 expands on this argument. It recounts the steps through which a number of leading biologists engaged with deep problems, such as the meaning of life and the nature of progress, and presented their views to the educated American public. It then focuses on the individual with which this book began—William Emerson Ritter. He linked progressive biological philosophy with public activism in a concrete program, centered on the Scopes trial, to get American people to see themselves as reasoning, improving organisms.

Chapter 9 shifts from speech to action and compares biologists' initiatives in the two prominent biological arenas of eugenics and sex. It examines why the former movement, led by Charles Davenport, was largely unsuccessful, while the latter, identified with Alfred Kinsey, had major consequences for the behavior and attitudes of Americans. It indicates to what extent, and in what ways, King was correct in his prediction that the twentieth century would be the age of biology.

The Development of High School Biology

DeWitt Clinton High School, the major academic public secondary school for boys in Manhattan during the first two decades of the twentieth century, was the central locus for the development of the American high school subject of biology. Beginning in 1902, nearly all freshmen—soon more than one thousand annually—enrolled in a year of coherent study of life. Clinton's teachers made their experience with these students the basis for national influence. They outlined and justified biology in science education journals; they participated on local, state, and national educational committees recommending the new subject; and, most significant, they published textbooks that reproduced their teaching throughout the country. By the 1920s biology had become a standard high school science and thus part of the common experience of middle-class Americans. In the spring of 1925, the teacher John T. Scopes of Dayton, Tennessee, was a typical performer within this new element of the nation's culture.

This chapter examines the reasons for high school biology's sudden genesis in early twentieth-century America and explores the implications of the subject's singular early developmental circumstances. It focuses on the structure of aims motivating high school biology's creators. As a subject that was both ostentatiously objective and intensely value laden, biology encompassed the many purposes that educators hoped to achieve from science teaching in the first half of this century.

The designers of high school biology, closely linked to the nation's leaders in both science and education, were immersed in the discourse of progressivism. In message and method, their subject was evolutionary to the core. Yet its focus was not on life's history, but on its present and its future. Clinton teachers designed biology for upwardly mobile, immigrant Manhattan boys between the ages of thirteen and sixteen. They taught students how to think about the cosmos, what to eat and drink, how to behave toward life-forms living and dead, what to consider clean and precious or dirty and dispensable, and how to react to intimate physiological functions in themselves and others. The ideal product of the biology course was a modern male—an individual whose physiological and intellectual development had converged, who understood his place in the world around him, and who could act intelligently to improve it.

This was a tall order. Many established science teachers did not believe that biology was an intellectually legitimate subject. Moreover, the realities of metropolitan life put formidable obstacles in educators' paths. Both teachers and students varied in their abilities and enthusiasms. The rapid expansion and increasing inclusiveness of the school system repeatedly threatened to overwhelm any established set of teaching practices. Manhattan seemed a particularly unpromising site for the study of living nature. Lastly, biology educators' pronouncements regarding important, but not easily discussed, subjects threatened to involve them in destructive controversy with parents and community groups. Problem areas included animal experimentation, dietary recommendations, and, above all, sex.

Teachers' responses to these problems epitomized the "luminous platitude" of progressivism: that advancement resulted from experimentation and adjustment rather than from confrontation. This perspective came particularly easily to biology teachers, since it was both the central biological principle they taught their students and the basic behavioral value they sought to convey to them. At the same time, commitment to experimentation and adjustment also expressed the most commonsensical sort of response to the realities of educational politics. Biology educators sought above all to establish their subject in the nation's schools; once it had become a natural stage in the development of Americans, it would—they hoped—open the way to new and broader experiences with life.

Examination of the origins of high school biology also opens a new perspective on the controversies over the teaching of evolution. Numerous writers, focusing on explicit textbook statements, have inferred that, sixty years after *The Origin of Species*, the subject was an established element within an otherwise innocuous life science curriculum. They have sought to explain why this seemingly unusual controversy arose in America in the 1920s and to determine whether the anti-evolution campaigns influenced the content of textbooks.[1] Viewing evolution as merely one among many contentious elements within a novel and fluctuating subject explains, among other things, why biology educators reacted with indifference to the conviction of Scopes. Recognizing that explicit discussion of the facts and theories of descent with modification formed only a small part of the function of evolutionism within biology education leads to reevaluation of what mattered and who won. Though Fundamentalists could censor explicit statements in textbooks through legislation, they had much less control over matters of perspective; educators continued to design biology to produce Americans who would be modern, secular, and humanistic.

LIFE IN HELL'S KITCHEN

Before 1900 there was no such experience as high school biology.[2] The vast majority of Americans gained their knowledge about the natural world from personal experience, local traditions, and primers or popular reading matter. The small percentage of the adolescent population who attended secondary schools (less than 4 percent in 1890) varied widely in the extent and nature of their exposure to life sciences. Students could study one or more of three subjects—physiology, botany, or zoology. Each was offered as an elective, for a variable number of hours, at any point in the curricular sequence.

The relations among the three life sciences were essentially competitive. Each had its own promoters and rationales. Physiology was sponsored by a wide spectrum of physicians and reformers (ranging from *Popular Science Monthly* editor W. J. Youmans to the Woman's Christian Temperance Union) who sought to provide young people basic information that would lead to healthy habits. Elementary botany, by contrast, was controlled by professors such as Asa Gray and Charles Bessey; it conveyed the order of creation, provided a model for systematic thinking, and trained students' faculties of observation and comparison. Zoology, the least widely taught subject, combined Louis Agassiz's emphasis on comparative anatomy with his romantic nature enthusiasm.[3] Though the American Society of Naturalists worked diligently in the years around 1890 to increase the amount of "natural science" in high schools, its members could not agree on what should be taught. The 1893 report of the prestigious Committee on Secondary School Studies (the Committee of Ten) of the National Education Association (NEA) both reflected and reinforced these competing perspectives on life. The subcommittee on natural history explicitly rejected the suggestion that subjects be blended. They recommended instead that schools choose between a full year of botany or zoology on the basis of teacher competency and student interest and that physiology, if offered, should be an additional, independent, course.[4] These disputes served to limit the visibility and status of the entire area.

Contrasting changes in the universities and the elementary schools during the 1890s held uncertain implications for high school life science instruction. In part 2, I sketched how proponents of biology, centered on Woods Hole, dominated such elite institutions as Johns Hopkins, Harvard, Columbia, and the University of Chicago. They emphasized laboratory research in the comparative anatomy and embryology of invertebrates, but they believed that their highly specialized work would ultimately lead to a general science of life. By comparison with botanists, they showed little direct interest in secondary education. In fact, a significant concern was that the high schools not take over the college course developed during the previous decade. Yet numerous teachers in normal schools and high schools, along with students interested in educational careers, began to train in the new university departments and at Woods Hole. They absorbed biologists' values, dicta, and hopes, and began to consider how to apply them at the secondary school level.[5]

Educational reformers working with elementary school-age children were also interested in the unified study of the natural world. But the form that interest took—"nature study"—represented tendencies quite contrary to university biology. Nature study advocates such as Ernest Thompson Seton argued that young children should be taught particular facts about the lives of common plants and animals in order to build sympathy with nature. They rejected biologists' abstract rationalism as well as their laboratory investigations of esoteric marine invertebrates. Nature study was a significant educational movement in the 1890s, and there was considerable interest in extending it beyond "the grades." But its promoters' antischolastic stance, as well as its popularity in the mass magazines (notably *St. Nicholas* and *Ladies' Home Journal*) and youth organizations (Woodcraft Indians and Boy Scouts of America), made its appropriateness for formal high school instruction doubtful.[6]

The expansion of high schools in the years around 1900 made them increasingly important centers for these intellectual and organizational crosscurrents. Total high school enrollment increased from 203,000 in 1890 to 915,000 in 1910, a period when the adolescent population grew less than 40 percent.[7] Expansion was even more dramatic in cities and towns in the Northeast, Midwest, and far West. Teachers and administrators who faced this larger, less affluent, and ethnically more differentiated student population were willing to experiment with new, potentially more efficient course arrangements.[8] In the mid-1890s, Chicago educators initiated the first substantial plan to coordinate instruction in the life sciences. Rejecting the perspective of the Committee of Ten, they established consecutive one-semester courses in botany and zoology under the rubric "biology." This administrative solution, however, apparently had little impact on the content of teaching. New York City was the place where bureaucratic standardization, metropolitan singularity, and special-interest politics combined to induce educators to design and propagate an integrated subject of biology.

The state of New York had long had the most centralized secondary school system in the country. Its board of regents shaped consensus among administrators, college professors, and teachers. In the natural sciences, the regent-sponsored New York State Science Teachers Association set and maintained standards for secondary schools by publishing syllabi for entrance examinations to state colleges. New York City had few public high schools and thus relatively little influence on state policies prior to the consolidation of Manhattan, Brooklyn, and the three other boroughs into greater New York in 1898. During the next decade, however, numerous schools were built, enrollment quadrupled to over 50,000, and the number of teachers increased proportionately.[9] The new subway system, together with organizations such as the city's High School Teachers Association, made New York City the only place in America where high school teachers could maintain regular contact with many others in their subject specialties. A number of younger faculty held advanced degrees from the elite universities, and some kept abreast of current intellectual developments through Columbia University's Teachers College, which began to act as a sponsor of innovation with the appointment of James Earl Russell as dean in 1897. Moreover, the most active New York City educators were uniquely situated to exert influence through writing textbooks. They could work on a personal basis with large metropolitan publishers such as Macmillan and the American Book Company (known as "the book trust"). They could use their teaching experience both to shape the state syllabus and to design books that would meet the resulting particular requirements of the nation's largest educational market. Publishers frequently promoted New York–oriented books throughout the country as part of their efforts to consolidate their industry.[10]

The specific stimulus for New York's leadership in developing biology, however, was an 1895 state law, sponsored by the Woman's Christian Temperance Union, that required every ninth-grade student to study "physiology and hygiene with special reference to the effects of alcoholic drinks and other narcotics."[11] New York City educational leaders, who considered the law political interference with their authority, sought to comply with its provisions in ways that minimized its impact and, if possible, furthered curricular rationalization. As a result, in 1902 the board of education combined physiology instruction with the botany and zoology previously offered as one-semester eleventh-grade electives into a required yearlong ninth-grade "biology" course.[12] This course initially consisted of separate units of botany, zoology, and physiology, often taught by different individuals. But as teachers repeated it annually over the next few years, they sought to integrate its parts and to justify its status as a school science.

Teachers and professors associated with a number of New York institutions, both in the city and upstate, participated in the extended discussions about biology that took place during the first two decades of this century. Yet from the beginning, the course's design center was at DeWitt Clinton High School.

This location for the origins of high school biology is particularly significant because Clinton represented—in its siting, student population, and faculty—the most novel and extreme characteristics of Progressive Era New York City.

Located at Fifty-eighth Street and Tenth Avenue, Clinton formed an island of gentility in one of the roughest areas of the city. The school began in 1897 in Greenwich Village and then expanded rapidly into four annexes scattered throughout Manhattan. In 1906 Columbia president Nicholas Murray Butler dedicated an expensive new six-story Dutch colonial structure (said to be the largest high school in the country) that included seventy-eight classrooms, fourteen laboratories, and an auditorium embellished with a pipe organ and two thirteen-by-sixteen-foot murals depicting the opening of the Erie Canal in 1825 by the school's namesake, the New York governor who had also pardoned naturalist Amos Eaton. The school's location in the center of Manhattan's West Side made it accessible, by subway and elevated train, from both ends of the island, as well as parts of Brooklyn and the Bronx. Yet it was blocks from any play area and was bordered by piers, railroad yards, and the Hell's Kitchen slum district. As Edward L. Bernays (class of 1908) recalled, students routinely endured the "carnage" of rock and snowball attacks from local gangs on the streets around the school; reaching the building meant "security and freedom from terror." Students hoped that the area would become increasingly institutional in years to come.[13]

Clinton's diverse yet selective student population epitomized both the potential for immigrants to overwhelm American civilization and the belief that meritocratic education offered the smoothest path to assimilation. Students noted proudly that their school was composed of "boys of all religious denominations, of all classes of social distinction and of every nationality," including "Cuba, Italy, Greece, and Armenia," thereby making it "the largest, most cosmopolitan, and most democratic boys' school in existence." Yet these students were not representative of New York City's adolescent population. All, of course, were male; they were, moreover, "those boys who wish to prepare for the greater universities and professional schools." Clinton was a democracy of opportunity, designed to appeal to some upper-middle-class elements and to provide an opening for the small percentage of families willing and able to commit sons to the uncertain path of upward mobility through professional training. Although many ethnic groups were represented in the early yearbooks, the vast majority came from German, Jewish, or older American backgrounds.[14]

Clinton's biology department, which included thirteen men by 1912, was led by three individuals who came from leading biology programs (Harvard, Columbia, and Chicago) and who worked at the forefront of the transformation of American education in the early twentieth century. Henry Linville, a

Figure 7–1

DeWitt Clinton High School, 1920, with power plant to the left rear. Photograph courtesy of the New York City Board of Education Archives.

rural midwesterner with a Harvard zoology Ph.D. (1897), founded the department in 1897 and chaired it until he transferred to Jamaica High School in Queens in 1908. Linville was a leader in educational progressivism: he organized the New York City Teachers League, one of the first American teachers' unions, and he established the left-wing educators' journal, *American Teacher*, dedicated to "democracy in education; education for democracy." Closely associated with John Dewey in both educational philosophy and in politics, he was the longtime head of the Teachers' Union of New York City and led the American Federation of Teachers from 1931 to 1935.[15] Benjamin Gruenberg, who taught at Clinton from 1902 to 1910, was closely associated with Linville, serving as managing editor of *American Teacher*. A Russian Jewish immigrant who grew up in Minneapolis, he worked evenings and summers to complete a Ph.D. at Columbia with T. H. Morgan in 1911. An active socialist, Gruenberg established the Political and Social Science Club at Clinton. After chairing biology programs at two other New York high schools, he became educational director of the U.S. Public Health Service in 1920 and later an editor for Viking Press.[16] George Hunter, the third significant member of the Clinton biology department, grew up in New York City and the commuter suburb of Mamaroneck. He pursued graduate study in zoology at the University of Chicago from 1896 to 1899 and spent seven summers at Woods Hole between 1895 and 1902. Hunter succeeded Linville as department head from 1908 to 1917, when he received an education doctorate from New York University and left the city to teach on the college level. Less politically engaged than Linville or Gruenberg, Hunter became a prolific textbook writer and through *Science Teaching at the Junior and Senior High School Levels*, an influential educational theorist.[17]

Between 1897 and 1917, Linville, Gruenberg, and Hunter guided the development of Clinton's biology course, and they modified and promoted it through the complex educational network that surrounded their base in mid-Manhattan. Contacts included other professionally active New York City biology teachers, most notably James Peabody, department chair at the coeducational Morris High School in the Bronx and Linville's fellow Harvard graduate student (A.M. 1896), and Maurice Bigelow (Harvard Ph.D. 1901), biology professor at Teachers College and coauthor of *The Teaching of Biology in Secondary Schools*, the first comprehensive, if tentative, guide to the subject.[18] Graduate school ties, local organizations, and the New York State Science Teachers Association continued to link them with university scientists. They disseminated their views through this association and other educational organizations, in popular magazines, and, above all, in *School Science and Mathematics*, the central forum for American science educators during the first decades of this century.[19]

As I have indicated, however, these teachers propagated their experiences most effectively in the form of textbooks. In 1906, when botany and zoology were still distinct curricular elements, Linville coauthored *A Text-book in General Zoology*.[20] A year later, Hunter produced his unified *Elements of Biology* and followed that with a series of revisions and variations that continued through *A Civic Biology* and *Problems in Biology* up to the posthumous publication of *Biology in Our Lives*.[21] Gruenberg published the widely used *Elementary Biology* and Linville contributed *The Biology of Man and Other Organisms*.[22] Other New York educators joined in textbook production with varying degrees of success. J. E. Peabody collaborated with botany teacher A. E. Hunt on *Elementary Biology* and *Biology and Human Welfare* (1924). M. A. Bigelow and his wife Anne produced *Introduction to Biology*. In 1916 W. M. Smallwood of Syracuse began a long-running series titled *Practical Biology* with I. L. Reveley and G. A. Bailey.[23]

Biology, as designed by these men, was extremely successful. According to one analysis, New York educators produced twelve of the eighteen "general biology" texts published between 1900 and 1925. Contemporary surveys indicated that from the early 1910s into the mid-1920s books by Hunter, Gruenberg, and Peabody and Hunt were by far the most popular biology texts nationwide. Based on these texts, general biology became a standard subject and the most widely taught scientific discipline in high schools. One year of coherent study of living nature thus became part of the normal experience of American youth; although what was learned is hard to pin down, what was taught became—like the content of American history—part of the public culture.[24]

BIOLOGY EDUCATION AND MENTAL DEVELOPMENT

The case for a year of biology early in high school was not self-evident. The course's origins, at least in New York, were largely bureaucratic; many teachers viewed it as a hodgepodge that conveyed little real science and diminished the possibility of ever establishing more than a year of life science in the curriculum. Biology's proponents could argue on the practical ground of enrollment that a required course in the earlier grades was preferable to a fluctuating combination of upper-class electives. But they also sought to show that their subject made sense intellectually and that it served their ideal student population—striving urban immigrant boys.[25]

The appeal of high school biology derived largely from its immersion within the richly circular currents of progressive thinking about education. G. Stanley Hall and John Dewey provided the best-known articulations of this perspective, but many of their ideas were common among American biologists, psychologists, and educators in the last two decades of the nineteenth century.[26] For these men the concept of development linked together embryology, natural and human history, psychology, and philosophy. Mental development was the most important aspect of this process; it resulted from efficient organization, successful adaptation to changing environments, and, in the case of humans if not of other organisms, the ability to alter surroundings "purposefully" over time. In both the individual and the race, mental development was thus a function of learning through experience. The immediate purpose of schooling was to structure students' experiences so that they would in fact learn; its broader goals were to adjust children to changing conditions and to develop, in at least some, "power"—the ability to confront novel circumstances creatively.[27]

High school freshmen were considered to be at a crucial developmental stage. As a result of stricter age grading in late nineteenth-century schools, nearly all were between thirteen and fifteen years old. The first elements of self-conscious awareness and the ability to use abstract concepts seemed to coincide with puberty. Study of biology could make these "young animals" aware of the cosmic processes in which they were enmeshed, and thereby facilitate their development from objects into "independent being[s]." In order to guide students to healthy, moral, rational maturity, biology educators provided instruction that included natural philosophy, training in the capacity to learn, and specific guidance on how to live.[28]

The Unity of Life

The basic distinction between biology and the courses with which it competed was obvious: it promised an overview of "life" instead of an acquaintance with some living things. As such, it held the paradoxical potential to be more abstract than botany or zoology and at the same time more elementary, able to draw from both of these subjects without adding them, and the time each took, together. This claim's plausibility depended on new stereotypes about the experience—or lack thereof—that city children were presumed to have of nature. Botany and zoology advocates had emphasized comparative anatomy in part to distinguish their subjects from common rural lore and to demonstrate that they were sufficiently difficult to be included in formal schooling. Nature study was conceptualized as a form of remedial education, filling gaps in a body of knowledge that intelligent people with sufficient country experience could acquire spontaneously. Metropolitan designers of biology, on the other hand, could reasonably assume "that the biological information of students, born and nurtured in a great city, would not be very great." This nearly complete ignorance about nature enabled teachers to feel justified in pulling together a variety of elementary, seemingly disparate topics and to present that mix to students within a theoretically informed structure as an academic science. Biology educators, even when dealing with enteric bacilli, were presenting a scholastic subject—a set of truths that could be treated with the same seriousness that sanctified instruction in mathematics, physics, or Latin.[29]

The unifying perspective that enabled advocates to fit all of life into one school year came directly from academic biologists. In the early 1890s, Charles O. Whitman in particular was highlighting the potential of "general physiology," a rubric whose origins could be traced to Herbert Spencer's *Principles of Biology* and back to German "vital materialism."[30] General physiology was too amorphous to get far in universities, but it held real attraction for educators. On the one hand it broke down the divisions separating botany, zoology, human physiology, and nature study, providing a basis for selecting organisms as diverse as beans and smut or perch and mosquitoes to illustrate particular topics. On the other hand, it presented the unifying theme of adaptive function. As early as 1904, Clifton Hodge of Clark University was urging teachers to ask living things, not "[W]hat are you? . . . Where did you come from?" but "[W]hat do you do in the economy of nature? What are the laws and conditions of your active life?" The 1910 New York state syllabus grounded "the unity of the biology course" in "the important fundamental functions performed by all living organisms" and "the interdependence of plant and animal life." Invocation of broad physiological principles became commonplace in textbook prefaces and programmatic statements.[31]

Translating these words into a practical course was more difficult. Peabody and Hunt's *Elementary Biology*, appropriately subtitled *Plant, Animal, Human*, strung its three separately paginated sections together with the hope that human physiology would "review, sum up, and give real significance to many facts learned earlier in the course." A few years later, however, books by the Bigelows, Hunter, and especially Gruenberg were structured around a handful of goal-directed processes by which life was maintained and advanced: the nature and conditions of plant growth, the roles of food, breathing, circulation, excretion, and sensation in human life (with hygienic applications), heredity, sex and reproduction, and the relations among organisms (including symbiosis and parasitism).[32]

The intellectual value of the new emphasis on function lay in its perceived ability to provide a philosophy of life for students who were uprooted yet transplantable. On a simple level, asking organisms what they did rather than where they came from expressed melting-pot egalitarianism. The deeper significance of this general physiology, however, lay in its sweeping picture of the universality and complexity of adaptive interaction between biological units and their physical and organic surroundings. One of Hunter's laboratory guides, for example, began the year with an account of the interactions among plants and insects in a patch of weeds. Linville presented the broader implications of this perspective, arguing that the study of biology would "give the student the knowledge whereby he can see himself in a setting." Instead of thinking of himself or "his race" in isolation, he would come to understand his life "in relation to all organic factors that affect the [human] race." Even if this "feeling" for biological interrelations remained unconscious, it provided "the added balance and sureness" that would enable the student to respond successfully to new circumstances.[33] While some interpreted the lesson of universal adaptation conservatively, arguing that the student should learn that "he, too, has an environment to which he must adapt himself," the more common view emphasized the mutuality of assimilation and progress. Gruenberg, for example, noted that his text "selected types of problems that best illustrate man's method of adapting himself or his surroundings to his needs," and Linville told the student that through the study of biology he would learn to be a "master of life" rather than a passive victim of his surroundings.[34]

Learning through Experience

Biology advocates believed that in addition to these matters of scientific content, their subject had, in Linville's words, the unique merit "that the practice of its method should promote the habit of thinking in men." As the student's first contact with a scientific discipline, biology functioned as a transition from the random and/or rote learning of childhood to mature analytic thinking. It was particularly suited to this pedagogical position because it could begin with the child's "natural" interest in novelty and raise it, inductively, to a more general enthusiasm for learning. George C. Wood of Brooklyn Boys' High School, for example, attested to the value of "a few minutes' discussion with any class in the presence of a growing plant or a kitten," and then described the payoff from laboratory exercises through which students learned the fundamental similarity of energy transformation in steam engines, plants, and boys: "Ah! that is the appeal to the child. He did it himself. How his eyes brighten! How his face shines! And he sneaks up after the recitation and tells you he likes biology."[35]

The physiological emphasis of biology made it particularly valuable as an introduction to science. Its target population, "boys of super-virile age," were considered to be interested in handling living animals and in understanding how systems worked, while they were said to view the drawing of plants or the memorizing of classification schemes as girlish.[36] The experience of handling organisms and analyzing and writing about their functions also led students to a much broader educational goal: to recognize that life was neither dirty nor mysterious. In biology class, squeamish or vulgar expressions about bodies would give way to neutral scientific description, and intellectual honesty and autonomy would replace superstition. Pointing to "ill directed" home training about the functions of the body, Frederic S. Lee jokingly exclaimed, "How much easier the work of the teacher would be if all parents were exterminated!" Benjamin Gruenberg hoped that learning the methods of biology would ultimately lead a new generation to "break from their dependence on hide-bound tradition and authority."[37]

Better Living through Biology

Like most arguments for liberal education, claims for biology's value in providing students a worldview and cognitive skills tended toward hopeful rhetoric. Teaching them how to live, on the other hand, involved concrete tasks. Noting that they lived "in an era when the pragmatist advocates his system of philosophy," New York biology educators were central participants in the movement during the 1910s to make secondary education more practical.[38] Yet they were less concerned with the vocational aims highlighted increasingly by physics and chemistry teachers than with preparing the student "to be a good animal" who would be ready for "efficient citizenship." In an environment that was alien to many students and their families and historically unprecedented in its congestion and artificiality, writers such as Peabody and Hunter equated citizenship with helping the student "to improve the surroundings in which he lives."[39]

Interpreted broadly, how to live included appreciation of fields and woods and recognition of the grandeur of the American continent, but the core of biology's "practical uses" revolved around more explicitly urban problems of nutrition, ventilation, and sanitation. All texts discussed the energy and protein contents of food in relation to their cost, detailed the causes and prevention of food spoilage, and, after 1920, emphasized the importance of vitamins. They condemned juvenile use of alcohol and tobacco and pointed to the dangers of other habit-forming drugs.[40] Hunter, among others, explained the importance of moderate exercise, good posture (including the value of adjustable school desks), and open bedroom windows for ensuring healthy respiration. The more radical Gruenberg discussed the need for dust-free air in factories.[41] Considerable space was devoted to the design and value of sanitary water and sewer systems, as well as to the importance of quarantine, clean eating utensils, and modern practices of manure and garbage disposal.[42]

The leading New York biology educators gradually came to realize that the most distinctive training in living that they could provide their students was to educate them about sex. As a prominent and universal biological function, sex was integral to the conceptual structure of their course. It was a subject about which students were deeply curious, and as such it could lead them to understand that science was meaningful. Lack of knowledge about sex was a source of many psychological, medical, and social problems. Initially, biology educators limited their coverage of sex to reproduction in plants and the embryology of nonplacental animals; in 1904 Maurice Bigelow argued that if a teacher would leave the student "alone with his thoughts in these deeper and more delicate questions" of human sexual phenomena, "he will arrive at the truth."[43] Within a decade, however, experience with students, along with the newly respectable social hygiene movement, led Bigelow and his colleagues to decide that "it is inadequate to leave students to make their own application of biological ideas [about sex] to human life. The demonstrated fact is that most of them do not, and so we need definite and applied teaching."[44]

Bigelow, Gruenberg, and Peabody participated in one of the first public American discussions of high school sex education, organized as a joint meeting in 1911 of the American Society of Sanitary and Moral Prophylaxis and the New York Association of Biology Teachers. These men, along with Hunter and other New York teachers, figured prominently in such discussions during the next decade.[45] Gruenberg ultimately resigned from the schools to direct a sex education program for the U.S. Public Health Service, and he and his wife Sidonie published a number of advice books for adolescents.[46] Sex was a

In the upper picture a little girl can be seen dumping garbage from the fire escape. She was a foreigner and knew no better. The picture below shows the result of such garbage disposal.

Figures 7–2 and 7–3

Examples of unscientific living, among immigrants and upper-class women. Messages of tolerance and cultural relativism could easily be woven into hygienic guidance. (Left, reprinted from George W. Hunter, *A Civic Biology*, 389; below, reprinted from Henry R. Linville, *Biology of Man and Other Organisms*, 441).

the weight of the body rests on the toes. We smile at the pictures representing the appearance of the skull of the Flathead Indians, and pity the Chinese women of an early day. But their age-long customs are now being altered rapidly, while our bone twisting habits are the results of shallow fashion under civilization, possibly to be replaced by some other whim equally bad. Although the waist-binding habit of women in the civilized world is being modified, the vice of

Fig. 216. Fashionable Heels and Army Shoes.

wearing high heels has become so firmly fixed in the early part of the Twentieth Century that it is impossible to buy ready-made certain qualities of women's dress shoes with sensible low heels.

natural part of biology, and teachers believed that their course could provide students with important information and a healthier, more rational perspective on the subject. Understanding the processes of reproduction, the nature and problems of sexual function, and the principles of heredity would help students not only to live better themselves but to produce children within marriage who would be as free as possible from defect and disease. The result, in a multitude of senses, would be better life for all.

Pedagogical Problems

High school biology was attractive to progressive science educators because it so fully integrated scientific content, educational theory, and social improvement through the evolutionary concepts of experience, learning, adaptation, and development. By the same token, however, it also shared in two of the most fundamental problems that have frequently been noted among Progressives—the "paradox" of simultaneous belief in natural progress and in reformist intervention, and seeming naivete regarding the ease with which ideas could change society.[47] Hell's Kitchen was a place where cosmic processes and moral imperatives seemed deeply out of synchrony. School overcrowding and underfunding, lack of aptitude or interest among many students, and pressures from a variety of social groups with contrary aims and unpredictable degrees of access to power all worked against the realization of reform through education—at DeWitt Clinton and more generally. These problems directly affected the efforts of biology teachers to show their students organisms in action, teach them scientific method, and alter their attitudes toward the environment, foods, and sex. Given the complexity of their situation and the sharp limits on their power, problems facing biology educators seemed often intractable. Yet they continued to advance their subject, relying on the intellectual position they already applied to both their organisms and their students: they combined advocacy of experimentation, willingness to accommodate, and faith in the future.

The central challenge for New York biology educators was to enable their students to experience the functional processes and web of interactions that made up the natural world. The most obvious places for such experiences were in American woods and streams, which Manhattan students such as Bernays considered as mysterious as "darkest Africa." But field trips posed insuperable problems for urban educators. Since the school calendar ran contrary to nature's spring-fall cycle, concepts seldom matured when the organisms did. The logistics of transporting hundreds of urban teenagers to a place where they might see something significant were daunting. Students who were hard to control in classrooms for one-hour periods could be impossible outdoors or could be destructive in their enthusiasm.[48]

Hunter sought to overcome these obstacles by proposing that teachers begin the school year with a trip to a vacant lot to study the interactions between insects and weeds, but the more original—and even spectacular—metropolitan response was the American Museum of Natural History. Supported by plutocrats and taxpayers, it provided "habitat groups" displaying a variety of woodland settings. Labels and cutaways showing the interiors of trees and below-ground phenomena enabled teachers to be sure that students could see what was important. Guards ensured order. Still, the museum was at best a partial solution to the problem of enabling urbanites to learn about nature through experience. There was little opportunity for active observation or the study of living organisms. Moreover, in their perfection, the museum's framed tableaux aestheticized nature and gave students the impression that the best place to experience nature was indoors on a rainy day.[49]

The compromises inherent in the museum were easily accepted because, despite its promoters' claims, it was never really central for learning about vital processes. School laboratories, although less dramatic, were perceived to provide the regular, analytically discrete exercises on organisms that would tie the pedagogical ideal of experiential learning to the biology course's physiological content. In the course of a school year, laboratory work could enable students to perceive the disparate elements involved in life.

There was a fundamental obstacle, however. Biologists at the turn of the century took for granted that the great researches of the previous decades in experimental physiology provided the paradigms for the acquisition of scientific knowledge. Yet American high school students simply could not learn about the nature of life by recreating the kinds of experiments pioneered by Claude Bernard and Ivan Pavlov. As early as 1894 the American Humane Association interrogated public figures about animal experimentation in schools. The universal response was that such activity could never be justified.[50] At the 1900 conference of the New York State Science Teachers' Association, Columbia physiologist Frederic Lee noted matter-of-factly that although "the method of vivisection" was of great value in physiology, "its employment with junior pupils is inadvisable." The next generation of biology educators never disputed this community limitation on realizing the experimental ideal. Instead they worked out ways to circumvent the absence of vivisection: students could make noninvasive experiments on themselves and each other, they could cut and poison plants without opposition (one reason for a growing interest in plant physiology), and they could experiment destructively on sentient but remote animals such as hydra. Comparative arguments would then close the circle of understanding.[51]

A corollary of these choices was a decline in the practice of dissection. Both college biology and high school zoology courses relied heavily on dissection of the animal series, both to convey the principles of comparative anatomy and to teach attention to detail. Problems arose as high school education expanded. The Humane Association argued that inducing students to be unemotional about dead animals blunted their moral and aesthetic sensibilities. A more immediate difficulty was the accusation that the common practice of dissecting cats fostered the theft and killing of pets. Teachers were also concerned that although "some classes of boys" were enthusiastic about dissection, many found it repugnant. Initially they argued that such objections could be overcome, but by the 1910s many decided that dissection was time-consuming, hard to coordinate, and not particularly important for realizing their aims. Hunter's 1916 laboratory manual included only the cursory dissection of a perch, and Gruenberg and Wheat's had none. Many schools continued the practice, but turned from fresh cats to the tidier embalmed specimens, mostly endothermic, sold by Chicago's General Biological Supply House.[52]

Uncertainty regarding officially sanctioned adolescent participation in the vivisection or dissection of various organisms encapsulated important, if poorly articulated, distinctions in modern culture: the unclear boundaries separating an ideal of mature rationality about biological material from the two poles of vulgar boyish cruelty and effeminate sentimentality. The tension between these poles—and the location of a middle ground—in fact extended beyond the school laboratory to touch on important questions of how students should relate to organisms around them while living in cities.

New York teachers, from their introductory discussions of the rich interactions among common organisms, through injunctions about park protection, to general messages about adaptation and balanced natural cycles, emphasized the harmony of natural processes. Another message coexisted with this, however, and threatened to overwhelm it: that the organisms close at hand— bacteria, worms, flies, mosquitoes, and rats—were not companions but competitors, to be exterminated in any way possible. Texts included, for example, long discussions of the life cycle and habits of what was called for a time the "typhoid fly," an animal whose monstrosity was reinforced with a giant model on display at the American Museum of Natural History. Hunter outlined methods for killing many pests and destroying their habitats. "A few boys" could eliminate mosquitoes in areas near marshes, which were still common

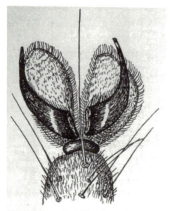

The foot of a fly, showing the hooks, hairs, and pads which collect and carry bacteria. The fly doesn't wipe his feet.

Figures 7–4 and 7–5

Dangers in the environment, and the value of collective action. Above, reprinted from George W. Hunter, *A Civic Biology*, 223; right, reprinted from J. E. Peabody and A. E. Hunt, *Elementary Biology*, 16.

Fig. 12.— Morris High School boys removing 63,020 eggs of tussock moth from four trees on school grounds. Work directed by Paul B. Mann. (Photographed by Lewis Enowitz.)

in New York's outer boroughs, by pouring oil into the water: "[W]hy not try it if there are mosquitoes in your neighborhood?" Like most Americans, biology teachers left the resolution of the tension between awareness of ecological complexity and the imperatives of hygiene to the future.[53]

More immediate difficulties in pedagogy arose regarding interactions between humans and the various natural products readily available to urban adolescents and their families. The law in New York, as well as many other states, required instruction in the dangers of alcohol and tobacco. Educators sought to mitigate the force of these explicitly political directives regarding the content of science education; yet despite their skepticism about the truth and value of such instruction, they followed the provisions of the laws. Similar if more subtle pressures influenced instruction about food. Depending on religion and ethnicity, immigrant children grew up with diets restricted in a

variety of ways. Texts seldom dealt with this question directly but instead presented general discussions of the importance of regular consumption of a
balanced diet chosen on the basis of economy; they offered extensive lists of
healthful food, prominently including shellfish and properly cooked pork.[54]

Sex

The interorganismal relations that generated the most serious pedagogical
difficulties were those associated with sex. Discussions of sex education thus
provide the clearest view of the range of self-imposed and external constraints
under which biology advocates worked and illustrate their approaches to re-
solving the conflicts in which they were enmeshed. Educators recognized
that sex was scientifically and socially fundamental, yet they faced a series of
dilemmas in introducing the subject to young adolescents in the public setting
of school science.[55]

Difficulties in teaching about sex began with the otherwise routine question
of pedagogical format. Many adolescents either did not know the basic facts of
human reproduction or could express their knowledge only in street language.
Textbooks were the usual means for providing students with specific scientific
information and terminology, but they could not include, for example, detailed diagrams of the reproductive tracts, out of fear that the books would fall
into inappropriate hands. The alternative of oral instruction, however, made
sex education a unique test of teaching ability. Insecurity in speaking about
the subject could destroy the classroom authority of an otherwise adequate
instructor. Oral presentations were particularly problematical in coeducational schools; the necessity of dividing classes in half and bringing in a teacher
who was the same sex as the students contradicted the ideal of bringing the
subject up "naturally" within the curriculum.[56]

Questions of content also posed problems. Biology advocates argued that
sex education should focus on healthy, normal phenomena rather than the
pathologies that visiting physicians frequently emphasized in order to scare
students about masturbation and venereal disease. Yet teachers could say little
about the nature and variety of sexual experience. Apart from reassurance that
occasional nocturnal emissions were physiologically innocuous, emphasis was
almost completely on the importance of and the means to restrain sexual
impulses.[57]

The underlying difficulty, as Clinton teacher E. F. Van Buskirk noted without irony, was that "the aim of sex instruction" was the "exact opposite" of the
rest of the biology course. "In most biology work, one of the most important

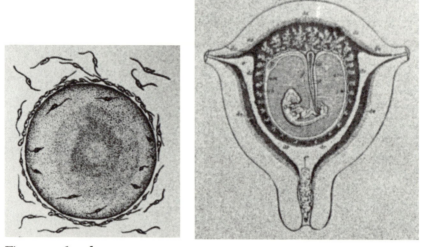

Figures 7–6 and 7–7

Possibilities and limitations in the representation of sex. Fertilization could be illustrated as effectively in algae as in humans; the human uterus, on the other hand, was completely detached from its anatomical context. Left, "Egg cell and sperm cell of rockweed," reprinted from Linville, *Biology of Man and Other Organisms*, 221; right, "Diagram of a mammalian uterus, showing attachment of an embryo to the lining," reprinted from M. A. Bigelow and A. N. Bigelow, *Introduction to Biology*, 405.

aims is to awaken interest and lead to experimentation." Student interest in sex, on the other hand, was already "very active," and the purpose of teaching was to respond only to necessary questions, without "exciting curiosity." No teacher claimed that mammalian sexual behavior was an appropriate subject for student experiments, either in the laboratory or in the field.[58]

How could these problems be resolved? As a *Journal of Education* editorial on New York City sex education debates noted, "It is easy for highly progressive persons to see that something should be done, easy to have a nightmare over conditions that need attention, but no one has a vision yet." Educators concerned about sex were uncertain about how to proceed and were faced with "many opposing forces." In 1913 the implementation by progressive Chicago school superintendent Ella Flagg Young of an official program of sex education by physicians raised a storm of controversy and lasted only a few months.[59] The New York City school board, by contrast, declined to approve an official program yet tolerated the initiatives of teachers such as Gruenberg, Hunter, and Peabody. Educators aimed to integrate sex into the curriculum, but they believed that success required pedagogical experimentation and careful testing. They hoped that through a gradual process of education, new generations of parents, teachers, and students would adjust to a more rational attitude about this central biological phenomenon.[60]

PRODUCING MODERN AMERICANS

In 1925, students in Dayton, Tennessee, were among the thousands nation-wide learning biology from Hunter's *Civic Biology*, a text featuring Long Island crustacea, trips to the American Museum of Natural History, and the advantages of sleeping on tenement balconies in hot weather. What relation did these features of the course have to the four pages on evolution that provided the basis for Scopes's indictment? Or, to put the question more broadly: how did the metropolitan origins of high school biology affect the teaching of evolution?

The influence of New York educators in this area was significant but ambiguous. On the one hand, evolution*ism* permeated thinking about the design of high school biology. Linville, moreover, played a direct role in defending the inclusion of evolution in secondary education: he was a member of six-person executive committee of the American Civil Liberties Union (ACLU) that initiated the challenge to Tennessee's anti-evolution law.[61] On the other hand, explicit presentation of evolutionary facts and theories was a minor element within biology instruction prior to 1925. It had no place in the New York state biology syllabus, and it was much less prominent in biology texts than in pre-1910 books on botany, zoology, or geology.[62]

This seeming inconsistency becomes comprehensible when one sees evolution as an issue similar to field study, vivisection, or sex. New York educators periodically and routinely noted that teaching evolution could disturb the classroom or the community, and they saw little value in antagonizing potentially vocal elements of the public. Although they experimented with different textbook treatments of evolution and participated in the ACLU project to quietly challenge anti-evolution laws, they compromised on such matters as the inclusion of the subject in state syllabi, the placement and extent of textbook discussion, and the degree to which evolution was presented as fact. Biology educators made few comments on the Scopes trial; it seems to have been a nonissue from the perspective of their journals.[63]

Compromise was easy because the topic of evolution was such a marginal part of the biology curriculum. The facts of evolution were historical in character and geological in substance, and hence outside the main lines of biology taught as an observational and experimental science. Elementary instruction on evolutionary mechanisms seemed impossible in the 1910s and 1920s given the degree of professional uncertainty about mutation, selection, and the inheritance of acquired characters. What was important was to convey an evolutionary *perspective*. This task, however, was largely independent of specific instruction about species change. Benjamin Gruenberg exemplified these attitudes: he stated that evolution was a very important subject, yet he devoted only ten pages to it in a five-hundred-page text and used much of that space to attack the significance of natural selection. He argued that the most important lesson of evolution was a "dynamic viewpoint." To convey this, "progressive change" was "indirectly suggested in the discussion of various processes and relations"; beyond that, he suggested that "a free discussion" would "ferret out lingering doubts and prejudices, not for the purpose of inculcating sound doctrine on the subject of evolution, but for the purpose of getting the students to feel the majesty of the larger concepts."[64]

Biology educators were more cunning than timid, maneuvering within limits set by their perception of the state of scientific knowledge, their experience in urban classrooms, and their assessment of their individual and collective power. They prevailed in the struggle that mattered to them most. High school biology became part of public education throughout the country, and its fundamental themes and images became part of middle-class culture. This view of life was coherently urban, male, and liberal. Among the millions of students exposed to biology, one of the earliest—Edward Bernays—can be seen as representative of the kind of person educators sought to produce. Bernays's first activity after graduating from Cornell's College of Agriculture was to produce the anti–venereal disease melodrama, *Damaged Goods*, on Broadway as an experiment in sex education. Learning rapidly about life while working as press agent for the Ballet Russe (which included a number of avid sexual experimenters) and as a war propagandist with the Committee on Public Information, he established himself in the metropolis as a founder of the key modern social technology of public relations. A model of pragmatic corporate liberalism, he viewed himself as an engineer correcting "maladjustments" between clients and their publics.[65]

Although neither the identity nor the importance of biology was ever seriously challenged after it became established nationwide in the 1920s, the content of the course responded to social and curricular changes. In the 1920s George Hunter settled in the suburbs of Los Angeles, and Clinton High School left its half-block in Hell's Kitchen for a twenty-one-acre campus in the most northerly part of the Bronx. These moves were emblematic of broader shifts in the ideal geographic locus of adolescence in America. Newer, more elementary courses in general science and health appropriated many of biology's simpler and more practical aspects. As a result, work in natural history became more prominent, and the academic level of the course rose. As with most educational innovations, its programmatic thrust became attenuated as it became an accepted part of the curriculum.[66] Still, the basic outlook that motivated the creators of biology remained. It was exemplified in the life and work of the next generation's major proponent of biology education—Hoboken-born evolutionary entomologist Alfred Kinsey. In the midst of the Great Depression Kinsey wrote a comprehensive guide to high school biology, arguing that the goal of the course was to teach the typical student "that it is an endlessly interesting world in which he is living."[67] Like his more senior colleagues in education, Kinsey then turned to the aspect of biology that most stimulated modern young Americans.

Big Questions

WHY THE SCOPES TRIAL MATTERED

Watson Davis was overwhelmed. As a journalist working for the scientific news agency Science Service, he was accustomed to meeting copy deadlines. His assignment in the summer of 1925 was much more complex, however, than simply covering a routine story. When the Science Service executive committee realized that the trial of John T. Scopes would become a major public event, they decided to do more than merely report the news. They took on the task of locating and paying for experts who would go to Tennessee to provide "competent and sane" advice to the Scopes defense team and, potentially, to testify.[1]

In early July, Davis scrambled to determine which scientists would be effective witnesses and available for service. Some declined—they already had plans, considered themselves too controversial, or were uninterested in participating in the Dayton circus. Ultimately, however, Davis telegraphed Scopes's lawyers a list of over a dozen names and shepherded to Dayton a cross-section of scientists, from the conspicuously Christian Harvard geologist Kirtley Mather, through respectable academic biologists Maynard Metcalfe (Johns Hopkins) and Winterton C. Curtis (Missouri), and on to the Jewish agricultural scientist Jacob Lipman (Rutgers). When Davis learned that the trial judge would not allow scientific experts to testify, he made sure that they provided statements to the media. Newspapers across the country featured efforts by both Davis and his scientists to explain evolution, to distinguish facts from theories, to argue for the compatibility of science and Christianity, and to assert the importance of science in civilization.[2]

In the 1950s, the drama *Inherit the Wind* presented the Scopes trial as a victory for intellectual freedom. More recently, scholars have viewed the trial as a largely successful attack by the religious right on education and on science or as, at best, a standoff.[3] In the previous chapter I argued that the trial had relatively little impact on school teaching. Here, without returning to the McCarthy-era celebration of freedom of conscience, I want to correct the view that scientists were either victims or bystanders. Leading academic biologists, while recognizing that the trial was hardly an ideal forum, believed that participation in it could help spread their scientific worldview—a perspective that was compatible with some interpretations of Christianity but was fundamentally secular.

Given the events described above in part 2, this activism seems surprising. The most notable characteristics of the generation of university biologists who matured between 1870 and 1900 were obsession with research and insulation from the culture around them. This was still the case for some leading figures in the 1920s: biology educator Benjamin Gruenberg complained that when he asked geneticist T. H. Morgan to help in the fight against the Fundamentalists, his old professor at Columbia "just laugh[ed] . . . and shrug[ged] his shoulder."[4] Yet others of the same generation were, by this time, much more broadly aware and activist. Between 1910 and 1925, they took up what can be called the "Big Questions" of biology: whether the variety of living things and the phenomena of life expressed some underlying order; whether progress was part of life's meaning; and what roles intelligence and free will had in biological change. These issues could be discussed in ways that could be appreciated by a significant segment of the American public. Biologists believed that such cultural activity was essential to progress within a modern democratic society. They took on roles as scientific sages; it was an easy transition, in the 1920s, for some to become biological evangelists.

Such activity was not new. It had been part of the cultural kit of naturalists in the Gilded Age. But for a generation raised on the mantra of academic research and accustomed to the now mature structure of the universities, speaking and advocating publicly required a significant shift in identity. What is amazing is, on the one hand, how many biologists did so and, on the other, that the role of sage, or even evangelist, became central to the identities of some leading life scientists.

In this chapter I explore how some early twentieth-century academic biologists became sages, outline the nature of their messages, and indicate the influence they had. I do this in two stages: first, through an overview of people and ideas during this period and then, through a more detailed examination of a single individual. Any of a large handful would do: George H. Parker, William M. Wheeler, or Lawrence J. Henderson at Harvard; Charles M. Child, Charles J. Herrick, or Warder C. Allee at Chicago; Herbert S. Jennings or Raymond Pearl at Hopkins; Edwin G. Conklin at Princeton; and others less prominently located, all shared many of the same ideas and expressed them in similar forums. I focus, however, on William Emerson Ritter, professor of zoology at the University of California from 1891 to 1923, and first director of the Scripps Institution for Biological Research, later the Scripps Institution of Oceanography. Ritter is worth detailed study, in part because he was a representative American academic biologist, both biographically and intellectually, and in part because he has not been examined in any depth. But the major value of focusing on him comes from the extent to which he merged ideas and actions. Ritter was a self-consciously philosophical biologist. He was also, as I indicated in the introduction, the scientist most deeply involved in creating Science Service, and he was working in Washington as its president when Watson Davis engaged in his semi-covert operation on behalf of Scopes and American biology.

THE ROUGH RIDER, AND OTHER SPOKESMEN FOR SCIENCE

Part 2 emphasized the degree to which academic biology developed around an obsession with research. Supported by endowments and hopeful medical reformers, men such as Whitman, Brooks, Mark, and Minot pursued the focused questions in basic science that had been framed by an international scientific leadership, with relatively little regard for the nature and degree of public interest in their work. The Marine Biological Laboratory embodied this emphasis.

The scientists of this founding generation immersed themselves so deeply in the creation of the academic discipline of biology, and in the particular programs of research they identified with that discipline, that they became temperamentally unfit for any other role. After passing fifty, Whitman, Minot, and Brooks each made an effort to function as a sage—to address the big questions of biology—but none was very successful. Whitman, after launching the MBL and the University of Chicago biology program, proposed a new vision of orthogenesis. But he planned fifteen years or more of experiments as a foundation for his views, and he died in 1910 with the project incomplete and unpublished.[5] Minot's problem was more in the content of his ideas. After completing a major textbook on human embryology in 1893, he turned to problems of broad interest, most notably aging and death. But his central assertion was that the significant changes occurred within the first year of human life, and what people experienced as aging was of minor biological significance. This approach gained him little attention.[6] Brooks focused on embryology and life-history studies from 1880 to around 1900. When he returned to big questions in his *Foundations of Zoology*, his arguments were so impenetrable that even his devoted students could make little sense of them. It is not surprising that he had little broader influence.[7]

Younger leaders of biology initially showed little interest in changing this situation. Jacques Loeb, for example, became nationally famous around 1900 through his invention of artificial parthenogenesis. He was profiled glowingly in newspapers and in the popular *McClure's Magazine,* but during the next decade was more likely to excoriate reporters for meddling, or to offer a cautionary lecture on "the limitations of biological research," than to explain to lay audiences the implications of new work. During the first decade of the 1900s, nearly all the presidential addresses of the ASN focused on intramural issues regarding the organization and coordination of the life sciences.[8]

When the last of the Gilded Age naturalists—men such as John Wesley Powell, Joseph LeConte, and Nathaniel Shaler—passed from the scene around 1900, positions as spokesmen for life science fell to a few university men who had avoided the "Ph.D. octopus" and to others on the margins of professional science. Stanford president David Starr Jordan and Cornell Agriculture School dean Liberty Hyde Bailey (both of whom completed formal education with bachelor's degrees around 1870) were the most visible life scientists in the years around 1900. Nature writers such as John Burroughs and John Muir, and the agricultural wizard Luther Burbank, also thrived in the new environment of newspaper chains and mass magazines. The most notable spokesman for life science in the early twentieth century was, however, sui generis. At the same time he worked as president of the United States, Theodore Roosevelt coauthored a monograph on the deer family, criticized writers such as William J. Long and Ernest Thompson Seton for being "nature-fakirs" (or fakers) who disregarded scientific method, pushed through the pure food laws, lectured the American people on their duty to reproduce, and launched the conservation movement. On leaving the presidency, he traveled to Africa to collect specimens for the American Museum of Natural History. In 1910 he was the first American to give the annual lecture at Oxford University endowed by Darwin's protégé, George J. Romanes. Roosevelt declared there that "he who would fully treat of man must know at least something of biology, of the science that treats of living, breathing things; and especially of that science of evolution which is inseparably connected with the great name of Darwin."[9]

The efforts of these men were remarkably successful. The 1900s marked a high point of public interest in both nature and organisms. However, their activities generated two mutually reinforcing problems. On the one hand, their ideas tended to be either vague or well tried and as a consequence had little impact on intellectual innovators. The aggressive young philosopher Arthur Lovejoy, for example, considered Jordan an intellectual fossil who was merely restating an old, naive version of materialism. Jordan in fact made use of his national status and his plentiful secretarial help to publish increasingly empty ruminations.[10] The other problem was that the intellectual authority that Americans perceived in men such as Roosevelt reinforced the marginality of academic biologists. Not only was a gentleman-amateur the best-known figure in their science, but this amateur worked actively to keep them down. In the early 1890s, Roosevelt had participated in a university review of the Harvard zoology department and circulated a minority report that complained bitingly about the program's neglect of large-animal work. A few years later he supported the attacks on the "new biology," and he continued in the new century, through essays and reviews, to question both the ideas and the status of science professors.[11]

ACADEMIC BIOLOGISTS ADDRESS THE PUBLIC

In the years after 1910, this situation changed. First tentatively and then, after 1915, with real intensity and variety, academic biologists spoke and wrote about Big Questions, in forums and in styles that the educated public could appreciate. Lectures, journal and magazine articles, textbooks, interdisciplinary collaborations, and, ultimately, testimony before public bodies brought both the discoveries and the reflections of leading professors before educated Americans.

The fundamental shift was generational. Brooks died in 1908, and Whitman two years later. Individuals such as Conklin, Jennings, Wheeler, Ritter, and Parker were then all nearing fifty. While the older men had devoted their lives to building a place for academic biology and continued to strive to attain status as researchers, the ones who came of age in the 1880s and early 1890s were accustomed to university positions and matured comfortably into roles as senior professors and influentials. They felt sufficiently secure to take on larger cultural tasks.

This transition coincided with changes in universities specific to the years around 1910. College students, spurred by the ferment of Progressivism and more specifically by reforming young college-educated author-lecturers such as Jack London and Upton Sinclair, were suddenly demanding relevance. Social scientists such as Edward A. Ross and John R. Commons became prominent as reformers. Biologists were swept up in this movement. At Berkeley, for example, an interdisciplinary group of professors that included Loeb produced a team-taught course on social problems in 1909; offerings on biology and society began to appear at a number of colleges a few years later.[12]

The most powerful stimulus for American biologists, however, was international. Around 1910 the publication and English translation of *Creative Evolution* made French philosopher Henri Bergson an intellectual star. Bergson's synthesis of evolutionary biology, perceptual psychology, and romantic metaphysics, expressed in an oracular yet accessible style, stimulated intellectuals across the arts and sciences. In the United States, discussion of Bergson built to a climax in early 1913 when the great man lectured to overflow crowds, in French, at Columbia University. The *New York Times* provided extensive daily coverage of Bergson's presentations. E. E. Slosson, a chemistry professor who had become editor of the middle-brow magazine *The Independent*, was a notable Bergson popularizer.[13]

Antiromantics such as Loeb were unrelievedly antagonistic to Bergson's "vitalism": Loeb titled his 1912 collection of essays *The Mechanistic Conception of Life* to make it "campaign literature" against Bergson. But many mainstream American biologists believed that Bergson's ideas needed at least to be taken seriously. In 1911, William E. Ritter published a long and positive assessment of Bergson in *Science*. Johns Hopkins biologist Herbert S. Jennings and Arthur Lovejoy, newly arrived as Jennings's philosophical colleague, both responded with efforts to refine the meanings of vitalism. All three were trying to articulate a perspective that would be resistant to accusations of mysticism but preserve what they considered Bergson's essential insight—that the individuality and unpredictability of organisms made biological phenomena different in important ways from events in the inanimate world.[14]

During these same years, presidential addresses at the ASN changed dramatically in topic and tone. In place of admonitions about interdisciplinary cooperation and sober discussions of problems such as adaptation, society presidents took advantage of their after-dinner settings to introduce more speculative themes. In 1911, in "Heredity and Personality," Jennings combined statistical levity with gestures toward the transcendent. He argued that his individual personality, although hereditarily determined, was the chance result of the combination of one of his mother's 17,000 different eggs with one of his father's 339 billion different sperm; that each of his parents was an equally fortuitous result and their mating highly unlikely; and that these considerations could then be taken back indefinitely through all his ancestors. "Gentlemen," he announced, "I must congratulate myself on my fortune in being with you this evening!" He drew from these incredible contingencies the conclusion that immortality was scientifically possible.[15] The following year, Edwin G. Conklin of Princeton brought Jennings's message down to earth in "Heredity and Responsibility," arguing that human development ultimately affirmed the importance of free will and moral responsibility. He roused his audience with a concluding quotation from Tennyson's *Ulysses*, urging them "to strive, to seek, to find, and not to yield."[16]

Although such addresses displayed new, more adventuresome attitudes, their direct impact extended only as far as the readership of *Science*. Philanthropists made biologists' ideas more visible. In the early 1910s, a few wealthy

Americans revived the idea, identified with the Lowell Lectures of the mid-nineteenth century, that the best way to encourage professors to apply their ideas to public concerns was by paying them to deliver elevated popular talks. David Starr Jordan presented the lectures published as *Call of the Nation* at the modernizing University of North Carolina in 1910. Conklin did better. In early 1914 he developed his ASN address into a series of lectures at Northwestern University. The resulting book, *Heredity and Environment in the Development of Men*, was widely used in undergraduate teaching for the next two decades. A few months later, Harvard zoologist George H. Parker lectured at an Amherst College series explicitly designed to provide "the basis of positive knowledge and experience" for successful "Social Control"—to enable the American people to cooperatively "seize their destiny and mold it to their conscious aims." His concise review of recent work on the nervous system, hormones, reproduction, and evolution appeared later that year as *Biology and Social Problems*.[17]

World War I intensified, and then channeled, this activity. Initially the range of arguments reflected both the intellectual and political diversity among American scientists. The pacifist Jordan emphasized that mass deaths among fit young men would weaken the germplasm of the belligerent nations; the more belligerent Raymond Pearl interpreted the dedication of the English upper classes as evidence of their good genetic qualities. William Ritter saw the war as a straightforward exemplification of the struggle for existence.[18] As the United States moved toward involvement, however, vocal American biologists formed a consensus around a liberal version of evolutionism. Vernon Kellogg's *Headquarters Nights* was the most prominent war tract by an American biologist. He described his own transformation from pacifist participant in Belgian war relief to active anti-German in relation to his contact with a German biologist-turned-officer. While Germans supposedly justified their nation's cause by appealing to the neo-Darwinian "*Allmacht* of natural selection applied rigorously to human life and society," Kellogg emphasized that the brotherhood of man was fundamental to evolutionary progress. The reconciliation of pacifists and interventionists—and of biologists and amateur evolutionists—around the identity of Allied victory, evolutionary progress, and love of humanity was reinforced through inclusion in Kellogg's book of a vigorous foreword by Colonel Roosevelt himself.[19]

This outlook, disconnected from its immediate wartime referents, became commonplace during the 1920s. One early and full elaboration of the perspective, Conklin's *Direction of Human Evolution*, opposed the idea that "our civilization, like other civilizations of the past, is showing signs of degeneration and decay." It argued instead that "the results of evolution may be summarized in three words: Diversity, Adaptation, Progress," and that the locus of progress had gradually shifted from bodily complexity, through intelligence, and finally to social organization. It looked forward, though admittedly with some trepidation, to a future "amalgamation of all races in all parts of the world" and optimistically anticipated rapid social evolution that would bring "the whole human race . . . together into a Society of Nations." Jennings breezily outlined a similar perspective in *Prometheus* and then developed it more seriously in *The Biological Basis of Human Nature*. Similar arguments were articulated by Kellogg, Parker, Ralph Lillie, Charles M. Child, Charles J. Herrick, Winterton C. Curtis, and others.[20]

It would be possible to explore the development and content of each of these scientists' ideas and to examine the extent of their agreement and disagreement.[21] But similarities are so predominant that the intellectual returns on such a project would diminish rapidly. A single case can provide a more concrete, and thus more vivid, illustration of this main stream of American biologists.

WILLIAM EMERSON RITTER AND THE GLORY OF LIFE

Like many academics of his generation, William E. Ritter traveled one of the faint and meandering paths that linked rural America with international science and national influence. Born in 1856 in central Wisconsin, the young Ritter commemorated in his self-chosen middle name the idealistic freethinker who had once passed through the area on a lecture tour. He worked on his family's farm and taught in local schools; he was passionately committed to "the original idea of American *equality*" and at the same time embraced such progressive ideas as the nebular hypothesis and evolution. Ritter was already in his mid-twenties when he enrolled in a teachers' college in Oshkosh; after graduation he pursued a longstanding plan to move to California, where he entered the state university in 1886 as a special student and received a bachelor's degree at age thirty-one. He benefited from an informal Berkeley policy to send promising graduates away for further training in the hope that they would be more likely than easterners to accept a job back in California. Thus subsidized, he enrolled in Harvard's zoology program in 1888.[22]

When Ritter returned to Berkeley three years later, he was a typical academic biologist. As an undergraduate he had absorbed the California geologist and evolutionary philosopher Joseph LeConte's expansive perspective on evolutionary progress. At Harvard, these broad concerns were transmuted into a tightly focused dissertation on the adaptations of a California blind fish. While

Figure 8–1

William E. Ritter, standing in
front of the Scripps laboratory
shown in figure I-1, about 1915.
Photograph courtesy of the
Scripps Institution of Oceanography Archives, University of California, San Diego.

in Cambridge, he prepared a report on "biological science at the University of California" that disparaged LeConte's "gigantic" chair of "geology and natural history" and proposed a new professorship and laboratory in biology. He emphasized the need for work in "morphology, embryology, adaptation to surrounding conditions, and heredity." Berkeley authorities supported this reform, appointing Ritter to an instructorship in biology and giving him the job of establishing a teaching laboratory.[23]

Ritter also envisioned a marine station. He had worked at Alexander Agassiz's Newport laboratory while a graduate student and had collected material for his dissertation in San Diego harbor during his honeymoon with physician Mary Bennett Ritter. A seaside laboratory for summer instruction and research seemed an integral part of the modern biology program he envisioned. In 1892 he occupied the ideal site. Pacific Grove, on Monterey Bay, was a faunally rich and unpolluted resort area accessible by train from Berkeley. Using a minuscule appropriation from the university, Ritter built a movable wood and canvas shelter and brought a dozen students, teachers, and "recreation seekers" to study the animals of the bay. This embryonic laboratory was immediately threatened, however, by a more developed competitor. David Starr Jordan, president of the new Stanford University, was using money from a wealthy San Francisco family to erect an "ample, well appointed laboratory" a short distance away. Ritter was so embarrassed at the "sorry spectacle" of "our little tent-house" next to Jordan's wooden building that he gave up the field. The next summer he pitched the university's tent three hundred miles south, on Santa Catalina, the isolated island off the coast of Los Angeles. He accomplished little there, however, and the marine laboratory project was abandoned.[24]

The arrival of the dynamic Benjamin I. Wheeler as University of California president reenergized Ritter's marine station plans in late 1900. The university provided funds to establish a temporary laboratory in the Long Beach section of Los Angeles, and Wheeler put Ritter into contact with wealthy Angelos who could support the endeavor. But fundraising efforts led nowhere: the city fathers wanted Long Beach to become a port, not a habitat for ocean science. The much smaller town of San Diego, by contrast, was more welcoming. The physician, amateur conchologist, and civic booster Fred Baker championed a project that would bring in interesting people and boost the town's status. In 1903, with money channeled through the chamber of commerce, Ritter established a laboratory in the Coronado Hotel's boathouse (continuing the hotel's tradition, going back to its 1887 contract with Henry Ward, of using science as a tourist attraction). At the end of the summer he gained local institutional recognition through the creation of the Marine Biological Association of San Diego.[25]

Initially Ritter thought that a "reconnaissance" of the fauna of the southern California coast would be "preparatory to selecting the most favorable locality for the permanent station."[26] But numerous factors—the hiring of ecologically trained Charles A. Kofoid, the influence of Ritter's mentor Alexander Agassiz, the lack of unpolluted bay habitats, a concern that the state university provide public service, and a desire to differentiate the new institution from other marine stations—led Ritter in 1903 to propose a continuing emphasis on "the survey idea." He explained to his San Diego supporters that his aim was "*a comprehensive investigation* of the *marine life and of the physical conditions under which it exists, of this immediate portion of the Pacific Ocean.*"[27] Two years later he elaborated, sharply distinguishing his emphasis on "*marine biology*" from others' focus on "*general biology prosecuted by researches on marine organisms.*"[28]

Ritter's most important activity during these years, however, was to interest the two wealthiest and most influential residents of the San Diego area—La Jolla newspaper magnate Edward W. Scripps and his unmarried half-sister and business partner, Ellen—in his work. Within a few months they became trustees, and nearly sole patrons, of the new Marine Biological Association.[29] Ritter's relationship with Edward and Ellen Scripps transformed him, over the next two decades, from a provincial biology professor to a scientific entrepreneur and spokesman with national influence. The initial money came from Ellen. Sixty-seven years old in 1903 with no direct heirs, she considered the association a good work. It was also a way to make her community a more refined and liveable place. By 1912 she had donated more than $200,000 to erect buildings for the institution in La Jolla and to endow Ritter's salary, enabling him in 1910 to give up teaching responsibilities in Berkeley and to live and work in La Jolla full time.[30]

Damned Old Crank

E. W. Scripps, though less free with cash for the institution, exerted a more profound influence on Ritter and on science. Scripps, along with Joseph Pulitzer and William Randolph Hearst, was one of the first generation of national mass media moguls. Born on an Illinois farm two years before Ritter, he escaped his rural backwater—not, like Ritter, through a slow process of formal education—but by riding the new wave of commercial culture. In 1872, at age eighteen, he moved to Detroit and began working on a newspaper run by his thirty-seven-year-old half-brother. He acquired a fortune and national influence over the next three decades by building up the first national newspaper chain, which extended to twenty-five cities by 1908.[31]

Scripps's success derived from (in addition to his ruthlessness as a businessman) his belief that his newspapers would speak to and for American working people. In contrast to Pulitzer and Hearst, Scripps did not seek to draw readers through screaming headlines and flashy graphics. Instead, he sold inexpensive, easy-to-read, plain-looking tabloids. As a populist, he attacked the rich, exposed municipal corruption, and supported unions. But the core of his newspaper philosophy was to provide his readers information—large numbers of stories each day (generally more than two hundred per issue)—on all kinds of matters of interest. His belief was that average people (the "ninety-five percent") actively wanted to learn about the world around them and that the daily newspaper was their basic (in most cases only) source of information.

Scripps was a self-described "damned old crank." As a young man he had been a womanizer and morphine addict, and he remained alcoholic, dyspeptic, and manic-depressive for much of his life. Yet he considered himself a clear-headed, original thinker, someone independent of both class allegiances and Sunday-school cant, a realist who maintained a core of idealism about

Figure 8-2

Edward W. Scripps, about 1915, at his estate north of San Diego. Photograph courtesy of the Ohio University Archives, Athens, Ohio.

the future of American democracy. In the 1890s he sought to reinforce his individuality by moving from suburban Cincinnati to an isolated mesa north of San Diego where, waking at noon and dressing in casual clothes, he coordinated the policies of his newspapers, dictated dozens of privately circulated "disquisitions" on journalism, politics, and the economy, and corresponded with such progressive intellectuals as Clarence Darrow and Lincoln Steffens.

Scripps first met Ritter in 1903 at the Coronado laboratory, which he termed, with typical irreverence, "Bugville." Initially he supported the laboratory out of a general belief in the value of science.[32] This positive attitude was gradually overlain with personal friendship: Scripps warmed to a neighbor who came from a similar midwestern rural background, who held congenial naturalistic attitudes toward the world, and who held status in a community that was independent of Scripps's newspaper empire. Scripps was a profane bully capable of writing Ritter that "your scientific friends are only lice after all, and . . . they stand more in need of being put out of existence than of being aroused to any greater energy." But he and the distinguished-looking, well-spoken Ritter began to be described as "David and Jonathan," as they began to visit each other for long discussions on a wide range of issues. In 1907 Scripps took the lead in pushing San Diego officials to provide the station 170 acres of oceanfront property just north of La Jolla.[33]

Gradually Scripps became engaged with Ritter's science. Ritter recalled that his friend repeatedly demanded that he explain "what kind of a thing this damned human animal is, anyway." Scripps noted that he himself was happy when he was acting "instinctually" but became "pessimistic" when he tried to reason, especially about such issues as property, sex, or the value of life. He wanted to be as optimistic and happy "as a reasoning being, as I [am] as an instinct controlled animal" and thought that biology might hold the key. It taught him that "most things that mankind generally considers to be wrong are really right and consistent with natural law and even with a universe ruling Deity." Some people reached "bliss" through dogmatism or pragmatism, but Scripps was willing to take a longer route. In particular, he wanted to gain insight into the tension that his different roles—as mouthpiece of American working men, as knowing manipulator of public opinion, as captain of industry, and as founder of a wealthy lineage—foregrounded: "On what data of finally determined knowledge is based the principle of the righteousness of democracy—the righteousness of the rule of the many weak over both themselves and the few strong members of the human family?"[34]

Scripps saw in Ritter's vision of biological research and theory some basis for understanding humans as political animals. He wanted biologists to provide Americans a naturalistic account of customs, ethics, religion, and philosophy that would increase their happiness and reinforce democracy. By 1909 he hoped that the marine station would develop into an interdisciplinary think tank where a small group of men would gather information and "make deductions which would be passed out to the world as authoritative and as the last word so far uttered concerning what is actually known, in order that the people might govern their conduct individually and as social organisms according to so much of nature's law as had been discerned."[35]

The Scripps-Ritter friendship endured for more than twenty years because Ritter, unlike Scripps's underlings and competitors, declined to be provoked by the old crank's bluster. Ritter was glad to have the support of both Edward and Ellen Scripps and was ready to listen sympathetically to E. W.'s speculations, but in most cases he continued to develop his ideas and enterprises as he believed they should. E. W. was a good enough manager to appreciate this attitude. However, by reinforcing Ritter's activities in some areas and not others, Scripps influenced the directions of Ritter's work.

Aristotle on the Beach

The initial effect of the Scripps siblings' patronage on Ritter was to confirm his inclination, evident in his 1901 plan for a biological survey of the waters off southern California, to reclaim the activities and identity of an American naturalist. During the 1890s Ritter had produced a mix of papers on the taxonomy, embryology, and adaptation of annelids that was quite typical of academic biologists. By 1905, however, he was arguing for the necessity of gaining a "speaking acquaintance" with the entire California marine fauna, and he was emphasizing the importance of collecting full life histories with the aim of understanding the distribution of marine organisms in space and time.[36] This project contained a substantial nostalgic element—especially through the contrast Ritter implicitly drew between his descriptive survey work and the focused experimentalism of his Berkeley colleague Jacques Loeb. Yet he was hardly returning to the projects of Agassiz or Baird. In place of their national inventories, Ritter pursued an explicitly local study, and he now justified it, not on grounds of utility or taxonomic completeness, but as the basis for abstract work. Influenced by his associate Charles Kofoid, he believed that the ultimate goal of the marine station could be a mathematically grounded ecology of the sea. Ritter was not engaged in a revolt against biology, but rather was offering a variation within that capacious framework.[37]

The crucial period in the Ritter-Scripps partnership extended from 1907 to 1910, as the Scrippses acquired property for the station, built its seaside laboratory, and agreed to pay Ritter's salary as resident director. In 1908 Ritter formally explained to E. W. and Ellen what they would be in for. Describing his activities thus far as merely "cranking the engine" of the institution, he abandoned his earlier emphasis on marine biology. He proposed instead what he admitted was a "grandiose not to say visionary" program, focusing on a "renovation of biological theory" that he believed would "turn out to be of the utmost consequence to human life and society" and would make the institution "as potent for general human welfare as any single institution of physical science that has hitherto existed." He challenged his backers to decide then and there whether he was "visionary and addle-headed and so not to be trusted with large responsibilities," or in fact had "a program vastly worth while, but one which if carried out will tax to the utmost the brain and treasury of the institution."[38]

E. W. and Ellen agreed, and the consequence was a rapid flowering of Ritter's interest in the foundations and implications of biology. His first writings in this broad and murky region were the essays on Bergson mentioned above and an article in *Popular Science Monthly* arguing that biology teachers should recognize and display more conspicuously to students their love for the animals they studied. Then, in 1912, Ritter used an official report on the history and prospects of the Marine Biological Station to offer what he called "a confession of faith about the larger meaning of science, of biology in particular."[39]

By this time, Ritter conceived his station as an institution for "pure science." This meant deemphasis on what he called "loaves-and-fishes problems," such as the declining lobster harvest.[40] But it did not mean restricting the station's mission to research. Ritter argued that "in a democratic country like ours" a biological research institution could only justify its existence by showing the significance of the knowledge gained "for the higher life of mankind." Instead of lobster, the station would provide "intellectual, spiritual sustenance" to the community—to the state and the nation to which it belonged. Ritter provided a menu of the ideas he hoped to convey to "generally but non-technically educated members of the community."[41] These included the conviction that life came only from other life; the universal applicability of the "principle of evolution"; the limited importance of the social darwinian "struggle-survival doctrine"; the orderliness of the organic world; the conception of the organism as a whole; the degree to which the capacities of organisms exceeded what was necessary for "ordinary life"; the material correlates of psychic phenomena; the laws of heredity; relations between the sexes; and *"innumerable bald, unphilosophized facts of living nature that would entertain and instruct, and consequently keenly interest thousands upon thousands of generally intelligent persons."*[42]

Ritter devoted the rest of his working life to the dual projects of transforming this litany of biological generalizations into a structured scientific faith and of proselytizing his outlook before both academic and popular audiences. I use the term "faith" here deliberately. Ritter claimed that for much of his life he had had a conventional, not very deep, interest in religion. By the 1910s, however, he was participating actively in the San Diego Unitarian Church and, more generally, in liberal religious circles.[43] He believed, on the one hand, that biology should explain the religious impulses of humankind; on the other, he wanted biology to form the basis for a truly modern religiosity.

Ritter's summa was the two-volume *Unity of the Organism: The Organismal Conception of Life*, published in 1919. The "organism" was, on the one hand, the phenomenon that Ritter confronted daily. His dredgers brought in a miscellany of animals and plants; describing and classifying these organisms was the basic activity that had defined him, over the last two decades, as a working scientist. On the other hand, "organism" placed Ritter in a venerable intellectual tradition. In his search for allies, Ritter acknowledged debts to his Berkeley philosophical colleague George Howison and his mentor Joseph LeConte, as well as Harvard philosopher Josiah Royce and the American pragmatists.

But he reached much further back, for the mantle of Aristotle. In a 1917 dialogue that Ritter imagined occurring one evening on the Scripps Institution pier, involving himself, LeConte, Darwin, William Wordsworth, Dante, Jesus, Aristotle, and Confucius, Aristotle had the last word.[44]

Ritter was proud that Aristotle had been particularly interested in his own favorite group of organisms, the tunicates. He was, in fact, the first naturalist-philosopher. Aristotle's method of describing particulars, classifying objects, and then moving toward comprehensive generalization represented a "natural history mode of philosophizing," according to Ritter. Moreover, Aristotle had placed particular emphasis on the category "organism." He recognized that living things were different from nonliving and that understanding different organisms required different sets of naturalistic categories. Animals had capacities plants lacked, and humans were rational and political. Ritter believed that he was following Aristotle in rejecting the existence of a transcendent spiritual or mental realm. He argued that "the chief end of science is to show in detail and literally how we live, move and have our being in nature. . . . Nature is man's maker as well as his sustainer." He pushed for consistently naturalistic explanations, but at the same time, he shared Aristotle's antagonism toward what Ritter called "elementalist" or "essentialist" explanations.[45]

For Ritter, the choice between "elementalism" and "organismalism" was not merely intellectual but also political. In the first months of World War I he toyed with the idea that population pressure could make aggressive war "scientifically justifiable" (and hence the United States should appease Japan by selling it Hawaii), but by the end of 1915 he was explicitly rejecting social darwinist arguments. He told colleagues at the AAAS that the dogmatic belief in "essence" shared by Nietzsche and Weismann was both unfounded and inherently "aristocratic" and argued instead for the existence of a democratic "web of life." With U.S. entry into the war in 1917, he began to attack the "philosophical brutism" that Germans supposedly drew from Weismann; at the same time, he sought to link his approach to biology with President Wilson's approach to the war. He telegraphed the president that his policies were "scientifically unassailable," and he urged federal officials to promote a "philosophy of life" that would be thought out as carefully as that of Germany, but would be stronger, "because our philosophy is far more nearly right than hers." The conclusion of *The Unity of the Organism*, composed in September 1918, exulted that the carefully prepared German military establishment, operating "under the guidance of a philosophy of mechanism and brutism," was losing to an American force "improvised in the course of a few months" but operating "under guidance of a philosophy of personality and humanism."[46]

Ritter's ultimate commitment was to the democratic implications of science. In the introduction to his book he declared forthrightly that "no faith of mine is greater because none is rooted more deeply in my scientific philosophy, than that in the ultimate triumph of popular, that is of democratic principles in all aspects of civilization. Indeed the *facts*—not the *theories*—of organic unity and integration which have dominated all my later work are the foundation of this faith."[47] Ritter's belief in popular government was based, not on a commitment to biological equality, but on the belief that people were diverse in their qualities and were autonomous. The actions of all these individuals formed the basis for progress.

Socio-biology and Americanism

Ritter's and Scripps's convictions lay behind the project in "socio-biology" that I described at the beginning of this book. The biologist Ritter and the journalist Scripps saw science popularization as a strategic salient where they could, together, improve the quality of the millions of human animals living in the United States. According to Ritter, Scripps had come "to regard his own newspaper organization, in common with all other newspapers, as a biological phenomenon"—a means through which humans acquired information and sometimes changed their beliefs. In spite of tendencies that Ritter later candidly characterized as "Fascist," Scripps concluded in 1919 that World War I had established popular rule as the predominant form of human political organization. Improving popular government would require "making all the people more intelligent," and this would consist largely in "intelligence about science," encompassing the physical, the biological, and the social. Since newspapers were the main vehicle through which the mass of Americans gained new knowledge, he proposed an organization to increase and improve newspaper science.[48]

Scripps and Ritter thus jointly conceived an American Association for the Dissemination of Science. Scripps proposed to provide $40,000 in seed money to create a journalistic bureau through which scientists could distribute stories to newspapers. The idea was that it would be self-sufficient. It would not, Scripps argued, "be a propaganda"; still, both he and Ritter expected that much of the association's activity would involve raising the level of discussion of political issues by providing relevant social and biological scientific perspectives. On broaching their scheme to others, Ritter and Scripps learned that leaders of the National Academy of Sciences had been making parallel plans (though without a source of funding) to influence the presentation of science in the press. The two groups joined forces, and in early 1920 they created

Science Service under the combined auspices of the Academy, the AAAS, and the Scripps organization, with a $500,000 endowment from Scripps.[49]

Ritter became president of the Science Service board, but real control was initially in the hands of its paid director, chemist-journalist E. E. Slosson. Scripps thought that Slosson, author of a book on the wonders of chemistry and a series of profiles of modern intellectuals, "sugar-coated" his science too much. Both Scripps and Ritter were unhappy with his emphasis on dramatic discoveries in physical science and his avoidance of the biological and social issues that had motivated the two of them. But Scripps had long before established a policy of not interfering with his managers, and Ritter was preoccupied with the process of retiring from the University of California; as a consequence, they did little to influence the organization's direction in its first years.[50]

The anti-evolution campaign refocused Ritter's attention on Science Service. William Jennings Bryan began to publicize the idea of legislative restrictions on teaching evolution in early 1922. The initial collective response by scientists was limited, and Science Service did little, in part because Slosson believed that such work would conflict with his bylaw forbidding "propaganda."[51] Ritter became involved because of local conflicts in California: he participated, along with Luther Burbank and David Starr Jordan, in a mass meeting organized in late 1924 to defend evolution teaching in that state. Ritter's contribution to this event was rhetorically less compelling than Burbank's paean to "science which has opened our eyes to the vastness of the universe and given us light, truth and freedom from fear where once was darkness, ignorance and superstition."[52] But it expressed his philosophical perspective succinctly. He argued that human progress depended, above all, on man's development of "a more logical and comprehensive philosophy of life, a more safety-insuring system of morals, and a more practically potent religious faith than he has ever yet evolved." These advances were possible "only on the basis of a knowledge of and confidence in the natural that will not have room for one jot or tittle of belief in the supernatural." The most important body of natural knowledge concerned the processes through which progress had occurred in the past: how thought and passion had enabled organisms "to solve more and more successfully the problems of their existence upon the earth." Such understanding of evolutionary progress gave people both a starting point "from which to attack human problems, and the courage of past achievements to hearten them for the advance." Outlawing evolution in the schools would thus take away from the next generation the guide that could enable them to attain a "fuller measure" of "the glory of life."[53]

For Ritter, both the problem and its solution were clear. Ignorant but vociferous believers in the "supernatural" were hindering human evolutionary progress by keeping young Americans from learning about evolution in the public schools. Science Service was uniquely situated to overcome this mal-function in the socio-biological process. Fostering progress depended on gaining the attention of masses of adult citizens and persuading them to be sympathetic toward, or at least tolerant of, the teaching of evolution in the schools. Formation of a body of public opinion opposed to Fundamentalism would prevent legislative action against science education.

In February 1925 Ritter went to Washington and took a direct role in Science Service for the first time. He brushed aside Slosson's worry that participating in the evolution controversy would violate the organization's antipropaganda policy. His initial ideas included supporting publication of modern translations of Genesis. When, in May, the ACLU arranged to challenge Tennessee's anti-evolution law, Ritter and the leadership of Science Service immediately moved to participate. They recognized that a well-publicized trial of a sympathetic individual would, like dozens of other legal confrontations over the previous generation, provide a human interest hook through which a national public could be taught about serious issues. They also wanted to be sure that the presentation of evolutionism was not solely in the hands of the notorious radical lawyer Clarence Darrow. They appropriated $1,000, from a fund Scripps had personally provided for special initiatives, to coordinate the participation of scientists in the trial. Their basic strategy was to enable moderate and responsible scientists to speak—in court, they hoped, but otherwise to the press. Science Service staffer Watson Davis was also there, both to make sure that communication between scientists and newsmen would occur and to provide copy on the broader issues at stake.[54]

At the 1927 annual meeting of the AAAS, held, pointedly, in Nashville, Tennessee, Ritter gave a general address that presented his perspective clearly. He pointed with pride to "the extensive . . . support given to science by the newspapers in the Scopes trial" as evidence for "the usefulness journalism may have to science," and he emphasized "the genuine effort by many papers" in 1925 "to get and publish data on the merits of the controversy." He argued that scientists and journalists shared a common conception of truth, and that their "union of interests and efforts" would expand if they deepened a common commitment to evolutionism. In particular, bolstered by his memory of the recently deceased Scripps, he emphasized that "belief in the supernatural" was "a gigantic error." It should be replaced by recognition of "the creative power of the natural order." If scientists and journalists would develop a partnership to spread such an evolutionary naturalism, they would enable Americans to adjust the "life of emotion" and the "life of reason"—not only with regard to religion, but in their dealings with economic issues and with sexuality. From Ritter's perspective, this project was fundamentally a national one.

In a stirring conclusion, he reminded his audience that enabling Americans to see themselves as passionate, reasoning, and cooperating organisms would be the fulfillment of "Americanism," as it had been sketched by "those two master builders of our Nation, Benjamin Franklin and Thomas Jefferson."[55]

The Scopes trial, like many a latter "trial of the century," was a messy event that, as Edward Larson has emphasized, had multiple and conflicting consequences. Few newspapers at the time went as far in their advocacy as Ritter envisioned. Yet they focused more Americans' attention on evolutionism than could dozens of Science Service press releases about new discoveries. They juxtaposed news coverage of the trial, educational features presenting the arguments and evidence for evolution, and, in some cases, discussions of the evolution of the Bible. Many presented both Scopes and the scientists as serious, even heroic, searchers for truth; more powerfully, they presented the trial as a contest between freedom and oppression and between civilization and ignorance. The most compelling argument that defenders of evolution mustered in legislatures after Scopes was that passage of an anti-evolution law would make them a "laughing-stock" around the country. Scripps and Ritter would have agreed that fear of ridicule was one of the characteristics of the "damned human animal," something that was both natural and a crucial marker distinguishing humans from other organisms. In his next book, Ritter emphasized that the basis for progress was maladaptation.[56]

Good Breeding in Modern America

Enlightenment, whether in the schools or in the public arena, was important work. However, as the high school teachers knew well, action ultimately mattered more than ideas—both for their objects of study and for themselves. Action came easily to scientists involved with focused and seemingly immediate problems, such as the USDA employees discussed in chapter 3. But what about professors concerned with American life more broadly? How could biologists committed to academic careers and research values make a difference regarding policy, norms, or behavior?

The lack of national significance in academic biology briefly became visible during World War I. The more practical aspects of the life sciences—agriculture, fisheries, forestry, and medicine—were well supported by the military and other branches of the federal government. University chemists, physicists, and even psychologists participated energetically in the war effort through committees organized by the new National Research Council (NRC). Biologists such as Edwin G. Conklin, however, worried in 1915 that their domain was being classed with "embroidery," as a useless, unmanly subject; the NRC committee on which Conklin served in fact drifted through the war, doing little more than organizing classes for college reservists. This embarrassment was forgotten with the return to normalcy: by 1920 the academic biologists who controlled the NRC Division of Biology and Agriculture were emphasizing that the greatest needs in their field were for graduate fellowships at private universities and increased support for the Marine Biological Laboratory.[1] But at least some professors sought to do work that would, in the long run, be important for the commonwealth. They wanted to participate in formulating the future for American people.

We can see this desire, the difficulties involved in realizing it, and the conditions that made for success by comparing the most visible areas of biologists' involvement with human problems in the first half of the twentieth century— eugenics and sex. These were hardly independent subjects. Their interconnections are in fact nicely encompassed in the multiple entendre of "good breeding"—a phrase that can allude, depending on context, to class and ethnic origins, home circumstances, schooling, high culture, the maintenance of favorable hereditary qualities, fertility, prenatal hygiene, or enjoyable coitus. At one end of the discussions of good breeding were evolution and heredity; at the other, sexual physiology and behavior. These poles were linked through a range of concerns, including hygiene, population, reproductive physiology, and birth control. The common interest was the biological improvement of humanity.

Yet eugenics and sex offer important contrasts, in both development and legacy. Eugenic biologists, for the most part, expressed mainstream values and operated within a framework of public discussion and advocacy. Sex biology was much more controversial and as a consequence was developed largely within settings that were comparatively private. The eugenics movement, so-called, rose and fell within a generation and left a confused legacy that scientists and historians are still trying to sort out. Work in the biology of sex, by contrast, had both definite and enduring consequences. Between the 1910s and the 1960s, in fact, the respectability of the two domains reversed: as eugenics became a dirty word, sex biology became a regular feature in mass magazines. A comparison of biologists' engagement with eugenics and with sex illuminates three issues in particular. It demonstrates how and how much American academic biologists sought to make human problems into biological problems during this period. It highlights the intellectual, organizational, and practical difficulties—but not the impossibility—of succeeding in such a transformation. Finally, it indicates what biologists concerned with Americans' problems actually accomplished in the first half of the twentieth century.

The Imperfect Amalgamation of Eugenics and Biology

Eugenics is the subject that has made the study of early twentieth-century biology important and meaningful within general history. In the last generation, at least a dozen books, including prize-winning contributions by Stephen J. Gould and Daniel J. Kevles, have examined aspects of American eugenics. Charles Benedict Davenport occupies a larger place in American history than Thomas Hunt Morgan, and the Carnegie Institution's support for the Eugenics Record Office has received much more attention than, for example, Scripps's patronage of Ritter's station. Three emphases have stood out. The origins of eugenics are traced to major scientists—to Davenport in the United States and then back to the English biometrician Karl Pearson and the polymath hereditarian Francis Galton. The core of the American movement is presented as old stock, upper class, racist, and coercive. Finally, though writers have disagreed over the causes of eugenics' apparent decline between 1925 and 1950 and over the degree of continuity between hereditarian movements at the beginning and the end of the twentieth century, there has been a continuing focus on the question (to use old eugenic terminology) of the nature and extent of the "taint" that eugenics has left within human biology, especially genetics, up to the present.[2]

Rather than inquiring directly into the origins and consequences of eugenics, I want to place biologists' involvement in that movement within the framework of their varied "socio-biological" pursuits. Such a perspective foregrounds some facts that are acknowledged but seldom emphasized. Eugenics, with or without the name, was a significant bandwagon well under way in the United States when academic biologists first joined it in the years around 1910. These scientists became involved, not because of developments in genetics, but as part of their heightened interest in social problems at the climax of the Progressive Era. Biologists' involvement in eugenics is notable, however, more for its diversity than for its direction. A biology-based eugenics, instead of rising around 1900 and then declining around 1930 for either intellectual or political reasons, was an enterprise that never really cohered. Its promoters were unable to integrate it into the disciplinary and work structure of the academic biology that had developed around the turn of the century. It was, on the one hand, too interdisciplinary and, on the other, insufficiently experimental. Biologists left eugenics less because of refutation or revulsion than from a dearth of innovation.

Interest in human biological improvement and the prevention and correction of "degenerative" tendencies were commonplace, if protean, concerns throughout the nineteenth century. Many people believed that heritage, home setting, and education combined to affect the quality of children; they worried that bad behavior—drinking, prostitution, and especially masturbation—could not only affect the individual but lead to weak offspring.[3] In the United States, these concerns first received pointed expression in the writings of Gilded Age reformers. The paradigmatic articulation of this perspective was Richard Dugdale's 1877 family survey, "*The Jukes*." Dugdale described how bad behavior and unfavorable conditions had intensified and been reproduced through generations, leading to a lineage of prostitutes, criminals, and idiots. He argued, however, that incarceration and education could reverse these trends and turn the next generation back toward health and respectability. Another example of the genre, "The Tribe of Ishmael," written by Indiana minister Oscar McCulloch, is notable primarily for clarifying the question of influence; it was Reverend McCulloch who introduced the naturalist David

Starr Jordan to the problem of the biological study of the poor when Jordan worked in Indianapolis in the 1870s.[4]

In the depressed first half of the 1890s, patrician scientists such as Nathaniel Shaler and Clarence King publicly worried that the apparent increase in paupers and defectives was due in part to civilization's relaxation of the rigors of natural selection.[5] A few years later, asylum physicians (including the aptly named Indiana doctor Harry Sharp) persuaded legislators in a number of states that the new surgical procedure of vasectomy would solve a range of problems among asylum and prison inmates, including masturbation, fathering of orphans, rape—and the decline of the race.[6] In 1906 Theodore Roosevelt was able to use the bully pulpit of the presidency to direct the attention of respectable Euro-Americans to the importance of maintaining their fecundity and to the danger of committing "race suicide."[7] Agricultural scientists, sociologists, and physicians united around Francis Galton's quarter-century-old neologism, "eugenics." The American Breeders' Association, a group of practical, government, and academic scientists led by USDA scientist and bureaucrat Willet M. Hays, established its Committee on Eugenics in 1907; the most prominent members were Jordan, Luther Burbank, Alexander Graham Bell, and sociologist Charles Henderson.[8] The most popular early twentieth-century American book on eugenics was Burbank's engagingly titled *Training of the Human Plant*, which argued for the inheritance of acquired characters and the positive value of racial hybridization.[9]

What is notable in this narrative, and comprehensible in light of previous chapters, is the lack of involvement of the founders of academic biology in these activities. Jordan—a Cornell-educated naturalist associated with federal science and, after 1891, an administrator working in the isolated setting of Stanford—was the exception proving the rule. He offered a general undergraduate course in evolution, including eugenics, from the early 1890s on; but despite his efforts to make his ideas available generally in the form of published lecture notes, his initiative was not widely emulated. Jordan's picture of the world was not rejected; most biologists, however, had more pressing research and teaching concerns.[10]

Academic biologists began to pay greater attention to eugenics as part of their broader effort around 1910 to become more relevant and influential within public culture. Preparing popularizations and textbooks were thus among eugenic biologists' first activities. Charles B. Davenport produced a pamphlet introducing the subject in 1910 and published a more detailed exposition a year later. William E. Kellicott, a biologist closely associated with Johns Hopkins, delivered talks at Oberlin College in 1910 that he published as *The*

Social Direction of Human Evolution. Biologists presented a series of eugenics lectures at the University of Chicago in 1911, and in 1912–13 a loose alliance of professors organized presentations on major college campuses around the country. Discussions of eugenics were incorporated into the more general books, such as George Parker's *Biology and Social Problems* and Conklin's *Heredity and Environment in the Development of Men,* which appeared around this time. By 1916 University of Wisconsin biologist Michael Guyer directed *Being Well-Born* at nonscience majors, and Harvard zoologist William E. Castle produced an advanced undergraduate textbook that combined genetics and eugenics.[11] American academic biologists were participants in, but not leaders of, major prewar eugenics organizations: the British-organized International Eugenics Congress (1912), the First National Conference on Race Betterment (1913), and the *Journal of Heredity* (founded in 1914).[12]

Biologists' early eugenic popularizations combined enthusiasm in general with uncertainty regarding particulars. They presented the subject as the climax of biology's potential to guide the development of humanity.[13] But they were divided and internally confused about both basic tenets and specific applications: on the one hand, the roles of Darwinian, Mendelian, and Lamarckian mechanisms and, on the other, such issues as the relative importance of "negative" and "positive" eugenics, the degree to which the American upper class was a hereditarily superior population, the eugenic consequences of war, and the advisability of either compulsory or voluntary sterilization. Guyer's textbook was representative: its concluding chapter began in fire-breathing fashion by comparing the increasing number of defectives with a flood of feces. It then proceeded through the hereditary excellence of the Darwins and Lowells, the decline in the birthrate among the best Americans, and the dangers of mongrelization, to a blistering dismissal of ethical objections to compulsory sterilization. Yet Guyer pulled back from nearly every specific positive conclusion. After demonstrating declining elite fertility through statistics on Harvard and Bryn Mawr graduates, he questioned whether "the woolly-witted son of opulence, so abundant in our colleges today," was in fact superior to the "alert eager boy—and his name is legion" unable to afford higher education. He described the arguments for sterilization and rejected those of major opponents, but then questioned its necessity, effectiveness, and moral and hygienic consequences.[14]

Eugenics, moreover, did not win the general support of biologists. Numerous prominent individuals remained unpersuaded, or rejected it after a brief period of support. William Ritter briefly expressed interest in eugenic goals in the late 1900s, but within five years was arguing against the main programs in the area. Jacques Loeb, while accepting in principle that family lineages might differ in talents, considered the science underlying eugenics to be suspect and its chief promoters to be racist. T. H. Morgan was criticizing eugenic arguments intensely by the late 1910s. As early as 1914, Wesleyan University evolutionist and bacteriologist Herbert W. Conn devoted a book to the argument that what seemed to him the core message of eugenic biology—that the only way to change humanity was to control marriage—was both depressing and false. He noted that many of the problems eugenicists emphasized, such as syphilis, were not hereditary in the true sense. He argued, in straightforwardly evolutionary and progressive language, that humans differed from other animals because they had evolved, through natural selection, a moral sense that provided the foundation for social feelings and social organization. This "artificial" construction was, and would continue to be, much more important in human progress than anything that could be done through the control of mating.[15]

The response voiced by Davenport, Guyer, and others to uncertainties and to criticism of eugenic enthusiasm was that the real need was for research so that scientific knowledge could catch up with reformist zeal.[16] But that response soon led to a deeper problem: it was very difficult to do eugenic research within the framework of the academic biology that had been established at the universities and at Woods Hole during the previous generation. Comparison with the German situation is illuminating. In the early twentieth century, basic biological work in Germany was subordinated to a highly developed system of academic medicine that had close ties to the nation-state. The German movement's medical and statist orientations were captured in its preference for the term "racial hygiene" rather than eugenics. Numerous scholars have shown how, between 1900 and 1940, German racial hygienists created a dynamic enterprise that both fostered research into human heredity and easily became integrated into Nazi efforts to eliminate the insane, retarded, deviant, and, ultimately, millions of Jews.[17] In the United States, by contrast, participation in eugenics ran contrary to the balance between specialization and organization that academic biologists had so laboriously established during the preceding generation. Biologists were institutionally and intellectually separated from clinical institutions—most completely from insane asylums, but also from public health offices and even from university hospitals. They were also increasingly isolated from psychologists, sociologists, and anthropologists. In addition, as we have seen, biology was dominated by an ethos of individual, focused, basic research. Biologists sought discrete discoveries of a fundamental nature that would be applicable across species.

It was difficult to reconcile eugenic biology with these structural elements and disciplinary values. We can see these problems in the work of Davenport, the most important American biologist involved with eugenics.

CHARLES B. DAVENPORT AND THE DIFFICULTY OF EUGENIC RESEARCH

For the first twenty years of his career, Charles Davenport was a typical academic biologist. He came to the subject as part of an adolescent rejection of modernity and search for personal autonomy. Growing up in Brooklyn, a few blocks from New York harbor and that symbol of America's metropolis, the Brooklyn Bridge, he was pushed by his father to forge a career as either a real estate developer or engineer. He completed a technical degree at the local Pratt Institute in 1886, but his real love was nature. For him that meant participation in the juvenile Agassiz Association, immersion in the Smithsonian literature that Spencer Baird distributed so generously, and summers on his family's Connecticut vacation farm.[18]

In 1888 Davenport rebelled against his father and enrolled in the Harvard zoology program. Initially he joined in the general interest in invertebrate embryology, with particular attention—appropriately—on the development of individuality. He summered in Woods Hole (at the Fish Commission) and in Newport, and he pursued a variety of projects, culminating in his encyclopedic *Experimental Morphology* in the late 1890s.[19] He then decided to focus on the quantitative study of variation and evolution. Having maintained his Brooklyn connections, he became director of the summer laboratory operated by the Brooklyn Institute of Arts and Sciences in Cold Spring Harbor, on the north shore of Long Island, in 1898. Davenport believed that this locale offered ideal conditions for work on the ecology of animal populations and their transformations: using his summer students as field collectors, he gathered and counted the animals on a cross-section of the half-mile long sand spit that projected out into the harbor. His aim was to assess the relative importance of local conditions and large-scale climatic or geographic factors in determining what organisms would thrive and how they would interact.[20]

In the early 1900s Davenport pursued one of the dream projects of American academic biology. He had left Harvard in 1899 to join the University of Chicago zoology department, and he soon identified himself as a protégé of Charles O. Whitman. He appropriated Whitman's idea for a "biological farm" on which evolution could be studied directly, and in 1902 he proposed that the new Carnegie Institution of Washington (CIW) organize a "station for

Figure 9–1

Charles B. Davenport, 1914.
Photograph courtesy of the Cold
Spring Harbor Laboratory Archives,
Cold Spring Harbor, New York.

experimental evolution" at Cold Spring Harbor. He was aware, as he explained to medical organizer and CIW board president John S. Billings, that
the program he proposed could aid in the "improvement of the human race"
and, more particularly, could illuminate the American problems of race mixture and the effect of climate "on 'blood,' " but he emphasized that this kind
of work would take a generation to launch. He pointed to himself, at thirty-
six, as a person who could devote twenty-five years to basic research.[21]

In early 1904, the Carnegie board endorsed Davenport's proposal; he immediately left Chicago and settled down to work, on a waterfront estate in
one of the wealthiest suburban enclaves in New York. He hired such promising young scientists as botanist George Shull and *Drosophila* pioneer Frank
Lutz and sought to shape a program that was responsive to rapidly changing
scientific views of the evolution problem. Initially he hoped to investigate
variation, mutation, and selection in order to uncover "the laws of evolution
of organic beings," but he soon came to focus on Hugo DeVries's mutation
theory and on the Mendelian "laws of inheritance of characteristics." He bred

poultry and canaries, and studied the inheritance of eye color and hair form in humans. Davenport participated with practical scientists in the new American Breeders' Association, but he advertised his academic identity by leading its Committee on Theoretical Research in Heredity.[22]

In 1909 Davenport suddenly embraced the eugenics movement. The immediate stimulus was probably contact at the Breeders' Association with Eugenics Committee chairman David Starr Jordan, who was interested in getting rid of his responsibilities.[23] He was energized, however, by his Long Island neighbor, Theodore Roosevelt, whose last major act as president was the convening of the North American Conservation Conference at the White House in February 1909. Less than a month later, Davenport emphasized that while conservation of coal reserves, forests, and domesticated animal breeds was certainly important, "there is one national asset to which too little attention is being paid . . . —the best of human protoplasm!" Within a year he took over the Eugenics Committee, began to investigate the heredity of feeblemindedness, and persuaded Mary Harriman, widow of the railroad magnate E. H. Harriman, to pledge the funds to create the Eugenics Record Office (ERO).[24]

Davenport believed that his key innovation was to link research in eugenics and advanced biology. He located his eugenics center on a large estate adjacent to the Station for Experimental Evolution, served as director of both organizations, and around 1920 merged them to form the Carnegie Institution Department of Genetics. He resisted the suggestion that he create an independent organization for "propaganda" on human problems, arguing in 1912 that he wanted to focus on research and that "the principles of heredity are the same in man and hogs and sun-flowers." He envisioned a long-lasting, integrated program of investigation.[25] Actions, however, did not match this rhetoric. During the course of the next decade the office became increasingly identified with propaganda for both compulsory sterilization and immigration restriction. And Department of Genetics staffer E. Carleton MacDowell asserted, in retrospect, that the biggest problem at Cold Spring Harbor under Davenport had in fact been the lack of "integration of the work of highly individualistic investigators," operating from different institutions and with "differing points of view." While MacDowell blamed this failure on Davenport's oversensitivity, secretiveness, and inability to provide leadership "on a philosophical plane," the underlying problem was that eugenic interests diverged too much from the values of academic biology for personal or "philosophical" resolution.[26]

Harriman, like most supporters of eugenics, wanted results. In the negotiations establishing the ERO, she advised Davenport that she was supporting his work, not building an institution. Her conception of scientific success derived from her participation, a decade earlier, on a collecting cruise to Alaska that her late husband had organized for a group of naturalists. She did not demand immediate biosocial change, but she was looking at least for visible accumulations of data.[27] Davenport already believed that collection of family pedigrees made up an important part of eugenic biology, and he realized that organizing and maintaining such information (in a fireproof vault) provided a rationale for a permanent institution. The consequence of this confluence of interests was that Davenport's eugenic biology research center took shape around a "record office" and was organized on the pattern of his earlier summer ecology project and, more broadly, on the Gray-Baird model of natural history investigation.

Davenport envisioned a grand collecting enterprise that would generate a comprehensive report in the indefinite future. His foundational document was *The Trait Book*, an extensive key to the classification of human variations from head to foot. He did some fieldwork—on Long Island and in Maine—himself, but he expanded his reach significantly by recruiting, training, and energizing dozens of low-paid or volunteer, mostly female, "eugenic field workers" through an annual summer school program. These agents, like Gray's or Baird's collectors, dispersed around the country to gather and input masses of semi-digested pedigree data. The ERO itself consisted of a small clerical staff, led by Davenport and his chosen subordinate, Harry Laughlin (a young midwesterner with a master's degree in agriculture, who originally came to Cold Spring Harbor to teach a summer course on farming), who, like herbarium and museum staffs, processed vast amounts of material—more than thirty-five thousand case histories by 1935, cross-referenced through nearly a million index cards.[28]

As data gradually accumulated, Davenport sought to demonstrate progress through a series of bulletins. Some dealt with paradigmatic cases of Mendelian heredity in humans. A 1916 study of Huntington's chorea traced this problem in a number of families and argued straightforwardly that this progressive neurological disorder was a Mendelian dominant. Davenport also identified cases of osteopsathyrosis and argued for a similar Mendelian basis. This work, however, involved essentially freak phenomena: disorders that were easy to identify because their effects were so dramatic but were limited in their significance because they affected only a minuscule portion of the population.[29]

Other bulletins were efforts to make a preliminary case for the implications that the ERO's data would have for biosocial policy. These projects, however, were deeply problematic. Historians have noted, sometimes with amazement, the naivete displayed in ERO reports such as *The Nam Family*, *The Hill Folk*, and other studies of "white trash" lineages; its studies on the heredity of feeble-mindedness; and Davenport's report on the supposed Mendelian basis for sailors' love of the sea.[30] The common problem in these studies was that by the scientific standards of the 1910s, they were overly interdisciplinary, involving a mix of what were recognizably biological, medical, psychological, and sociological assertions. Such work would be difficult for anyone trained only in biology, and particularly so for a group that was working in isolation from both social scientists and physicians. Davenport initially made efforts at outreach, collaborating in 1909–10 with child psychologist Henry H. Goddard and some superintendents of schools for feebleminded children, but such partnerships were extremely time-consuming; most projects after 1911 were done in-house. The consequence, however, was that by the mid-1910s, psychiatrists were attacking Davenport for poaching on their territory, psychologists such as J. E. Wallace Wallin ridiculed the low standards in his work, sociologists dismissed him as a naive pedant, and anthropologists attacked his efforts to control the NRC division of psychology and anthropology. Rather than taking on the difficult task of responding to these criticisms, Davenport shifted his ground: from families to occupational groups, and finally, in the 1920s, to race.[31]

The deeper difficulty for the ERO, however, was not its publications but its structure. Its isolation, identification with its unique database, and emphasis on natural history work all meant that the office's only products were its bulletins. Junior scientists at the adjacent Station for Experimental Evolution contributed to the journal literature, participated in disciplinary associations, and, in some cases, ultimately integrated themselves into academic science. Frank Lutz, for example, moved to the American Museum in 1909, and George Shull became a professor at Princeton in 1916. ERO staffers, by contrast, were either professionally marginal or permanently rooted in the ERO files. Women such as Florence Danielson and Anna Finlayson lacked advanced degrees. Laughlin obtained a doctorate from Princeton essentially by courtesy, and Arthur Estabrook and Howard Banker devoted years to the preparation of a few reports. No one moved from the ERO to a significant university chair.[32]

Davenport apparently recognized this problem. In the mid-1910s he sought to develop a new track of experimental eugenics, attracting young biologists to projects that would be both academically significant and eugenically meaningful. The most notable figure in this effort was the later-disillusioned Mac-Dowell. He came to Cold Spring Harbor from Harvard in 1915 with a mandate to develop an animal model for eugenic problems. He was drawn, however, by his interests, audiences, and results, back into biology.

In the 1910s, the most significant intersection of experimental biology, eugenic concerns, and political controversy involved alcohol. At Cornell Medical College in New York City, biologist Charles Stockard was reporting that alcohol acted as a "germinal taint" in guinea pigs, producing hereditary degeneration in the offspring of both female and male animals intoxicated repeatedly prior to mating. These findings became part of the accelerating arguments over Prohibition. Davenport, like most eugenicists, supported temperance and believed that the offspring of both male and female drinkers suffered from their parents' indulgences. He hired MacDowell, who had acquired expertise with the newly popular stand-in for humanity, the white rat, from the mammalian geneticist and eugenicist William Castle, to see if "alcoholized" animals had lower fecundity or produced mentally inferior offspring.[33]

In less than a decade, however, a confluence of forces transformed this experimental breeding project from a model for eugenic reform into a failed contribution to basic biology. The rapid change in the political climate following the adoption of Prohibition lowered interest in results that were pro-temperance; the complexity of the experimental problem, which involved genetics, physiology, and behavior, as well as extrapolation from rats to humans, was such that results were equivocal; MacDowell was a young man trying to produce academically significant papers. The breeding program was disrupted when staff scattered during World War I. When MacDowell analyzed his data in 1921, he presented the results as indications that acquired characters were inherited and claimed that their relevance to human alcohol problems was only "incidental." As support for neo-Lamarckism rapidly faded in the 1920s, MacDowell's prospects outside Cold Spring Harbor diminished; he remained there, studying growth and then cancer. The promise of eugenic biology was unfulfilled. With resources essentially static, Davenport and his staff continued to emphasize the survey approach.[34]

If we look more widely at American eugenic biologists, what is notable is how little eugenics research they did. In the 1920s, biologists prominent in eugenic organizations included William E. Castle and Edward M. East at Harvard, Edwin G. Conklin at Princeton, Henry F. Osborn at the American Museum of Natural History, and Frank Lillie at Chicago. They spoke positively about eugenic goals and acknowledged the need for better data and clearer intellectual foundations, but neither they nor their students did much to remedy this situation. Senior men pursued their ongoing noneugenic research programs, and younger biologists went in other directions. The exception that proved the rule was Charles Stockard. In the mid-1920s he obtained a major grant from the Rockefeller philanthropies to study the inheritance and development of temperament in different breeds of dogs. This project was clearly designed to provide an animal model for human racial differences. He built a dog farm in Peekskill, New York, and worked on the project, essentially in isolation, for fifteen years. At his death in 1939 he had published almost nothing, and experts brought in to assess the data declared the project a total loss. The ideas and skills that worked in biology were hard to transfer to eugenics.[35]

From this perspective, we can see why the most prominent innovations in eugenics in the United States came from groups other than academic biologists. Nativist politicians promoted ethnically discriminatory immigration restriction policies, psychologists developed intelligence testing into a tool for asserting that races differed in cognitive capacity, state fair promoters devised "fitter family" contests, asylum physicians pushed programs for the sterilization of the insane and retarded, and at least one physician practiced and promoted euthanasia of defective infants.[36] These initiatives expressed the priorities, methods, and standards of their primary backers. Biologists joined these movements to varying degrees, but they easily abandoned them when they slowed down or came under fire. Raymond Pearl, who had argued boldly when working on poultry at the Maine Agricultural Experiment Station in the 1910s that alcohol was a eugenic plus because drinking kept the hereditarily weak from reproducing, had no compunctions about criticizing mainstream eugenics brutally in 1927, a few years after he moved to the Johns Hopkins Medical School; he shifted his primary concern to the broader problem of overpopulation. East, who was arguing strenuously for the exclusion of the Irish in 1920, made a similar transition. Herbert S. Jennings held on to eugenic ideals into the 1930s, but at the same time he was arguing that little progress could ever be made in eliminating undesirable genes. A few years after Davenport's retirement in 1934, Carnegie Institution leaders and academic geneticists agreed that the ERO was a failure, and they shut it down.[37]

Daniel Kevles and Pauline Mazumdar have explained how, in both England and the United States in the 1930s and 1940s, the predominant interest in problems of human heredity shifted from a eugenic to a medical context. Medical geneticists were willing to generate data much more slowly than biologists, and the questions they asked were more limited. The payoff they sought was knowledge that could be applied to the particular difficulties of the patients before them. It was a very unacademic attitude, as well as a retreat from the earlier belief that understanding human breeding would be the foundation for a major improvement in the composition and quality of the American people.[38]

The new self-denial on the part of biologists regarding eugenics did, however, resolve the kinds of conflict between scientific beliefs and human desires that had put Clarence King under psychiatric care in 1892. Working slightly more than a half-century later on the site of King's confinement, Columbia University biologists Leslie Dunn and Theodosius Dobzhansky prepared *Heredity, Race, and Society* for the new Anglo-American Penguin/Pelican paperback series "on the sciences, arts and skills of man." Though the series advertisement heralded "man . . . as integrator, governor and administrator of the world we live in," Dunn and Dobzhansky's aim was a negative one, as far as eugenic biology was concerned. They emphasized the futility of nearly all efforts to improve the genetic makeup of human populations, and they argued that the belief that "human race hybrids are inferior ... must be counted among the superstitions." While anticipating ultimate "race fusion," they celebrated the "variety of human cultures"; instead of worrying about the intellectually inferior, they honored "the simple kindness of heart of a plain man."[39] From their perspective, Americans could breed with whom they wished; they did not presume to guide or to censure.

SOLVING THE PROBLEMS OF SEX

Sex was a different matter. Biologists became increasingly visible sources of information and guidance on human sexual phenomena during the first half of the twentieth century. The history of this activity is still murky. A pair of general surveys of the history of sex research include American work. There have been notable histories of reproductive science, examinations of birth control and venereal disease, and two recent biographies of Alfred Kinsey.[40] These works have demonstrated both the importance of the private enthusiasms and fears of sex researchers as well as the intricate interactions among investigators and the groups that financed and protected their work—most notably the Rockefeller philanthropies and the National Research Council's Committee for Research on Problems of Sex (CRPS). We still need, however, a comprehensive and detailed study that will explain how scientists assessed the sexual problems of Americans, including themselves, and will show where and how scientists influenced patterns of behavior.

My aim is the more modest one of comparing biologists' involvement in sex problems with their participation in eugenics. Though unable to integrate eugenics issues with the structure and values of their academic discipline, biologists were able to generate major research programs around sex. Eugenics occupied a no-man's-land among the biological, social, and medical sciences, but between the 1910s and the 1940s, sex was increasingly understood as a part of biology. And although eugenic biologists had little enduring influence, sex biologists were able to have a major impact on the thinking and behavior of large numbers of Americans. They participated deeply in, to use Paul Robinson's phrase, the "modernization of sex."

Most biologists who came to Woods Hole each summer in the early twentieth century were there for sex. Marine invertebrates were scientifically interesting in large part because they were so open in their efforts to reproduce; they would spawn in tidal pools at the feet of collectors, or would even unconsciously enlist scientists as agents of fertilization. Biologists' behavior at Woods Hole was determined by the breeding habits of their organisms of choice. Ernest E. Just was typical in spending evenings on a pier with a lantern, waiting to net specimens of the marine worm *Nereis* at the moment of spawning. Many scientists spent their days hunched over microscopes observing fertilization and the early stages of development, or examining the structures found in sperm, ova, and embryos.[41]

The specificity of these studies was such, however, that sex acts were, in an important sense, lost from scientific view. To biologists, the spawning event was usually a mere detail of life history, insignificant by comparison with spermatogenesis, ovigenesis, fertilization, and development. The fundamental problem of sex, as perceived by biologists, was not one of activity, but of difference—not the behavioral interactions of males and females, but the existence and characteristics of the two sexes. The sexual behavior of invertebrates, and its resonance with the lives of biologists themselves, appeared largely in stray moments of off-color humor, such as Jacques Loeb's joke that he prevented inadvertent fertilization in his experiments on artificial parthenogenesis in sea urchins by using only "ladies' handkerchiefs" to wipe out the bowls in which he put eggs.[42]

It is not surprising, therefore, that university biologists were nowhere to be found when, in the early 1910s, it suddenly turned "sex o'clock in America." In New York City, campaigners against prostitution and venereal disease joined with enthusiasts for the ideas of Sigmund Freud to make sex a subject that could be discussed in public. Their initial cause was *Damaged Goods*, a French melodrama that dealt with the spread and consequences of syphilis. As chapter 7 indicated, biology educators had some involvement with the antisyphilis Society for Sanitary and Moral Prophylaxis (soon renamed the American Social Hygiene Association, or ASHA), and former DeWitt Clinton student Edward Bernays began his career in public relations by promoting *Damaged Goods*. Still, leadership in the social hygiene movement was composed largely of conservative reformers working under the patronage of John D. Rockefeller, Jr. Efforts expanded after American entry into World War I, with a focus on limiting the sexual activities of young lower-class Americans, especially their participation in prostitution. At the end of the war, the federal Public Health Service planned to expand these propaganda efforts into a permanent nationwide program of sex education, hiring New York biology teacher Benjamin Gruenberg as coordinator.[43]

The apparent success of these efforts led university biologists finally to become involved with "sex problems." In 1920, in the wake of the government's entry into antisyphilis education, ASHA was searching for a new mission. Max

Exner, the association's sex education director, and Earl Zinn, a staffer trained in psychology, proposed a new focus: fostering research. They hoped to advance knowledge that would both strengthen the scientific credibility of sex educators' pronouncements and enable them to take a less monolithically negative tone in their teaching. They gained Rockefeller's approval to initiate a program of investigations dealing with normal aspects of sex, with the understanding that it would be managed through the National Research Council (NRC), an independent scientific organization, rather than through ASHA, which was closely associated with the Rockefellers and admittedly devoted to "propaganda." Because, as Adele Clarke has emphasized, human sex was widely considered an "illegitimate" research area, the small group of scientists and foundation staffers who participated in what became the CRPS would determine the directions of work for more than a generation. The activity of this group forms the starting point for understanding the development of sex biology.[44]

Exner and Zinn understood that their first tasks were to interest respectable scientists in their program and to outline a workable plan of action. At least some of the information they wanted was straightforward, at least in principle: basic scientific facts about the physiology of orgasm, good data on the sexual activities of adolescents, and scientifically authoritative guidance regarding the effects of masturbation on physical and mental health. Zinn and a Carnegie Institution consultant agreed at the outset, however, that no American scientist at that time could publicly report that he had performed "rigid experimental investigation of the basic physiological and psychological factors in relation to sex and sex expression" (that is, measurements of respiration and blood pressure during coitus, experiments on conditions influencing orgasm, or controlled studies of adolescent masturbation). Zinn believed that it would be both necessary and, in the long run, more desirable, to answer such questions as part of a broader "genetic account of sex tendencies."[45]

The meaning of this phrase became evident at a conference Zinn organized in October 1921 to build support for the program at the NRC. The twelve participants included medical scientists Victor Vaughan and Walter Cannon, biologists Edwin G. Conklin and Michael Guyer, and psychologists Robert Yerkes, Carl Seashore, and Helen Thompson Woolley. When Woolley, a feminist, called for studies of birth control, she was told that the NRC would consider such work "propaganda" and hence outside its domain. Participants agreed that human "sex behavior" was important, but it was only part of what they should study. Vaughan argued, for example, that "the internal secretions of the sex glands" should be the "center, warp and woof" of both sex education and the proposed research program. The conference report listed a wide variety of subjects for investigation. These included infra-human sexual physiology and behavior, differences in sexual habits among people of different ages, races, and degrees of civilization, the nature of the "sex impulse," the health effects of masturbation, intercourse, and abstinence, and eugenic concerns about population and race. Seashore asserted hopefully that all these proposed research areas could be subsumed under the umbrella term, "biology of sex."[46]

A more powerful notion of sex biology emerged, however, in the course of the next year. Yerkes, Cannon, and Conklin became the nucleus of the new NRC Committee for Research on Sex Problems.[47] The initial grant proposals, divided among psychology, physiology, and biology, included surveys of the sex lives of Harvard students and of a comparable sample from the "quite different racial group" (Jewish immigrants) at City College, investigations of the estrus cycle in guinea pigs, and studies of sex determination in marine invertebrates and fungi.[48] The committee apparently decided to deemphasize socio-racial problems, at one end of the evolutionary ladder, and plants and lower animals, at the other. In December 1922, they assessed two competing program statements: one by Yerkes that emphasized relevance and prioritized physiological and psychological projects, and one by Frank Lillie (who had replaced Conklin) that focused on "fundamental research" and on biological problems of sex determination, differentiation, and interrelations. They then agreed to divide their domain officially into four parts: biology of sex, physiology of reproduction, "psychobiology of sex (infra-human)," and human psychobiology, including social science.[49]

The committee's dual use of "biology"—as both a general term describing all the committee's work and as one of the subgroups within that framework—was reminiscent of the ambiguity that Charles O. Whitman had used a generation earlier in promoting his new academic disciplinary structure. Both the natural science emphasis at the NRC and the initial domination of the committee by influential biological scientists Conklin and Lillie, as well as Cannon, made such an approach seem natural. It was strongly reinforced by the smoothly operating, well-formulated research programs that applicants in biology and physiology submitted. By March 1922, Lillie, Herbert Evans at Berkeley, Stockard and his assistant George Papanicolaou at Cornell, and George Corner at Johns Hopkins had each presented detailed requests for aid on projects dealing with hormones and the ovarian cycle, and they were able to report regular progress over the next years.[50]

This aura of competence in endocrinology contrasted with the vague ideas and abortive projects that psychologists, sociologists, and medical sexologists presented. In the first round of applications, for example, Minnesota psychologist Karl Lashley presented a schizoid proposal that combined a detailed outline of work on sexual maturation and behavior in rats with a sweeping plan to survey and interview unspecified individuals concerning their sexual knowledge and premarital sexual experiences. (Only the first part was funded.) Iowa psychologist Bird Baldwin requested funds to determine whether boys began to produce sperm before they grew pubic hair, a project whose rationale the committee considered obscure at best. At Harvard, psychologist William McDougall and physician Roger Lee proposed competing questionnaires on the sex lives of Harvard students; Lee's plan was funded, but ended without significant results when university administrators withdrew their support. Yerkes himself had elaborate plans for studying sexuality in primates, but could do nothing to realize them until after 1924, when he left his staff position at the NRC for a professorship at Yale.[51]

During the course of the 1920s, the biological emphasis of the committee grew. The most important influence was the creation of the Social Science Research Council (SSRC) in 1923 and the subsequent decision of the NRC to limit itself to "natural science." When committee attempts to gain joint NRC-SSRC sponsorship for human behavior projects, or to become independent, both failed, social scientist William F. Ogburn resigned; the committee came to consist almost completely of biological and medical scientists. The shift in focus was expressed in the 1925 change in the committee's name from "sex problems" to "problems of sex."[52]

At this same time, both foundation decisions and the dynamics of research were increasing the importance of sex work within biology. At least four times between 1924 and 1934, Frank Lillie sought to gain Rockefeller support for an institute of "racial" or "genetic" biology at the University of Chicago. While money for sex biology continued to flow, this plan was deflected, and then finally submerged, in a general endowment of Chicago zoology timed to coincide with Lillie's retirement.[53] Charles Stockard did receive a big grant for his eugenic dog farm in 1925; as a result, however, CRPS money went in a straighter path to his long-suffering immigrant assistant, George Papanicolaou, the individual who actually did the work on female cycles and who, through his development of the eponymous Pap smear, would ultimately become more famous than his boss.[54]

Initially the committee discussed the creation of a single central research institute, and Lillie pushed himself forward as the most likely candidate. But both self-interest and a desire for openness induced the committee members to divide the available money among about a dozen recipients each year. In the first twenty years of its existence the CRPS distributed some $1 million among more than eighty grantees. This cohort of scientists worked on a variety of projects, including the largely unsuccessful surveys of human behavior, and such technological endeavors as an assessment of the use of X-rays for sterilization. But the committee's central interest was and remained, until 1940, the support of experimental investigation of sexuality. Following Lillie's scheme, the CRPS focused, in a developmental framework, on maturation and divergence of the two sexes, the cellular and hormonal products of the gonads, the female ovulatory cycle and pregnancy, and male-female relations. The emphasis throughout was on hormones as determiners of both structure and behavior.[55]

Sex biologists produced a mix of real successes and overreaching claims. Papanicolaou, Evans, Corner, and Edgar Allen worked out the details of the human ovulatory cycle. Lillie, biochemist Frederick Koch, and others at Chicago isolated and explored the properties of the "male hormone" testosterone. Numerous scientists supported by the CRPS showed that hormones produced by the gonads had significant effects on the body and behavior and that they offered some therapeutic possibilities. On the other hand, widespread expectations regarding the existence and complementarity of single "male" and "female" hormones were never realized, nor were beliefs about the effects that hormones had in each sex. Sex hormones did not have the clear and safe therapeutic effects that researchers hoped for. Psychobiologists such as Yerkes developed sweeping and transitory theories about the physiological and evolutionary bases for human sex differences and interactions. The most controversial claim was that of Gregory Pincus, who erroneously asserted in 1936 that he had cloned rabbits. Looking back two years later, Lillie pointed to only a few definite accomplishments in sex biology and then called for more research.[56]

Still, as Clarke and Nelly Oudshoorn have shown, by the late 1930s sex biologists formed themselves into a mainstream American research community, which had established a track record of discoveries. They forged a wealth of formal and informal relations with chemists, pharmaceutical companies, physicians, and even some social scientists. Their most important achievements, however, were to make the case that sexuality was a subject that properly stood within the domain of biologists, to show that it could be discussed openly both among scientists and with the public, and, finally, to educate Americans to think of sexuality as fundamentally a biological issue. George Corner began his 1938 advice book for boys by arguing that adolescents should be able to get the same "clear, accurate and practical" information about sex as they were able to do "on every other important branch of science, . . . from botany to radio construction."[57]

From the perspective of the millennium, these achievements may not seem like much. In the context of the 1930s, however, they were both difficult and important. The things that sex biologists accomplished were precisely what eugenic biologists were not able to do. Eugenic biology, centered in a single, isolated institution, never became an active, ramified research network. Its leaders produced few significant factual discoveries. Many biological, social, and medical scientists were either oblivious or antagonistic to eugenic biology;

and, conversely, the entire eugenics movement was so broad and shallow that it easily dissipated between 1930 and 1945, its remains scattered among demography, medical genetics, and public health. Biologists never became dominant within eugenics, and few Americans thought seriously about their future mates from a eugenic biological perspective.

ALFRED KINSEY'S AMERICA

The success of sex biologists over the two interwar decades is important because their achievements provided the foundation and protection for Alfred Kinsey, a biologist who, paradoxically, was closer to eugenics, in both his methods and his aims, than he was to endocrinology. Kinsey presented an intensely biological perspective on sex; and he pointed toward a modern and liberatory view of sexuality. He provided a scientific reference point for groups as varied as *Playboy* hedonists, feminists, and homosexuals, influencing at least two generations of Americans.

Kinsey was a member, though a marginal one, of the biologists' "club." His scientific breeding was quite respectable: he received a Harvard Ph.D. in 1919 for work with William M. Wheeler, who had been Charles O. Whitman's student and colleague from 1886 to 1899. As a significant figure in biology education, Kinsey was widely known as an interpreter of nature and a guide for teachers. On the other hand, he was a loner, had built his career at Indiana University (in the 1920s and 1930s an undistinguished institution), and, working as an entomologist and taxonomist, occupied one of the edges of biological science. James Jones's recent comprehensive biography has emphasized how Kinsey's private passions—in particular, his desire to overcome his sexually repressed adolescence and his interest in justifying his homosexual desires— led him to investigate human sexual behavior beginning in the late 1930s. Stephen Gould has explained that the methods he developed were drawn to a significant extent from his prior experience collecting and processing information on variability in his chosen animals, the gall wasps. In the present context, however, the best way to think about Kinsey's work is through comparison with that of Charles Davenport. Kinsey's project was, in structure and methods, similar to Davenport's eugenic field studies.[58]

The two enterprises were not, to be sure, identical. Heredity and sexual behavior were different subjects. Davenport embodied the kind of puritan repression that Kinsey was determined to overcome. As a participant in the automobile age, Kinsey engaged personally in fieldwork to an extent that was difficult prior to the 1920s. He also devoted much more attention than Davenport to methods of extracting specific and internally consistent answers from his interview subjects. But there were underlying similarities. Kinsey, like Davenport, left behind New York City, a domineering father, and engineering school for invertebrate biology, the Harvard graduate program, and quantitative approaches to evolutionary problems. Both began work on humans only at around the age of forty. They both entered problem areas that were already occupied by anthropologists, sociologists, psychologists, and psychiatrists. Both took an omnium-gatherum attitude to their material, combining data collected by themselves, informally trained assistants, and outsiders, and through hearsay. They accumulated huge quantities of data, which they hoped to process into knowledge at a central office. Faced with so much material, they both relied on a young assistant acquired through personal contacts: in Davenport's case, Harry Laughlin; in Kinsey's, Indiana undergraduate Clyde Martin. Though claiming to study humanity in general, they both were most interested in those Americans who lived outside respectable society: criminals, prostitutes, hobos, and, in Kinsey's case, homosexuals. They both had major difficulties in their methods of sampling.

Kinsey developed his project to gather case histories of sexual behavior essentially on his own between 1938 and 1941. Had it remained such, it would have been another individual effort in sexology and would probably either have been stopped by local opponents or supplanted by war work in the aftermath of Pearl Harbor. In 1941, however, Kinsey interested the CRPS in his activity. Jones has explored the details surrounding this decision, emphasizing shifts in the relations between the committee and the Rockefeller Foundation staffers, the interest on the part of the aging Robert Yerkes to leave a significant legacy, and the considerable charisma of Kinsey, described by the mild-mannered George Corner as "the most intense scientist I ever knew."[59]

As with Davenport, however, underlying elements of patronage and institutional relations were significant. In 1941, the CRPS was more than a dozen years away from its last involvement with sociological or clinical studies of human sexual activity. For the previous five years no anthropologist or sociologist had participated on the committee. Kinsey came to this group of biologists, animal psychologists, physiologists, and psychiatrists as a biologist, and he argued strenuously that he was working on the biology of sex, building on the outlook, techniques, and reputation he had established in taxonomy and biometry during the preceding two decades. The committee evaluated Kinsey's project from a biological perspective.

Between 1941 and 1954, Kinsey received more than $400,000 in Rockefeller Foundation money through the NRC and CRPS. These funds enabled him to hire a small group of assistants (none of whom was a biologist) and,

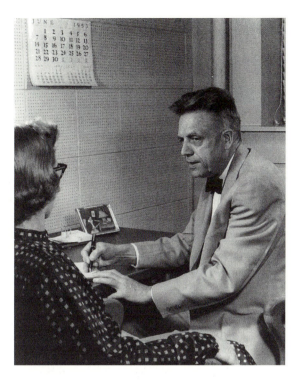

Figure 9–2

Alfred C. Kinsey, recording responses in a mock interview, 1953. The framed photograph on the table is of Mrs. Kinsey. Reproduced by permission of the Kinsey Institute for Research in Sex, Gender and Reproduction, Inc., Bloomington, Indiana. Photo by Bill Dellenback. The most frequently reproduced photograph, distributed by United Press (see Jones, *Kinsey*, following 460), shows Kinsey, with three assistants at an open filing cabinet, examining a file folder; a calculator stands in the rear.

in 1947, to establish an institute legally independent of Indiana University. Equally important was the prestige and protection that came from sponsorship by the most prominent scientific and philanthropic organizations in the country. By 1948 Kinsey had collected, either personally, through his assistants, from other sex researchers, or from individuals such as the record-keeping pedophile "Mr. X," data on more than 12,000 Americans. *Sexual Behavior in the Human Male* was based on the records of the activities of 5,300 white American men.[60]

Kinsey presented, on the one hand, the facts about "the human animal." He described the study of human sexual behavior as a problem in taxonomy—the characterization and measurement of variation among individuals. He focused on the physiologically defined and easily quantified act of ejaculation. On the other hand, he pointed with pride to the fact that he had consulted with specialists in thirty-three different areas, ranging from anatomy and endocrinology to penology and public-opinion polling. Moreover, the framework of the project was fundamentally that of a social survey: the collection, tabulation, and analysis of responses to a questionnaire. His first undertaking was to map the frequency of orgasms as a function of age, marital status, class and occupation, urban vs. rural, and religion; the second was to classify orgasms by "source of sexual outlet," including masturbation, premarital and marital intercourse, homosexual relations, and contacts with animals.

Kinsey's purposes were clear. He wanted first to demonstrate that men in twentieth-century America had more orgasms than was commonly assumed. His second concern was to show how variable the amount of sexual activity in healthy individuals was: from one orgasm in thirty years to nearly fifty thousand over the same period. This was, he noted without puzzlement, thousands of times more variability than in the morphological characters he had previously examined in insects.[61] His third emphasis was on the range of outlets, especially the incidence of homosexual contacts. His general point was that many men who had lots of sex of different kinds were successful individuals and good Americans; freed from (largely religious) repression, more people would be able to enjoy themselves and others. This was biologically grounded sex education with a vengeance.

The scientific criticisms of Kinsey recapitulated those of Davenport. The ERO's field workers had traveled the countryside, opportunistically interviewing idiosyncratic populations about unverifiable, often long-past events, and then shoehorning responses into a biological framework whose intellectual limitations were papered over by the quantity of data. Social scientists and psychiatrists scored Kinsey's animalistic conception of human nature and his identification of sexual expression with the narrowly biological phenomenon of ejaculation; Margaret Mead and others were particularly unhappy with the excretorial connotations of his key term, "outlet." They questioned whether one-time interviews would elicit either accurate or full accounts of sometimes decades-old events. The most pointed objection was that Kinsey's huge accumulation of data had been skewed, by sampling biases, to heighten the prominence of "deviant" behavior. They thus cast doubt on Kinsey's most consequential assertion: that he was presenting a true picture of American life. These attacks, raised immediately after publication of Kinsey's book, were restated and refined repeatedly over the next decade; Jones emphasizes that Kinsey devoted his energy to deflecting, rather than addressing, these arguments.[62]

While Davenport was losing credibility rapidly in the 1920s, within fifteen years of the creation of the ERO and ten years before his retirement, Kinsey remained at the center of American discussions of sexuality from late 1940s to the 1970s, and beyond. Insofar as any individual scientist could, he altered, through his writings, the way millions of people behaved in a fundamental aspect of their lives. Why was he so successful? Some reasons are straightforward. What Kinsey was arguing was consistent with many of his readers' desires, in ways that the strictures that Davenport and other eugenicists offered were not. Kinsey's arguments reinforced social forces that were already loosening Americans' sexual inhibitions—the rise in the standard of living, the lengthening of adolescence and the postponement of wage-earning, and increasing geographic and social mobility. Particular events, such as women's increasing autonomy, the social mixing that occurred in World War II, the dominance of Hollywood in establishing mores, and the emergence of middle-class gay enclaves in major cities, also reinforced the plausibility of Kinsey's views.

The factors more specific to biology, however, are what matter in the present context. Kinsey collected data and produced books that, despite their flaws, described sexuality with arresting clarity and made a prima facie case that variation in frequency and sources of sexual outlet were significant issues. The figure of "Dr. Kinsey" was also important: his public identity as a modern academic biologist gave him a kind of authority that combined scientific objectivity, professorial ordinariness, and clinical intensity. Kinsey generally posed, after 1945, either with interview subjects, his family, files, calculators, or colleagues (see figure 9–2); all these attributes were different from that of the stereotypical nineteenth-century naturalist Charles V. Riley (see figure 2–7).

Moreover, as Kinsey knew well, his arguments were persuasive because they drew on a foundation of biology teaching that he and other science educators had been providing for a generation. American adolescents had been trained to be enthusiastic about animal activities, to overcome their squeamishness in the face of the organic, and to experiment; they were urged to think of themselves as adapting and behaving organisms. *Sexual Behavior in the Human Male* was the chapter that had been missing from their high school textbooks.

Lastly, other biologists in the 1940s and 1950s enabled Americans to overcome the chief nonreligious fears that kept them from increased sexual activity: pregnancy and infectious diseases. The likelihood of pregnancy stood as the greatest barrier to increased heterosexual contact, certainly outside of marriage, and, to a considerable degree, within it. Mechanical contraceptives had existed for decades but were limited in availability, and, equally important, in cultural visibility. The big change occurred in the 1950s and early 1960s, largely as the result of the work of CRPS alumnus Gregory Pincus. Supported by female birth control advocates, he utilized the core findings of sex biologists regarding the effects of hormones on the female cycle to develop the oral contraceptive. The Pill put millions of women into new physiological states of hormonally induced reversible sterility, enabling them to think of their bodies and of sexual behavior in new ways, much closer to those of men; it formed an integral element of the sexual revolution of mid-century.[63]

The second barrier to increased sexual activity, infection, was particularly important for Kinsey's "high-rating" individuals, who generally interacted with multiple partners. As with contraception, some remedies for syphilis had existed prior to the 1940s; but the real biotechnological breakthrough, penicillin, reached the public just prior to the appearance of *Sexual Behavior in the Human Male*. For the next thirty years, infectious diseases were minor interruptions in the activities of sexual athletes. Antibiotics were particularly important in supporting the intensely sexual gay subcultures that emerged in large cities during the 1960s. In the 1980s the AIDS virus formed a new and much more formidable barrier to sexual athleticism, but not before gay people became an organized and self-assured minority who continued to challenge tradition in the pursuit of liberty and happiness.

The weaknesses in Kinsey's conception of sexuality and the incompleteness of his data were insignificant by comparison with the strength he drew from his participation in the community of sexual biologists, as well as his engagement with the biologists' project to culture Americans. Lester Ward, near the end of his life, privately imagined the possibility that civilized people could free sexual activities from "that slavery to function which belongs to a pain economy" and treat them instead as "truly aesthetic forces and the most effective of all sources of happiness." The Kinsey Report, the Pill, and penicillin provided Americans the tools to experiment and to determine what new sexual possibilities might work. Reviewing *Sexual Behavior in the Human Male*, Columbia University literary critic Lionel Trilling was amazed by Kinsey's "democratic pluralism of sexuality" and by the extent to which other biologists, and university and foundation leaders more broadly, promoted "acceptance and liberation." To the europhilic Trilling, the Kinsey Report was shallow and confused, but it was, he admitted, "characteristically *American*."[64]

For American scientists, the world changed on 23 September 1941. It was, from most perspectives, an ordinary day in the early mobilization of scientists for the war anticipated in the near future. It had particularly long and deep reverberations, however, because an official in the federal Office of Scientific Research and Development prepared a memo stating that universities could charge the government a percentage for overhead on all their research contracts. This workaday bureaucratic decision established the contract as the solid basis for a new set of relations between academic scientists and the government. It would assure the place of universities in the cold war–era Establishment and provide a new status for researchers within American society.[1]

The new order for American science gradually incorporated biologists. During World War II they worked on problems that included the development of penicillin, control of tropical insects, and nutrition; yet they were, as in World War I, relatively marginal participants. In the immediate postwar setting they struggled for an identity independent of medical scientists and for a recognized place at the federal table. Gradually, however, they became visible and successful. Geneticists were significant participants in Atomic Energy Commission research programs and in the public debates that grew out of them. The National Institutes of Health (NIH) broadened its medical mandate to include cellular and molecular studies pursued "extramurally" at universities. In the early 1950s biologists acquired their own division within the new National Science Foundation. The National Defense Education Act (1958) provided money for students in biology together with the other sciences. By 1960, biologists were full participants in the increasingly systematized American system of scientific research.[2]

The fundamental change in the quarter-century after 1941 was in scale. Any of a range of statistics tell the same story of relative stasis in the 1930s and then a long-term boom. The number of doctorates in all life sciences was essentially static from 1931 to 1948; by the late 1950s the number more than tripled, and then it tripled again in the next decade. The number of entries annually in *Biological Abstracts* quadrupled between 1940 and 1960. Between 1946 and 1964, NIH research grants grew from $780,000 to more than $529 million; federal funds for "basic research" in biological sciences (not including medical and agricultural research) swelled from $8 million in 1953 to $189 million in 1964.[3]

This increase in the number of people, publications, and dollars was accompanied by growing diversity. Research communities formed around such new model organisms as neurospora and phage. New disciplinary identities appeared. Scientists were geographically more dispersed as the percentage of doctorates from New England universities declined and those from the western states increased; the National Science Foundation encouraged this trend after 1964 through its Centers of Excellence program, designed to build up academic research at hitherto second-tier universities.[4]

During this period, biologists established, for the first time, close ties with physicists, who themselves were newly preeminent in the wake of the war. From the turn of the century and earlier, life scientists had communicated with chemists through the hybrid discipline of biochemistry. "Biophysics," by contrast, had been a word in search of a subject-matter, rather than an active science, for the half-century prior to 1945. This changed with the efforts, internationally, of such leading physicists as Erwin Schroedinger and Neils Bohr, and with the activity in the United States of men such as Max Delbruck, Leo Szilard, and George Gamow. As numerous scholars have emphasized, the penetration of physicists into biology was facilitated by Rockefeller Foundation research manager Warren Weaver. Physicists' interest gave biology a new visibility and status.[5]

Harder to pin down, but important, was the emergence of a new work ethos in biology. Physical and medical scientists established an increasingly normative emphasis on the continued production of basic research results by the early 1900s. T. H. Morgan's Drosophila Group had displayed such values during the interwar years, but Robert Kohler has shown how small-scale and tenuous drosophila research in fact was prior to 1940. This emphasis on production of discrete results diffused through biology after World War II, along with the physical scientists and the research contract framework they had worked out with the government.[6]

The cold war liberal consensus about science provided the frame for these new developments. Wartime science leader Vannevar Bush had used penicillin as a lead example in his argument that university scientists, given freedom and support to do what they wished, would ultimately produce specific and noncontroversial results. This perspective was institutionalized at the NIH, which distributed money widely to biologists on the grounds that the advancement of knowledge would lead to new ways to combat and prevent disease.

The best known, and perhaps fullest, personification of all these changes was James D. Watson. As an undergraduate during the war years he had studied zoology, with particular interest in ornithology, at the evolution-dominated University of Chicago. He then went to Indiana University for graduate work.

Missing Kinsey's evolution course by a few years, he joined the university's new cutting-edge genetics program. The Italian Jewish refugee Salvador Luria, brought to southern Indiana with the aid of the Rockefeller Foundation, initiated him into the phage group. This physicist-led study of bacterial viruses was intellectually fundamental, an exciting set of puzzles, and, plausibly, a long-term foundation for more effective control of major diseases. In 1950 Watson began a postdoctoral European tour, sampling laboratories and meetings on genetics, biochemistry, and polio. Disdaining older biologists who seemed to waste their time discussing "the role of the geneticist in this transitional age of changing values" on radio, he reached, with English physicist Francis Crick, for the fundamental breakthrough discovery of the structure of DNA. Back in the United States, he participated in the rise of molecular biology, and then transformed the Cold Spring Harbor Laboratory into a major biomedical research center.[7]

This story of the burgeoning of biology, especially in its molecular elements, in the decades after World War II is well known and has been well told. More important in the present context, however, is what faded away as those changes occurred.

The decade around 1940 marked an abrupt and extensive generational transition in academic biology. The cohort that entered the emerging area in the 1880s and early 1890s was remarkable for its cohesion and stamina. Nearly all the individuals mentioned in the second half of this book—Morgan, Lillie, Wilson, Conklin, Davenport, Parker, Wheeler, Ritter, and Jennings, among others—knew each other, communicated, and collaborated for half a century. They continued to lead both research and disciplinary affairs into the 1930s. Morgan, for example, led the Drosophila Group from 1910 until his (unofficial) retirement, at seventy-three, in 1940. Lillie was the central figure at the MBL and University of Chicago from the early 1900s up to the mid-1930s, and he concluded his career in the late 1930s with the simultaneous presidency of both the National Research Council and the National Academy of Sciences. When these men finally faded from the scene, their places were taken by a much more miscellaneous group of much younger men. Some were Europeans, some came from physical sciences, some rose suddenly from obscurity to fill the leadership vacuum. They had relatively few connections among each other and little psychological investment in the world that existed before the war.

The singular community cohesion that had existed in biology ended around this time, not only because of the generational shift but also as a result of intellectual, geographic, and economic changes. The importance of the MBL as the American biologists' club gradually declined after 1925, as marine embryos became less important research objects, the automobile opened up new possibilities for summer work, and home ownership became an important element of Woods Hole life. A variety of groups—most notably those interested in phage, who summered in Cold Spring Harbor beginning in 1945—established their own communal centers. No single circle could believe, rightly or wrongly, that it encompassed either the national scene or biology as a whole. Biologists adopted, largely from chemists and physicists, the practice of annual focused gatherings (such as the Gordon Research Conferences and the Cold Spring Harbor Symposia); this practice was an acknowledgment of both the increased numbers of life scientists and the diversity of their interests.[8]

An additional shift was a decline in biologists' efforts to explain their ideas and to justify the significance of their activities to the public. In chapter 8, I argued that in the 1920s disciplinary leaders were quite visible as interpreters and advocates. Over the next two decades, however, journalists increasingly took over the work of popularization. They emphasized such spectacular outcomes of research as penicillin and the polio vaccine, and affirmed that the general advance of science would necessarily result in the conquest of disease. After 1945, biologists followed Vannevar Bush in accepting such broad-brush arguments and having their serious discussions about priorities within closed offices. Scientists who sought to address the public not only lost status among such members of the new generation as Watson, but could really suffer. Kinsey ultimately paid for his best-seller notoriety with invective from a congressional committee and loss of foundation support. His departmental colleague, the vocal geneticist Hermann Muller, was intermittently excluded from international scientific meetings by the Atomic Energy Commission. The chemist-turned-biologist Linus Pauling went through a number of interrogations, lost his passport, and was browbeaten on national television by Mike Wallace after he became involved in antinuclear and other leftist causes.[9]

The diversity and extent of the changes that biologists experienced in the decades around 1950 were so significant that they provide a natural stopping point. Continuing this story would require another book, a different set of tools, and a different author. In this work I have sought to make six interrelated historical points. American biology, broadly defined, developed within the context of efforts to create a continental and democratic American nation. Particular American conditions—geographic, communal, and institutional— were necessary for the shaping of biological science that occurred. "Biology" itself became a key to imagining a national community of life scientists. Tension existed between individual and academic interests, on the one hand, and public and national ones, on the other. Contrary to usual expectations, biology was as much if not more about culture, in its specific biotechnological sense, than about nature. Finally, biologists did affect American development, imperfectly but positively.

Though focused on the nineteenth and early twentieth centuries, this history, like any except the most antiquarian ventures, does offer some indications about the pathways leading to the present, and it opens up questions about the contours of the road ahead. In particular, it emphasizes how recent and unstable was the emphasis in biology on basic research, pure science, and the search for the nature of life. I remember discussions in the late 1970s with young molecular biologists at both Johns Hopkins and the University of Georgia (a Center of Excellence) about the then new enterprise of genetic engineering. Most downplayed outsiders' concerns about the creation of new pathogens or monsters. They were disturbed, however, by the realization that some of their professors were as interested in profits as in papers and by the idea that products could be as important as fundamental knowledge about nature.[10]

What this book indicates is that the transition that has occurred in biology since the 1970s is in many respects a reconnection with an older tradition. Both recombinant DNA technology and the more varied innovations in the manipulation of reproduction (such as egg transplantation, sperm sifting, and cloning) have placed culture, in its biotechnological meaning, back at the center of biological science. Experimenters in these areas have oriented their activity not toward understanding nature, but toward its manipulation and improvement. Their science is linked to technology, industry, and profits.[11]

On the other hand, the controversies of the last two decades over biotechology and scientists' ongoing responses have reaffirmed that biology differs from physics and chemistry in both the depth and variety of its public components. Biologists have dealt with the general problems of reconciling the public funding of research with the corporate utilization of results, as well as the difficulties of ensuring the safety of their products. But they have also faced the unique responsibilities arising from the fact that the products of their laboratories are intentionally designed to affect the bodies of American humans (capable of filing suits) in certain ways, or from the fact that engineered bacteria, plants, and animals may reproduce and become independent actors within the public realm. There are good reasons why bioethics is the one element of the science studies movement that has truly become established, both institutionally and legally, in the last two decades, and why biologists— James Watson in particular—made the ethical, legal, and social implications of genomics an integral part of the Human Genome Project. In the last two decades biologists have established an enduring public presence.[12]

In 1994, Democratic senator Barbara Mikulski, a former social worker who had grown up in the shadow of Johns Hopkins, lectured the National Academy of Sciences that scientists should stop treating public funding as an "entitlement" program to support individual "curiosity" and instead think of themselves as part of one American "community" and organize their activities around "important national goals." These statements made scientific leaders nervous, and at least some were relieved when Mikulski lost power with the Republican victory later that year. Her speech is less significant for its impact on federal appropriations, however, than as a contemporary articulation of an enduring perspective among American leaders.[13]

For biologists, this perspective is less a challenge than an opportunity. Mikulski's paradigm of a scientific enterprise that had public support because it was oriented toward "national needs" was the Human Genome Project. She explicitly contrasted this positive relationship with the inability of physicists to articulate the national significance of their grand project, the Superconducting Supercollider, within a post–cold war context. This book has shown how American biologists have long worked toward national goals (or, stated more modestly, desirable policies and practices) and, in so doing, have advanced science. The more important point, however, is that biologists have not only implemented, but shaped, those policies. Through their networks and choices of work, they have influenced, quite out of proportion to their numbers, what American goals would be.

The termination of the supercollider and the contemporaneous rise of the genome project together mark the transition, a century after Clarence King's vision, from the Age of Energy to the Age of Biology. Within this new context, both scientists and other Americans who think about the aims with which American biologists might be concerned in the coming decades can gain through awareness of what their desires have, over the long term, been. The interests of individuals and moments are varied, but underneath has lain a continued coherent desire to culture American life. This has encompassed both agriculture (including modern biotechology) and environmental management (including wilderness and species preservation). It has included not only the preservation of health and combating of disease, but also the enhancement of possibilities for both present and future generations associated with sexual, reproductive, and genetic technologies. Finally, it has entailed efforts more narrowly "cultural"—to present Americans with a naturalistic perspective on life and to persuade them that they are organisms who have the capability to use intelligence to deal with the modern world. These desires have formed foundations for articulating and realizing the promise of American life.

PREFACE AND ACKNOWLEDGMENTS

1. Philip J. Pauly, *Controlling Life: Jacques Loeb and the Engineering Ideal in Biology* (New York: Oxford University Press, 1987).

2. Dorothy Ross, ed. *Modernist Impulses in the Human Sciences, 1870–1930* (Baltimore: Johns Hopkins University Press, 1994).

3. Robert E. Kohler, *Lords of the Fly: Drosophila Genetics and the Experimental Life* (Chicago: University of Chicago Press, 1994).

INTRODUCTION
TOWARD A CULTURAL HISTORY OF AMERICAN BIOLOGY

1. William E. Ritter diary, 28 June 1919, carton 9, William E. Ritter Papers, Bancroft Library, University of California, Berkeley. The Scripps Institution for Biological Research was renamed the Scripps Institution of Oceanography in 1926.

2. William E. Ritter, "The Philosophy of E. W. Scripps," 78–83, typescript, 1943–44, box 4, folder 95, Ritter Papers, Scripps Institution of Oceanography, La Jolla, California.

3. *The Century Dictionary and Cyclopedia*, 10 vols. (New York: Century Company, 1889–1900), 2:1393.

4. Herbert Croly, *The Promise of American Life* (New York: Macmillan, 1909), 12–13; David W. Levy, *Herbert Croly of the New Republic* (Princeton: Princeton University Press, 1985).

5. Croly, *Promise of American Life*, 429–48.

6. Ibid., 400. Most immediately, Croly was a student at Harvard in the late 1880s when Ritter was a graduate student in zoology; both considered themselves protégés of the evolutionary philosopher Josiah Royce.

7. David A. Hollinger, *In the American Province* (Bloomington: Indiana University Press, 1985), 3–43; Hollinger, *Science, Jews, and Secular Culture* (Princeton: Princeton University Press, 1996); Robert B. Westbrook, *John Dewey and American Democracy* (Ithaca: Cornell University Press, 1991); James Livingston, *Pragmatism and the Political Economy of Cultural Revolution, 1850–1940* (Chapel Hill: University of North Carolina Press, 1994).

8. Richard Rorty, *Achieving Our Country* (Cambridge, MA: Harvard University Press, 1998); Thomas Geoghegan, *The Secret Lives of Citizens: Pursuing the Promise of American Life* (New York: Pantheon, 1998); Michael Lind, *The Next American Nation: The New Nationalism and the Fourth American Revolution* (New York: Free Press, 1995).

CHAPTER ONE
NATURAL HISTORY AND MANIFEST DESTINY, 1800–1865

1. Jefferson took some Indian artifacts to Monticello. There were in fact two Lewis and Clark collections: a large quantity of material (including the prairie dog) shipped by flatboat from Fort Mandan (near modern Bismarck, North Dakota) in spring 1805 during the outbound phase of the expedition, and a smaller amount gathered beyond the Rockies and on the return trip. On the Lewis and Clark expedition, see, most recently, Steven Ambrose, *Undaunted Courage: Meriwether Lewis, Thomas Jefferson, and the Opening of the American West* (New York: Simon and Schuster, 1996); on the explorers' "booty," see Paul Cutright, *Lewis and Clark: Pioneering Naturalists* (Urbana: University of Illinois Press, 1969), 349–92.

2. Cutright, *Lewis and Clark*, 358–63; on Barton, see Joseph Ewan, "Barton, Benjamin Smith," *Dictionary of Scientific Biography*, 1:484–86; Jeanette E. Graustein, "The Eminent Benjamin Smith Barton," *Pennsylvania Magazine of History and Biography* 85 (1961): 423–38.

3. Cutright, *Lewis and Clark*, 359–63; Frederick Pursh, *Flora Americae Septentrionalis; or, a Systematic Arrangement and Description of the Plants of North America*, 2 vols. (London, 1814; reprinted, with introduction by Joseph Ewan, Vaduz, Liechtenstein: J. Cramer, 1979).

4. The specimens not described by Pursh lay in the storerooms of the American Philosophical Society until 1896; they included at least seventy species unknown to botanists at the time Lewis collected them. Twenty-one new animals were published, but the expedition gained little credit for them because four authors dispersed the descriptions among six publications between 1811 and 1823. Cutright, *Lewis and Clark*, 365–68, 386–87.

5. On early naturalists in North America, see Brooke Hindle, *The Pursuit of Science in Revolutionary America, 1735–1789* (Chapel Hill: University of North Carolina Press, 1956), 11–35, 302–26; Joseph Kastner, *A Species of Eternity* (New York: Knopf, 1977); William Cronon, *Changes in the Land: Indians, Colonists, and the Ecology of New England* (New York: Hill and Wang, 1983); Thomas P. Slaughter, *The Natures of John and William Bartram* (New York: Knopf, 1996); Amy R.W. Meyers and Margaret Beck Pritchard, eds., *Empire's Nature: Mark Catesby's New World Vision* (Chapel Hill: University of North Carolina Press, 1998). On science in the early republic, see John Greene, *American Science in the Age of Jefferson* (Ames: Iowa State University Press, 1984); Charlotte Porter, *The Eagle's Nest: Natural History and American Ideas, 1812–1842* (Tuscaloosa: University of Alabama Press, 1986); Christoph Irmscher, *The Poetics of Natural History: From John Bartram to William James* (New Brunswick, NJ: Rutgers University Press, 1999).

6. Thomas Jefferson, *Notes on the State of Virginia* (Paris, 1785). On Jefferson as a political naturalist, see Daniel Boorstin, *The Lost World of Thomas Jefferson* (New York: Henry Holt, 1948); Edwin T. Martin, *Thomas Jefferson, Scientist* (New York:

H. Schuman, 1952); Alexander O. Boulton, "The American Paradox: Jeffersonian Equality and Racial Science," *American Quarterly* 47 (1995): 467–92.

7. Toby A. Appel, "Science, Popular Culture, and Profit: Peale's Philadelphia Museum," *Journal of the Society for the Bibliography of Natural History* 9 (1980): 619–34; Joel J. Orosz, *Curators and Culture: The Museum Movement in America, 1740–1870* (Tuscaloosa: University of Alabama Press, 1990), 44–57; David R. Brigham, *Public Culture in the Early Republic: Peale's Museum and Its Audience* (Washington, DC: Smithsonian Institution Press, 1995).

8. Alexander Wilson, *American Ornithology; or, the Natural History of the Birds of the United States*, 9 vols. (Philadelphia, 1808–14); John James Audubon, *The Birds of America*, 4 vols. (London, 1827–38); Audubon, *Ornithological Biography; or an Account of the Birds of the United States of America . . .*, 5 vols. (Edinburgh, 1831–38); Thomas Say, *American Entomology, or Description of the Insects of North America*, 3 vols. (Philadelphia, 1824–28); Clark Hunter, ed., *The Life and Letters of Alexander Wilson* (Philadelphia: American Philosophical Society, 1983); Alice Ford, *John James Audubon* (Norman: University of Oklahoma Press, 1964); Patricia Tyson Stroud, *Thomas Say: New World Naturalist* (Philadelphia: University of Pennsylvania Press, 1992). More generally, see Porter, *Eagle's Nest*, and Ann Shelby Blum, *Picturing Nature: American Nineteenth-Century Zoological Illustration* (Princeton: Princeton University Press, 1993), 3–118.

9. Jeannette E. Graustein, *Thomas Nuttall, Naturalist: Explorations in America 1808–1841* (Cambridge, MA: Harvard University Press, 1967).

10. Constantine Rafinesque, review of *Flora Americae Septentrionalis*, by Frederick Pursh, *American Monthly Magazine and Critical Review* 2 (1817): 171.

11. On the European situation, see N. Jardine, J. A. Secord, and E. C. Spray, eds., *Cultures of Natural History* (Cambridge: Cambridge University Press, 1996); David Philip Miller and Peter Hanns Reill, eds., *Visions of Empire* (Cambridge: Cambridge University Press, 1996). On the United States, see Slaughter, *Bartram*; Greene, *American Science*; Porter, *Eagle's Nest*.

12. Charles C. Sellers, *Charles Willson Peale* (New York: Scribner, 1969).

13. Hunter, *Alexander Wilson*, 45–61, 409–28.

14. Ethel M. McAllister, *Amos Eaton, Scientist and Educator, 1776–1842* (Philadelphia: University of Pennsylvania Press, 1941). Some naturalists continued with dubious activities: the French botanist André Michaux assisted Citizen Genêt in his disruptive activities in the 1790s; two decades later, the Englishman Charles Hamilton Smith used his interest in American fauna and flora as a cover for spying. Henry Savage, Jr., and Elizabeth J. Savage, *André and François André Michaux* (Charlottesville: University of Virginia Press, 1986); Simon Baatz, *Knowledge, Culture, and Science in the Metropolis* (Annals of the New York Academy of Sciences, vol. 584, 1990): 22.

15. Baatz, *Knowledge, Culture, and Science*, 33; Keir Sterling, introduction to Constantine Rafinesque, *Rafinesque: Autobiography and Lives* (New York: Arno Press, 1978), ix.

16. Patrick O'Brien, *Joseph Banks: A Life* (Boston: David R. Godine, 1993), 193–242; Miller and Reill, *Visions of Empire*; John Gascoigne, *Science in the Service of Empire* (Cambridge: Cambridge University Press, 1998).

17. Richard Bushman, *The Refinement of America: Persons, Houses, Cities* (New York: Knopf, 1992).

18. Orosz, *Curators and Culture*, 54; Greene, *American Science*, 257.

19. Porter, *Eagle's Nest*, 144, 147.

20. David M. Oestricher, "Unraveling the Walam Olum," *Natural History* 105 (October 1996): 14–21.

21. Stroud, *Thomas Say*, 31.

22. Baatz, *Knowledge, Culture, and Science*, 24, 30–31; see also Baatz, "Philadelphia Patronage: The Institutional Structure of Natural History in the New Republic, 1800–1833," *Journal of the Early Republic* 8 (1988): 111–38.

23. Graustein, *Thomas Nuttall*, 67; Washington Irving, *Astoria*, vol. 15 of *Complete Works of Washington Irving* (Boston: Twayne, 1976), 119.

24. James Fenimore Cooper, *The Prairie: A Tale*, ed. James P. Elliott (Albany: State University of New York Press, 1985), 67–72, 101.

25. Andrew Denny Rodgers III, *John Torrey: A Story of North American Botany* (Princeton: Princeton University Press, 1942), 6–12, 72; Christine Chapman Robbins, "John Torrey, 1796–1873: His Life and Times," *Bulletin of the Torrey Botanical Club* 95 (1968): 519–645. Had the Revolution been suppressed, William Torrey would have been singled out as a criminal for carrying out the execution of a British officer, and more particularly for subjecting him to the ignominy of hanging.

26. As a consequence Eaton gained support for deliverance from what he considered a frame-up. In 1815 his sentence was commuted, and two years later he was pardoned by William Torrey's political patron, DeWitt Clinton. He went on to write a series of elementary books in botany and geology, and moved upstate to Troy to organize the Rensselaer Polytechnic Institute. Rodgers, *Torrey*, 12–17; McAllister, *Eaton*, 149–56.

27. John Torrey, *A Catalogue of Plants Growing Spontaneously within Thirty Miles of the City of New York* (New York, 1819).

28. Torrey to Eaton, 21 March 1818, in Rodgers, *Torrey*, 27.

29. Jane Loring Gray, quoted in Rodgers, *Torrey*, 69; Torrey to Gray, 29 December 1842, Asa Gray Papers, Gray Herbarium Archives, Harvard University (hereafter Gray Papers).

30. Hooker to Torrey, 2 December 1823, John Torrey Papers, New York Botanical Garden, New York (hereafter Torrey Papers).

31. Hooker to Torrey, 9 September 1824, Torrey Papers.

32. Hooker to Torrey, 8 July 1825, Torrey Papers. See also Torrey to Hooker, 3 February and 20 August 1824, 28 April and 3 August 1825, 27 October 1826, Director's Correspondence, 44:174–78, Royal Botanic Gardens, Kew, England.

33. William Morwood, *Traveller in a Vanished Landscape: The Life and Times of David Douglas* (New York: Clarkson N. Potter, 1974), 128–34; William Jackson

Hooker, *Flora Boreali-Americana; or, the Botany of the Northern Parts of British America*, 2 vols. (London, 1829–40).

34. Hooker to Torrey, 8 July 1825, Torrey Papers.

35. Asa Gray, "Autobiography," in Asa Gray, *Letters of Asa Gray*, ed. Jane Loring Gray, 2 vols. (Boston: Houghton Mifflin, 1893), 1:4. On Gray, see A. Hunter Dupree's classic biography, *Asa Gray* (Cambridge, MA: Belknap Press of Harvard University Press, 1959). My interpretation is also influenced by Graustein, *Thomas Nuttall*, 327–29, 352–60.

36. This setting has been investigated most thoroughly in conjunction with the origins of Mormonism: see, e.g., Whitney R. Cross, *The Burned-Over District* (Ithaca: Cornell University Press, 1950), and Richard L. Bushman, *Joseph Smith and the Beginnings of Mormonism* (Urbana: University of Illinois Press, 1984).

37. Dupree, *Asa Gray*, 16–17, 20, 22 (quotation, 35).

38. Eaton complained impotently to Torrey that "I will not call [Gray] a conceited 'upstart;' because he has obtained your patronage," but then declared his "contempt" for the young man. Eaton to Torrey, 4 November 1835, in McAllister, *Eaton*, 238–39.

39. Gray to N. Wright Folwell, 3 May 1836, quoted in Dupree, *Asa Gray*, 55.

40. Torrey to Gray, 18 and 28 July 1836, Gray Papers.

41. On Gray's roles, see Graustein, *Thomas Nuttall*, 338–39; on contributors, see Dupree, *Asa Gray*, 95–96.

42. Asa Gray, "The Flora of North America" (1882), in Gray, *Scientific Papers*, 2 vols. (Boston: Houghton Mifflin, 1889), 2:252.

43. John Torrey and Asa Gray, *A Flora of North America: Containing Abridged Descriptions of All the Known Indigenous and Naturalized Plants Growing North of Mexico; Arranged According to the Natural System*, 2 vols. (New York: Wiley and Putnam, 1838–43).

44. John A. Lowell to Asa Gray, 27 and 28 February 1846, Gray Papers. On Harvard during this period, see Ronald Story, *The Forging of an Aristocracy: Harvard and the Boston Upper Class, 1800–1870* (Middletown, CT: Wesleyan University Press, 1980).

45. Theophilus Parsons, "Charles Greely Loring," *Proceedings of the Massachusetts Historical Society* 11 (1870): 263–91, esp. 267; on the centrality of the Massachusetts Hospital Life Insurance Company within Brahmin society, see Robert F. Dalzell, Jr., *Enterprising Elite: The Boston Associates and the World They Made* (Cambridge, MA: Harvard University Press, 1987), 103–10.

46. "Record of persons employed at the Gray Herbarium," Records A-1, and "Assistants," Records A-5, both in Gray Herbarium Archives, Harvard University.

47. Asa Gray, *The Botanical Text-book* (New York: Wiley and Putnam, 1842); Gray, *A Manual of the Botany of the Northern United States* (Boston: J. Munroe, 1848); Gray, *First Lessons in Botany and Vegetable Physiology* (New York: Ivison and Phinney, 1857); Gray, *Botany for Young People and Common Schools* (New York: American Book Company, 1858).

48. Dupree, *Asa Gray*, 204–10; Elizabeth B. Keeney, *The Botanizers: Amateur Scientists in Nineteenth-Century America* (Chapel Hill: University of North Carolina Press, 1992), 32–36.

49. Gray, *Letters*, 1:272n; Wayne E. Fuller, *The American Mail: Enlarger of the Common Life* (Chicago: University of Chicago Press, 1972), 65–67, 164–65; Richard R. John, *Spreading the News: The American Postal System from Franklin to Morse* (Cambridge, MA: Harvard University Press, 1995), 159–61. On the importance of postal services in nineteenth-century science, see Martin Rudwick, *The Great Devonian Controversy* (Chicago: University of Chicago Press, 1985), 36–37.

50. Dupree, *Asa Gray*, 110, 187; Asa Gray, *Botany. Phanerogamia*, vol. 15 of *Reports of the United States Exploring Expedition* (Philadelphia: C. Sherman, 1854–56); Asa Gray, ed., *Botany. Cryptogamia*, vol. 17 of *Reports of the United States Exploring Expedition* (Philadelphia: C. Sherman, 1874).

51. William Stanton, *The Great United States Exploring Expedition of 1838–1842* (Berkeley: University of California Press, 1975), 330–38.

52. Emory to Torrey, 28 September 1853, Torrey Papers.

53. Dupree, *Asa Gray*, passim.

54. Robbins, "Torrey," 597.

55. Rodgers, *Torrey*, 220–27.

56. Asa Gray, "A rapid glance at the botanical regions, chiefly N. Hemisphere," lecture notes, 27 pp., box AT, file 43, Gray Papers. Both the comment about "annexation," and a letter to Torrey on 28 March 1848 (Torrey Papers) mentioning that he was then lecturing on "geographical botany" make the 1848 date most probable.

57. On antebellum landscape painting, see Angela Miller, *The Empire of the Eye: Landscape Representation and American Cultural Politics, 1825–1875* (Ithaca: Cornell University Press, 1993).

58. Dupree, *Asa Gray*; Edward Lurie, *Louis Agassiz: A Life in Science* (Chicago: University of Chicago Press, 1960).

59. C. C. Felton, "Professor Agassiz," *Boston Advertiser*, 22 May 1855, quoted in Lurie, *Agassiz*, 217.

60. Lurie, *Agassiz*, 174–75; Louis Agassiz and Augustus A. Gould, *Principles of Zoölogy: Touching the Structure, Development, Distribution, and Natural Arrangement of the Races of Animals, Living and Extinct . . .* (Boston: Gould and Lincoln, 1848).

61. Lurie, *Agassiz*, 196–98.

62. Geologist Arnold Guyot established himself at Princeton. Others were unsuccessful: artist Jacques Burkhardt remained on staff; Leo Lesquereux, who was deaf, pursued paleobotany while working as a watchmaker in Ohio; Daniel Girard fought with Agassiz after moving to Washington and ultimately returned to France; and Eduard Desor, the one person who sought to compete with Agassiz directly, was driven out of town by the Boston elite after impugning Agassiz's morals. See Lurie, *Agassiz*, passim; James R. Jackson and William C. Kimler, "Taxonomy and the Per-

sonal Equation: The Historical Fates of Charles Girard and Louis Agassiz," *Journal of the History of Biology*, 32 (1999):509–555.

63. Lurie, *Agassiz*, 131, 178.

64. Circular letter by Louis Agassiz, 3 pp., 1853, and "Prospectus. Contributions to the Natural History of the United States, in Ten Vols. Quarto, by Louis Agassiz," 2 pp., 28 May 1855, both in Box 13, RU 7002, Spencer F. Baird Collection, Smithsonian Institution Archives, Washington, D.C.

65. "Museum of Natural History, and Comparative Zoology. Speech by Prof. Agassiz," *Boston Courier*, 26 February 1859.

66. Lurie, *Agassiz*, 220, 232.

67. Dorothy G. Wayman, *Edward Sylvester Morse: A Biography* (Cambridge, MA: Harvard University Press, 1942), 137.

68. Joel J. Orosz, "Disloyalty, Dismissal, and a Deal: The Development of the National Museum at the Smithsonian Institution, 1846–1855," *Museum Journal* 2 (1986): 22–33.

69. For Agassiz's position in the discussions of the natural compatibility among Africans, the southern climate, and the slave system, see George M. Frederickson, *The Black Image in the White Mind* (New York: Harper and Row, 1971; reprint, Hanover, NH: University Press of New England, 1987), 130–64, esp. 137, 145, 161.

70. "Museum of Natural History,"; on Agassiz's plans for his museum, see Mary P. Winsor, "Louis Agassiz's Ideas of Museum Arrangement," California Academy of Science, *Memoirs*, in press.

71. American Association for the Advancement of Science, *Proceedings*, 1850, p. 107; Louis Agassiz, "Sketch of the Natural Provinces of the Animal World and Their Relation to the Different Types of Man," in *Types of Mankind*, 8th ed., ed. Josiah Nott and George Glidden (Philadelphia: Lippincott, 1857), lxxv.

72. Editorial, *Boston Courier*, 24 March 1859.

73. Gray's father-in-law had already taken that position: in 1851, as attorney for the escaped slave Thomas Sims, he defended resistance to the Fugitive Slave Law on the grounds that the constitution applied to all races of men. See *Trial of Thomas Sims, on an Issue of Personal Liberty . . . Arguments of Robert Rantoul, Jr., and Charles G. Loring, with the Decision of George T. Curtis . . .* (Boston: W. S. Damrell, 1851), 24–32.

74. Gray to Torrey, 8 December 1858, Torrey Papers. Ultimately Torrey held on to his position; see Robbins, "Torrey," 604–5.

75. Gray to Torrey, 7 January 1859, Torrey Papers.

76. Dupree, *Asa Gray*, 252–59.

77. Asa Gray, "Diagnostic Characters of New Species of Phaenogamous Plants, Collected in Japan by Charles Wright, Botanist of the U.S. North Pacific Exploring Expedition," American Academy of Arts and Sciences, *Memoirs*, n.s., 6 (1859): 437–49. See also American Academy of Arts and Sciences, *Proceedings* 4 (1859): 130–35, 171–79; Dupree, *Asa Gray*, 249–63.

78. Dupree, *Asa Gray*, 144–48, 266.

79. Gray, "Diagnostic Characters," 443.

80. Gray to J. D. Hooker, 31 March 1860, and Gray to Francis Boott, 16 January 1860, both quoted in Dupree, *Asa Gray*, 287, 270.

81. Gray to George Engelmann, 11 November 1861, in Gray, *Letters*, 2:469; Dupree, *Asa Gray*, 306–8.

82. Gray to J. D. Hooker, 18 February 1861, in Dupree, *Gray*, 307; Gray to Darwin, [January] 1862, 10 November 1862, 25 July 1865, 18 February 1862, in Gray, *Letters*, 2:472–77, 490, 538.

83. Lurie, *Agassiz*, 305–6, 318; Mary P. Winsor, *Reading the Shape of Nature: Comparative Zoology at the Agassiz Museum* (Chicago: University of Chicago Press, 1991), 46. In contrast to Gray's racial optimism, Agassiz doubted that blacks could become socially equal to whites within the foreseeable future. Lurie, *Agassiz*, 305–9.

84. Dupree, *Asa Gray*, 313–19; Lurie, *Agassiz*, 331–34.

85. Dupree, *Asa Gray*, 322–23.

CHAPTER TWO
CULTURING FISH, CULTURING PEOPLE:
FEDERAL NATURALISTS IN THE GILDED AGE, 1865–1893

1. Mark Twain, *The Innocents Abroad, or the New Pilgrims' Progress* (Hartford, CT: American Publishing Company, 1869), 26–27.

2. Alternatively, he made it up: I could find no record of a collector sent to southern Europe or the Middle East in either the 1867 or 1868 Smithsonian *Annual Reports*. Mark Twain and Charles Dudley Warner, *The Gilded Age: A Tale of To-day* (Hartford: American Publishing Company, 1873).

3. On antebellum Washington, see Constance M. Green, *Washington: Village and Capital, 1800–1878* (Princeton: Princeton University Press, 1962); James S. Young, *The Washington Community, 1800–1828* (New York: Columbia University Press, 1966).

4. The basic source on the beginnings of federal science remains A. Hunter Dupree, *Science in the Federal Government: A History of Policies and Activities* (Cambridge, MA: Belknap Press of Harvard University Press, 1957; reprint, Baltimore: Johns Hopkins University Press, 1986). On this period, see also Robert V. Bruce, *The Launching of Modern American Science, 1846–1876* (New York: Knopf, 1987); James Rodger Fleming, *Meteorology in America, 1800–1870* (Baltimore: Johns Hopkins University Press, 1990); Hugh Richard Slotten, *Patronage, Practice, and the Culture of American Science: Alexander Dallas Bache and the U.S. Coast Survey* (New York: Cambridge University Press, 1994); Joel J. Orosz, *Curators and Culture: The Museum Movement in America, 1740–1870* (Tuscaloosa: University of Alabama Press, 1990), 155–65.

5. Quotation in E. F. Rivinus and E. M. Youssef, *Spencer Baird of the Smithsonian* (Washington, DC: Smithsonian Institution Press, 1992), 74 (this is the most modern and complete biography of Baird). See also Joel J. Orosz, "Disloyalty, Dismissal,

and a Deal: The Development of the National Museum at the Smithsonian Institution, 1846–1855," *Museum Journal* 2 (1986): 22–33: Pamela M. Henson, "Spencer Baird's Dream: A U. S. National Museum," California Academy of Science, *Memoirs*, in press.

6. Smithsonian Institution, *Directions for Collecting, Preserving, and Transporting Specimens of Natural History*, 2nd ed. (Washington, DC: Smithsonian Institution, 1854).

7. See, e.g., Spencer F. Baird, *Catalogue of North American Birds* (vol. 9) and *Reptiles and Amphibians of North America* (vol. 10) of *Report of Explorations and Surveys to Ascertain the Most Practical and Economical Route for a Railroad from the Mississippi Revier to the Pacific Ocean* (Washington, DC: A. D. P. Nicholson, 1858). On antebellum western surveys, see William H. Goetzmann, *Exploration and Empire* (New York: Knopf, 1966), 265–302. Early National Museum appropriations are in George Brown Goode, ed., *The Smithsonian Institution, 1846–1896* (1897; reprint, New York: Arno Press, 1980), 322. Most Smithsonian tenants were collectors who were either preparing to go out into the field or were working to catalog the specimens they had obtained. They humorously organized themselves as the "Megatherium Club," named after the extinct American giant sloth. See Rivinus and Youssef, *Baird*, 94.

8. On the development of Washington and the federal bureaucracy after the Civil War, see Alan Lessoff, *The Nation and Its City: Politics, "Corruption," and Progress in Washington, D.C., 1861–1902* (Baltimore: Johns Hopkins University Press, 1994), with statistics on pp. 18–19; also Stephen Skowroneck, *Building a New American State: The Expansion of National Administrative Capacities, 1877–1920* (Cambridge: Cambridge University Press, 1982), 47–84; Cindy Aron, *Ladies and Gentlemen of the Civil Service: Middle-Class Workers in Victorian America* (New York: Oxford University Press, 1987).

9. David Starr Jordan, *Days of a Man*, 2 vols. (Yonkers: World Book Company, 1922), 1:126.

10. My argument in this chapter draws on Michael James Lacey, "The Mysteries of Earth-Making Dissolve: A Study of Washington's Intellectual Community and the Origins of American Environmentalism in the Late Nineteenth Century" (Ph.D. diss., George Washington University, 1979); also Lacey, "The World of the Bureaus," in *The State and Social Investigation in Britain and the United States*, ed. Michael J. Lacey and Mary O. Furner (Washington, DC: Woodrow Wilson Center Press, 1993), 127–70. See also James K. Flack, *Desideratum in Washington* (Cambridge, MA: Schenkman Publishing, 1975), 58–98.

11. Rivinus and Youssef, *Baird*, 116–27; Orosz, *Curators and Culture*, 201–12; Smithsonian Institution, *Annual Report*, 1875, p. 8; Goode, *Smithsonian Institution*, 325–29.

12. Dean Conrad Allard, Jr., *Spencer Fullerton Baird and the U.S. Fish Commission* (New York: Arno Press, 1978), 76–85; Tim D. Smith, *Scaling Fisheries: The Sci-*

ence of Measuring the Effects of Fishing, 1855–1955 (Cambridge: Cambridge University Press, 1994), 38–51.

13. Smithsonian Institution, *Annual Report,* 1876, pp. 11–13; also Joseph Henry to Alexander Agassiz, 18 October 1877, and Agassiz to Henry, 22 November 1877, Alexander Agassiz Papers, Museum of Comparative Zoology Library, Harvard University.

14. Goode, *Smithsonian Institution,* 329–30.

15. Z. L. Tanner, "Report on the Construction and Outfit of the United States Fish Commission Steamer Albatross," U.S. Fish Commission, *Report,* 1883, pp. 3–5; Allard, *Baird,* 304–6.

16. Mike Foster, *Strange Genius: The Life of Ferdinand Vandeveer Hayden* (Niwot, CO: Roberts Rinehart Publishers, 1994); William Culp Darrah, *Powell of the Colorado* (Princeton: Princeton University Press, 1951); Wallace Stegner, *Beyond the Hundredth Meridian: John Wesley Powell and the Second Opening of the West* (Boston: Houghton Mifflin, 1954); Willis Conner Sorensen, *Brethren of the Net: American Entomology, 1840–1880* (Tuscaloosa: University of Alabama Press, 1995), 133–38.

17. *Index-Catalogue of the Library of the Surgeon-General's Office, United States Army* (Washington, DC: GPO, 1880–); Carleton B. Chapman, *Order Out of Chaos: John Shaw Billings and America's Coming of Age* (Boston: Boston Medical Library, 1994), 79–80, 159–65.

18. On Baird's influence, see, e.g., Charles Schuchert and Clara Mae LeVene, *O. C. Marsh: Pioneer in Paleontology* (1940; reprint, New York: Arno Press, 1978), 266.

19. The Biological Society of Washington had 161 local members in early 1885: see *Proceedings of the Biological Society of Washington* 3 (1886): xi–xxiv.

20. On Coues, see Paul R. Cutright and Michael J. Broadhead, *Elliott Coues: Naturalist and Frontier Historian* (Urbana: University of Illinois Press, 1981).

21. Smithsonian Institution, *Annual Report,* 1887–88, p. xxvi; Smithsonian Institution salary list, ca. 1880, box 10, RU 7081, Smithsonian Institution Archives.

22. Aron, *Ladies and Gentlemen,* 94, 129–30.

23. Darrah, *Powell,* 323; B. T. Galloway, "50 Years Ago in the Department of Agriculture," typescript, 14 pp., Galloway file, National Fungus Collection, Systematic Botany and Mycology Laboratory, USDA Agricultural Research Service, Beltsville, Maryland.

24. Bureau reports kept the government printers busy when Congress was not in session. Major publication series in the 1880s included the Smithsonian's *Annual Report, Contributions to Knowledge,* and *Miscellaneous Collections;* the National Museum's *Annual Report, Bulletin, Proceedings,* and *Journal;* and the Fish Commission's *Annual Report* and *Bulletin.* The Department of Agriculture's subdivisions nearly all began separate, sometimes multiple, sets of bulletins and circulars in the 1880s, in addition to the department's *Annual Report.* Page totals are from *United States Government Publications. A Monthly Catalog* 2 (1886).

25. Allard, *Baird*, 322–41; Allard, "The Fish Commission Laboratory and Its Influence on the Founding of the Marine Biological Laboratory," *Journal of the History of Biology* 23 (1990): 251–70; Paul C. Galtsoff, *The Story of the Bureau of Commercial Fisheries Biological Laboratory, Woods Hole, Massachusetts,* U.S. Department of the Interior, circular #145, 1962.

26. Flack, *Desideratum*, 61–69.

27. Wilcomb Washburn, *The Cosmos Club of Washington: A Centennial History, 1878–1978* (Washington, DC: Cosmos Club, 1978), 15–27; Flack, *Desideratum*, 82–93. Initially the club rented space in the Corcoran Building at Fifteenth Street and Pennsylvania Avenue. In late 1882 it occupied 23 Lafayette Square; three years later it moved into the Dolly Madison House, a few doors away. It remained there until the federal government claimed the land in the 1940s.

28. John W. Aldrich, "The Biological Society of Washington: A Centennial History, 1880–1980," *Bulletin of the Biological Society of Washington* 4 (1980): 1–40; Flack, *Desideratum*, 143–48.

29. For the details and fate of these plans, see chapters 4 and 6.

30. George Brown Goode et al., *The Fisheries and Fishery Industries of the United States*, 7 vols. (Washington, DC: GPO, 1884–87); J. W. Powell to W. B. Allison, 26 February 1886, in *Testimony before the Joint Commission to Consider the Present Organization of the Signal Service. . .*, Senate Misc. Doc. 82, 49th Cong., 1st Sess., 1886, p. 1077.

31. J. W. Powell, "The Personal Characteristics of Professor Baird," *Bulletin of the Philosophical Society of Washington* 10 (1887): 73; Norman Klose, *America's Crop Heritage: The History of Foreign Plant Introduction by the Federal Government* (Ames: Iowa State College Press, 1950); Allard, *Baird*, 262–95; Joseph E. Taylor III, "Making Salmon: The Political Economy of Fishery Science and the Road Not Taken," *Journal of the History of Biology* 31 (1998): 33–59.

32. William T. Hornaday, "The Extermination of the American Bison, with a Sketch of Its Discovery and Life History," U.S. National Museum, *Annual Report*, 1887, pp. 367–548; David A. Dary, *The Buffalo Book: The Full Saga of the American Animal* (Chicago: Swallow Press, 1974), 198–200; Helen Lefkowitz Horowitz, "The National Zoological Park: 'City of Refuge' or Zoo?" in *New Worlds, New Animals*, ed. R. J. Hoage and William A. Deiss (Baltimore: Johns Hopkins University Press, 1996), 128.

33. United States Entomological Commission, *First Annual Report . . . Relating to the Rocky Mountain Locust and the Best Methods of Preventing Its Injuries and of Guarding against Its Invasions* (Washington, DC: GPO, 1878).

34. Smithsonian Institution, *Annual Report*, 1867, p. 43; Morgan B. Sherwood, *The Exploration of Alaska* (New Haven: Yale University Press, 1965), 32–33.

35. Allard, *Baird*, 180–238.

36. Kurk Dorsey, *The Dawn of Conservation Diplomacy: U.S.–Canadian Wildlife Protection Treaties in the Progressive Era* (Seattle: University of Washington Press, 1998).

37. Theodore Gill, "The Proper Use of the Term Biology," *Proceedings of the Biological Society of Washington* 1 (1882): 102–4.

38. "Darwin Memorial Meeting," *Proceedings of the Biological Society of Washington* 1 (1882): 43–100; Elliott Coues, *The Daemon of Darwin* (Boston: Estes and Lauriat, 1885); Lester F. Ward, *Glimpses of the Cosmos*, 6 vols. (New York: G. P. Putnam, 1913–18), 5:69. On the range of interpretations of evolution by American naturalists, see Ronald L. Numbers, *Darwinism Comes to America* (Cambridge, MA: Harvard University Press, 1998), 24–48.

39. For more extensive discussion of the intellectual perspectives of federal naturalists, see Lacey, "Mysteries of Earth-Making Dissolve"; also Curtis M. Hinsley, Jr., *Savages and Scientists: The Smithsonian Institution and the Development of American Anthropology, 1846–1910* (Washington, DC: Smithsonian Institution Press, 1981). Pamela M. Henson, "'Objects of Curious Research': The History of Science and Technology at the Smithsonian," *Isis* 90 (1999): S249–S270.

40. Lester F. Ward, *Dynamic Sociology*, 2 vols. (New York: D. Appleton, 1883). On Ward, see Samuel Chugerman, *Lester F. Ward: The American Aristotle* (Durham: Duke University Press, 1939); Clifford H. Scott, *Lester Frank Ward* (Boston: G. K. Hall, 1976).

41. Scott, *Ward*, 19–26; Lester F. Ward, "Our Better Halves," *Forum* 6 (1888): 266–75.

42. The publication, U.S. Treasury Department, Bureau of Statistics, "Special Report on Immigration" (1871), is reprinted in Ward, *Glimpses of the Cosmos*, 1:183–201.

43. Lester F. Ward, *A Guide to the Flora of Washington and Vicinity*, U.S. National Museum bulletin #22 (Washington, DC: GPO, 1881); Scott, *Ward*, 27–28.

44. Henry N. Andrews, *The Fossil Hunters: In Search of Ancient Plants* (Ithaca: Cornell University Press, 1980), 208–13; Lester F. Ward, "Sketch of Paleobotany," *Fifth Annual Report of the United States Geological Survey, 1883–1884* (Washington, DC: GPO, 1885), 363–453; Ward, "The Geographical Distribution of Fossil Plants," *Eighth Annual Report of the United States Geological Survey, 1886–87*, 2 vols. (Washington, DC: GPO, 1889), 2:663–960.

45. Lester F. Ward, *Types of the Laramie Flora* (U.S. Geological Survey, bulletin #37, 1887). Ward was also involved in the debates over dating some major east coast strata: see Lester F. Ward, "The Potomac Formation," *Fifteenth Annual Report of the United States Geological Survey, 1893–1894* (Washington, DC: GPO, 1895), 307–97, 741–55; Schuchert and LeVene, *O. C. Marsh*, 323–24.

46. See esp. Lester F. Ward, "Incomplete Adaptation as Illustrated by the History of Sex in Plants," *American Naturalist* 15 (1881): 89–95; Ward, "Evolution in the Vegetable Kingdom," *American Naturalist* 19 (1885): 637–44, 745–53; Ward, "The Course of Biologic Evolution" (1890), in Ward, *Glimpses of the Cosmos*, 4:198–219.

47. Lester F. Ward, *The Psychic Factors of Civilization* (Boston: Ginn, 1893), 250.

48. Ward, "Broadening the Way to Success," *Forum* 2 (1886): 340–50, reprinted with preface and notes in Ward, *Glimpses of the Cosmos*, 4:31–43 (quotation, 36).

49. Ward, *Glimpses of the Cosmos*, 5:74–80; on Goode, see George Brown Goode, *The Origins of Natural Science in America*, ed. Sally Gregory Kohlstedt (Washington, DC: Smithsonian Institution Press, 1991).

50. G. Brown Goode and Tarleton H. Bean, *Oceanic Ichthyology*, 2 vols. (Washington, DC: GPO, 1895).

51. Goode et al., *Fisheries and Fishery Industries*, 1:vii.

52. See, e.g., William J. Rhees, ed., *Visitor's Guide to the Smithsonian Institution, National Museum, and Fish Ponds* (Washington, DC: Judd and Detweiler, 1881), 70–71.

53. Smithsonian Institution, *Annual Report*, 1881, p. 91.

54. G. Brown Goode, "Outline of a Scheme of Museum Classification," *Proceedings of the United States National Museum* 3 (1880): 597–600 (appendix 13, dated 10 April 1882). He quietly acknowledged that this system was not in fact consistent or complete; the number of zoological specimens, and the interests of the curators (including Goode himself), entailed a separate arrangement for plants, animals, and minerals, extending beyond the "industrial and economic" displays. See Smithsonian Institution, *Annual Report*, 1881, p. 91.

55. Smithsonian Institution, *Annual Report*, 1881, p. 86.

56. Goode, "Outline of a Scheme of Museum Classification," 600.

57. G. Brown Goode, "First Draft of a System of Classification for the World's Columbian Exposition," U.S. National Museum, *Annual Report*, 1891, p. 649f.; U.S. National Museum, *Annual Report*, 1893, p. 110; Robert W. Rydell, *All the World's a Fair: Visions of Empire at American International Expositions* (Chicago: University of Chicago Press, 1984), 43–46.

58. E.g., Neil Harris, *Cultural Excursions: Marketing Appetites and Cultural Tastes in Modern America* (Chicago: University of Chicago Press, 1990), 68; Margaretta M. Lovell, "Picturing 'A City for a Single Summer': Paintings of the World's Columbian Exposition," *Art Bulletin* 78 (1996): 41; Rydell, *All the World's a Fair*; James Gilbert, *Perfect Cities: Chicago's Utopias of 1893* (Chicago: University of Chicago Press, 1991).

59. Mark Twain, *A Connecticut Yankee in King Arthur's Court* (New York: Charles L. Webster and Co., 1889); Samuel Langhorne Clemens, *Letters from the Earth*, ed. Bernard De Voto (New York: Harper and Row, 1962).

CHAPTER THREE
CONFLICTING VISIONS OF AMERICAN ECOLOGICAL INDEPENDENCE

1. Roland M. Jefferson and Alan E. Fusonie, *The Japanese Flowering Cherry Trees of Washington, D.C.: A Living Symbol of Friendship*, USDA Agricultural Research Service, National Arboretum Contribution #4, 1977, 9. Memos are reprinted on pp. 49–54. Roger Daniels, *The Politics of Prejudice: The Anti-Japanese Movement in California and the Struggle for Japanese Exclusion* (Berkeley: University of California Press, 1962), 31–45. This chapter is drawn from Philip J. Pauly, "The Beauty and

Menace of the Japanese Cherry Trees: Conflicting Visions of American Ecological Independence," *Isis* 87 (1996): 51–73, © 1996 by the History of Science Society. All rights reserved.

2. "Destroy Tokio Gift Trees," *New York Times*, 29 January 1910; [Editorial], *New York Times*, 30 January 1910; Jefferson and Fusonie, *Cherry Trees*, 11–15.

3. Alfred W. Crosby, *Ecological Imperialism: The Biological Expansion of Europe, 900–1900* (Cambridge: Cambridge University Press, 1986); see also William H. McNeill, *Plagues and Peoples* (Garden City, NY: Doubleday, 1976); Richard Grove, *Green Imperialism* (Cambridge: Cambridge University Press, 1995), 230–44.

4. On USDA research, see Thomas S. Harding, *Two Blades of Grass: A History of Scientific Development in the U.S. Department of Agriculture* (Norman: University of Oklahoma Press, 1947); for individuals and bureaus, see references below. On experiment stations, see Charles E. Rosenberg, *No Other Gods: On Science and American Social Thought* (Baltimore: Johns Hopkins University Press, 1976); Alan I. Marcus, *Agricultural Science and the Quest for Legitimacy* (Ames: Iowa State University Press, 1985).

5. On American cosmopolitanism in the late nineteenth century, see Howard Mumford Jones, *The Age of Energy: Varieties of American Experience, 1865–1915* (New York: Viking Press, 1971); on nativism, see John Higham, *Strangers in the Land: Patterns of American Nativism, 1860–1925* (New Brunswick, NJ: Rutgers University Press, 1955). The term "nativist" was coined in the 1840s to designate American-born opponents of (mostly Irish) immigration.

6. William Cronon, *Changes in the Land: Indians, Colonists, and the Ecology of New England* (New York: Hill and Wang, 1983); also Carolyn Merchant, *Ecological Revolutions: Nature, Gender, and Science in New England* (Chapel Hill: University of North Carolina Press, 1989).

7. Jack Ralph Kloppenburg, Jr., *First the Seed: The Political Economy of Plant Biotechnology, 1492–2000* (New York: Cambridge University Press, 1988), 50–57; Stephen A. Spongberg, *A Reunion of Trees* (Cambridge, MA: Harvard University Press, 1990).

8. Leland O. Howard, "Danger of Importing Insect Pests," USDA, *Yearbook*, 1897, pp. 529–39; Lyster H. Dewey, "Migration of Weeds," USDA, *Yearbook*, 1896, pp. 274–79.

9. Theodore S. Palmer, "The Danger of Introducing Noxious Animals and Birds," USDA, *Yearbook*, 1898, pp. 98–106; Palmer, "A Review of Economic Ornithology in the United States," USDA, *Yearbook*, 1899, p. 289. On the acclimatization movement, see Michael A. Osborne, *Nature, the Exotic, and the Science of French Colonialism* (Bloomington: Indiana University Press, 1994); Linden Gillbank, "A Paradox of Purposes: Acclimatization Origins of the Melbourne Zoo," in *New Worlds, New Animals*, ed. R. J. Hoage and William A. Deiss (Baltimore: Johns Hopkins University Press, 1996), 73–85; Ian R. Tyrrell, *True Gardens of the Gods* (Berkeley: University of California Press, 1999).

10. Edward H. Forbush and Charles H. Fernald, *The Gypsy Moth. Porthetria dispar (Linn.): A Report of the Work of Destroying the Insect in the Commonwealth of Massachusetts, Together with an Account of Its History and Habits Both in Massachusetts and Europe* (Boston: Wright and Potter, 1896), 3–23; Edward Tenner, *Why Things Bite Back: Technology and the Revenge of Unintended Consequences* (New York: Knopf, 1996), 122–27.

11. Norman Klose, *America's Crop Heritage: The History of Foreign Plant Introduction by the Federal Government* (Ames: Iowa State College Press, 1950); Kloppenburg, *First the Seed*, 50–61.

12. Dean Conrad Allard, Jr., *Spencer Fullerton Baird and the U.S. Fish Commission* (New York: Arno Press, 1978), 271–81.

13. Quarantine measures dated back to colonial times, and the federal government began to get involved in the 1810s. Such procedures were extended to livestock around 1880. See Alan M. Kraut, *Silent Travellers: Germs, Genes, and the "Immigrant Menace"* (New York: Basic Books, 1994), 24–25; Ulysses G. Houck, *The Bureau of Animal Industry of the United States Department of Agriculture* (Washington, DC: Hayworth Printing Company, 1924).

14. Asa Gray, "The Pertinacity and Predominance of Weeds," *American Journal of Science*, 3rd ser., 18 (1879): 161–67; Liberty Hyde Bailey, "Coxey's Army and the Russian Thistle: An Essay on the Philosophy of Weediness," in Bailey, *The Survival of the Unlike*, 4th ed. (New York: Macmillan, 1901), 200–201. Bailey delivered this address to the AAAS in the nadir of the great depression following the Panic of 1893. Coxey's army was a group of unemployed men who had marched on Washington to demand public road-repair jobs (which would have included weed clearing). Bailey smugly contrasted the poor marchers themselves to weeds: while the former had been eliminated by federal police, the latter were "beyond the reach of the sheriff."

15. Lyster H. Dewey, *Legislation against Weeds*, USDA Division of Botany, bulletin #17, 1896; similarly, see Leland O. Howard, *Legislation against Injurious Insects*, USDA Division of Entomology, bulletin #33 (old series), 1895.

16. "Report of the Secretary," USDA, *Annual Report*, 1888, p. 12; "Report of the Secretary," USDA, *Annual Report*, 1889, pp. 8–9.

17. Other divisions included microscopy, gardens and grounds, chemistry, soils, and statistics, along with the essentially autonomous weather bureau.

18. Entomologist Leland Howard recalled the world of the USDA in the 1880s in two autobiographies: *History of Applied Entomology (Somewhat Anecdotal)*, Smithsonian Miscellaneous Collections, vol. 84, 1930, and *Fighting the Insects: The Story of an Entomologist* (New York: Macmillan, 1933). See also two manuscripts by Beverly T. Galloway: "50 Years Ago in the Department of Agriculture," and "The Genesis of the Bureau of Plant Industry," Galloway file, National Fungus Collection, Systematic Botany and Mycology Laboratory, USDA Agricultural Research Service, Beltsville, Maryland (hereafter National Fungus Collection). On the importance of scientific entrepreneurship and interbureau competition, see A. Hunter Dupree, *Science in the*

Federal Government (New York: Harper and Row, 1964), 176–83; William A. Niskanen, *Bureaucracy and Representative Government* (Chicago: Aldine, 1971).

19. C. Hart Merriam, "The Geographic Distribution of Animals and Plants in North America," USDA, *Yearbook*, 1894, pp. 203–14; Keir B. Sterling, *The Last of the Naturalists: The Career of C. Hart Merriam* (New York: Arno Press, 1974), 257–314; Palmer, "Review of Economic Ornithology"; Leland O. Howard, *Chinch Bug: General Summary of Its History, Habits, Enemies, and of Remedies and Preventives to Be Used against It*, USDA Division of Entomology, bulletin #17 (old series), 1888; Leland Howard and Charles L. Marlatt, *San Jose Scale, Its Occurrences in the U.S., With Full Account of Its Life History and Remedies to Be Used against It*, USDA Division of Entomology, bulletin #3 (new series), 1896, pp. 33–35; *Condensed Information Concerning Some of the More Important Insecticides*, USDA Division of Entomology, circular #1 (2nd ser.), 1891; David G. Fairchild, *Bordeaux Mixture as a Fungicide*, USDA Division of Vegetable Pathology, bulletin #6, 1894.

20. Richard C. Sawyer, *To Make a Spotless Orange: Biological Control in California* (Ames: Iowa State University Press, 1996).

21. Gustavus Weber, *The Bureau of Entomology* (Washington, DC: Brookings Institution, 1930), 27–31, 46–47; Howard, *Fighting the Insects*, 53; Howard, *The Mexican Cotton-Boll Weevil*, USDA Division of Entomology, circular #6 (2nd ser.), 1895; Howard and Marlatt, *San Jose Scale*; on discontent, see Lawrence Goodwyn, *Democratic Promise: The Populist Movement in America* (New York: Oxford University Press, 1976); Robert C. McMath, Jr., *American Populism: A Social History, 1877–1898* (New York: Hill and Wang, 1993). The extent of disruption in the hinterlands was brought home to Washington staffers when Kansas State College of Agriculture, the academic home of a number of them, was taken over by Populists, and vegetable pathology assistant David Fairchild's father was forced out as president: see James C. Carey, *Kansas State University: The Quest for Identity* (Lawrence: Regents Press of Kansas, 1977), 67–86.

22. Leland O. Howard, *Some Mexican and Japanese Injurious Insects Liable to Be Introduced into the United States*, USDA Division of Entomology, technical bulletin #4, 1896; Howard, *The Mexican Cotton-Boll Weevil*, USDA Division of Entomology, circular #14 (2n ser.), 1896; Douglas Helms, "Technological Methods for Boll Weevil Control," *Agricultural History* 53 (1979): 287; Howard, *The Gipsy Moth in America*, USDA Division of Entomology, bulletin #11 (n.s.), 1897; Beverly T. Galloway, ed., *Proceedings of the National Convention for the Suppression of Insect Pests and Plant Diseases by Legislation, Held at Washington, D.C., March 5 and 6, 1897* (Washington, DC: GPO, 1897); Howard, "Danger of Importing Insect Pests."

23. On Palmer, see W. L. McAttee, "In Memoriam: Theodore Sherman Palmer," *The Auk* 73 (1956): 367–77; Sterling, *Last of the Naturalists*, 129, 149. The Division of Economic Ornithology and Mammalogy was renamed the Division of Biological Survey in 1896.

24. "Report of the Division of Biological Survey," USDA, *Annual Report*, 1899, pp. 67–68; Palmer, "Danger."

25. Palmer, "Danger," 96.

26. Ibid., 105.

27. Ibid., 106.

28. Ibid., 107–8.

29. James Wilson to John Fletcher Lacey, 15 January 1900, John Fletcher Lacey Papers, State Historical Society of Iowa, Des Moines. Environmental historians have emphasized the Lacey Act's provisions regulating interstate commerce in game. This entry of the federal government into wildlife conservation was less controversial and less radical in principle, however, than its prohibitions on importations. For the legislative history, see Jenks Cameron, *The Bureau of Biological Survey: Its History, Activities and Organization* (Baltimore: Johns Hopkins University Press, 1929), 70–83; Theodore Whaley Cart, "The Struggle for Wildlife Protection in the United States, 1870–1900: Attitudes and Events Leading to the Lacey Act" (Ph.D. diss., University of North Carolina at Chapel Hill, 1971), 158–66, 188–89. More generally, see John F. Reiger, *American Sportsmen and the Origins of Conservation*, rev. ed. (Norman: University of Oklahoma Press, 1986); Thomas R. Dunlap, *Saving America's Wildlife* (Princeton: Princeton University Press, 1988); Dunlap, "Remaking the Land: The Acclimatization Movement and Anglo Ideas of Nature," *Journal of World History* 8 (1997): 303–19.

30. On Wilson's perception of the farm problem, see "Report of the Secretary," USDA, *Yearbook*, 1897, p. 10; Barbara Kimmelman, "A Progressive Era Discipline: Genetics at American Agricultural Colleges and Experiment Stations, 1900–1920" (Ph.D. diss., University of Pennsylvania, 1987), 25–41.

31. Oscar E. Anderson, Jr., *The Health of Nation: Harvey W. Wiley and the Fight for Pure Food* (Chicago: University of Chicago Press, 1958), 99; on Galloway, see John A. Stevenson, "Galloway, Beverly Thomas," *Dictionary of American Biography*, 10:217; Galloway, "50 Years Ago"; Galloway, *Commercial Violet Culture* (New York: A. T. de la Mare, 1899); David G. Fairchild, *The World Was My Garden: Travels of a Plant Explorer* (New York: Charles Scribner's Sons, 1938), 18–29; John A. Stevenson, "Plants, Problems, and Personalities: The Genesis of the Bureau of Plant Industry," *Agricultural History* 28 (1954): 155–62. Being a "hermaphrodite" was not equivalent to being homosexual—manners were more important than sexual partners. Galloway was the married father of three.

32. "Report of the Secretary," USDA, *Annual Report*, 1900, pp. xxxiii–xxxiv. Galloway was already skeptical about inspection for plant diseases and had broadened his division's domain to include environmental influences on disease susceptibility and the search for disease-resistant varieties. See USDA, *Annual Report*, 1893, p. 258; Galloway, "Plant Diseases and the Possibility of Lessening Their Spread by Legislation," *Convention for Suppression of Insect Pests*, 8–11; Galloway to Walter T. Swingle, 10 February 1897, entry 160, vol. 17, RG 54, National Archives, Washington, D.C. In 1895 Galloway broadened his division's name from Vegetable Pathology to Vegetable Physiology and Pathology.

33. Klose, *America's Crop Heritage*; David G. Fairchild, "Report of the Special Agent in Charge of Seed and Plant Introduction," USDA, *Annual Report,* 1898, pp. 3–4; USDA, *Annual Report,* 1899, pp. lxix–lxx; Beverly T. Galloway, "Searching the World for New Crops," typescript, 1920, Beverly T. Galloway Papers, Special Collections, National Agricultural Library, Beltsville, Maryland (hereafter Galloway Papers).

34. Fairchild, *World,* 30–37, 79–84; Marjory Stoneman Douglas, *Adventures in a Green World* (Miami: Field Research Projects, 1973).

35. Klose, *America's Crop Heritage,* 109–19; Isabel S. Cunningham, *Frank N. Meyer: Plant Hunter in Asia* (Ames: Iowa State University Press, 1984); Rose Schuster Taylor, *To Plant the Prairies and the Plains: The Life and Work of Niels Ebbesen Hansen* (Mt. Vernon, IA: Bios, 1941); Paul De Kruif, *Hunger Fighters* (New York: Harcourt, Brace, 1928). The Iowan Hansen had in fact been sent out independently by Wilson in early 1897. After setting up the SPI Section, Fairchild left for a second multiyear tour with Lathrop. He reclaimed his government position in 1903, and he soon solidified his social standing by marrying Alexander Graham Bell's daughter. On hearing of the marriage, Lathrop wrote a deeply felt letter to "dear old Fairy" professing his acceptance of this choice; after some initial tension, Marian Bell Fairchild established cordial relations with the gentleman she and her husband called "Uncle Barbour." See Barbour Lathrop to David Fairchild, 6 May 1905, David Fairchild Papers, Fairchild Tropical Garden, Miami (hereafter Fairchild Papers); Fairchild, *World,* 119, 457–58.

36. David G. Fairchild, "The Plant Introduction Work of the Department of Agriculture, Memorandum for Conference in Secretary's Office, March 1915," typescript, file "Plant Introduction Reports," Fairchild Papers; Fairchild, *Systematic Plant Introduction: Its Purposes and Methods,* USDA Division of Forestry, bulletin #21, 1898; Fairchild, *World,* passim; Harvey Levenstein, *The Paradox of Plenty: A Social History of Eating in Modern America* (New York: Oxford University Press, 1993), 29.

37. Beverly T. Galloway, "Division of Vegetable Physiology and Pathology," USDA, *Yearbook,* 1897, pp. 106–7; Herbert J. Webber and Ernst A. Bessey, "Progress of Plant Breeding in the United States," USDA, *Yearbook,* 1899, p. 468; Kimmelman, "A Progressive Era Discipline."

38. See, e.g., Walter Swingle to David Fairchild, 8 January 1899, Swingle file, Fairchild Papers; David Fairchild, "Notes Prepared for an Informal Address to the Washington Staff of the Office of Foreign Seed and Plant Introduction . . .," typescript, 9 October 1922, file "Plant Introduction Reports," Fairchild Papers.

39. "Report of the Secretary," USDA, *Annual Report,* 1900, pp. lxxv–lxxvi; "Report of the Secretary," USDA, *Annual Report,* 1901, p. xxiii. On the politics of consolidation, see Beverly T. Galloway, "The Genesis of the Bureau of Plant Industry," typescript, 30 June 1926, National Fungus Collection; also Stevenson, "Plants, Problems, and Personalities." On the structure and personnel of the new bureau, see "Report of the Chief of the Bureau of Plant Industry," USDA, *Annual Report,* 1901, pp. 43–45. The Bureau combined the Divisions of Agrostology, Botany, Pomology, Vegetable Physiology and Pathology, and Gardens and Grounds.

40. Beverly T. Galloway, "Applied Botany, Retrospective and Prospective," *Science* 16 (1902): 49–59; Galloway, "The Twentieth Century Botany," *Science* 19 (1904): 11–18.

41. Galloway, "Twentieth Century Botany," 12–13; Galloway, "Applied Botany," 55–57, 59.

42. Fred Wilbur Powell, *The Bureau of Plant Industry: Its History, Activities, and Organization* (Baltimore: Johns Hopkins University Press, 1927), 101; Weber, *Bureau of Entomology*, 157; Fairchild, *Systematic Plant Introduction*, 20–21; Fairchild to Swingle, 28 April and 13 October 1898, Walter T. Swingle Papers, Special Collections Department, Otto G. Richter Library, University of Miami, Coral Gables, Florida; Cunningham, *Frank N. Meyer*; Fairchild, *World*, 289, 410–11; Jefferson and Fusonie, *Cherry Trees*, 6–7.

43. Leland O. Howard, *The Insect Menace* (New York: Century, 1931); Howard, *Fighting the Insects*, 118–37; W. D. Hunter and W. E. Hinds, *The Mexican Cotton Boll Weevil*, USDA Division of Entomology, bulletin #145, 1904, pp. 13–14. On the basis of their work against these enemies, the entomologists acquired bureau status in 1904.

44. Charles L. Marlatt, *An Entomologist's Quest: The Story of the San Jose Scale* (Baltimore: Monumental Printing, 1953); "Charles Lester Marlatt," *Proceedings of the Entomological Society of Washington* 57 (1955): 37–43. In 1906 Marlatt married a second time, to Helen McKey-Smith, daughter of the Episcopal bishop of Philadelphia. Money from his parents, the estate of his first wife, and his second wife combined to make Marlatt a wealthy man. He built a large house on fashionable Sixteenth Street and was prominent in country club circles, joining foursomes with Franklin Roosevelt when Roosevelt was assistant secretary of the navy. (Author's telephone conversation with Florence Marlatt, daughter of Charles and Helen Marlatt, 11 July 1995.)

45. In 1908 the annual meeting of the American Association of Nurserymen voted to terminate negotiations about such a law. See Richard P. White, *A Century of Service: A History of the Nursery Industry Associations of the United States* (Washington, DC: American Association of Nurserymen, 1975), 168–71; Marlatt, *Entomologist's Quest*, 328–29.

46. Fairchild, *World*, 316–18, 326–27; Jefferson and Fusonie, *Cherry Trees*, 5–6.

47. Gustavus A. Weber, *The Plant Quarantine and Control Administration: Its History, Activities and Organization* (Washington: Brookings Institution, 1930), 3.

48. Jefferson and Fusonie, *Cherry Trees*, 16–21.

49. Weber, *Plant Quarantine*, 3.

50. See Marlatt's testimony, 19 February 1912, "A Bill to Regulate the Importation and Interstate Transportation of Nursery Stock (H. R. 18000)," *Hearings before the Committee on Agriculture on Miscellaneous Bills and Other Matters*, 51–106; Weber, *Plant Quarantine*, 10–29.

51. Weber, *Plant Quarantine*, 87, 61; for Marlatt's views, see his testimony, 27 April 1910, in "Inspection of Nursery Stock," *Hearings before the Committee on Agri-*

culture during the Second Session of the 61st Congress, 3 vols., 3:494–501; Haven Metcalfe, "The Chestnut Bark Disease," *Journal of Heredity* 5 (1914): 8–17; Charles L. Marlatt, "Plant Quarantine No. 37," *American Florist* 53 (1919): 411.

52. Galloway succeeded Liberty Hyde Bailey as dean of the College of Agriculture. His collapse was precipitated by attacks on his rationalizing policies by faculty and alumni, but it probably had deeper causes: he intermittently suffered from deep depression and ultimately took his own life. See Gould P. Colman, *Education and Agriculture: A History of the New York State College of Agriculture at Cornell University* (Ithaca: Cornell University Press, 1963), 250–60; Beverly T. Galloway to David Fairchild, 16 May and 31 October 1916, and Agnes Galloway to Fairchild, 20 May 1921, Fairchild Papers; Stevenson, "Galloway." Klose, *America's Crop Heritage,* 123–24.

53. Charles L. Marlatt, "Losses Caused by Imported Tree and Plant Pests," *American Forestry* 23 (1917): 75–80; Weber, *Plant Quarantine,* 29–42.

54. David G. Fairchild, "The Independence of American Nurseries," *American Forestry* 23 (1917): 213–16 (quotation, 216).

55. Weber, *Plant Quarantine,* 29; Galloway, "Searching the World for New Crops," 14; Beverly T. Galloway, "The Beginnings of the Bell, Maryland Plant Introduction Garden," typescript, n.d., Galloway Papers, National Fungus Collection; Fairchild, *World,* 425, 471–75; Fairchild, *Exploring for Plants* (New York: Macmillan 1930); Galloway, "Rational Plant Exclusion," typescript, 1925, Galloway Papers, National Agricultural Library.

56. Weber, *Plant Quarantine,* 108–16. Marlatt complained wistfully to Fairchild that given the American governmental system, he was not a dictator—"all we can do is to come as near to [absolute power] as we can." Marlatt to Fairchild, 10 September 1929, Fairchild Papers. Their personal relations finally soured when, in his autobiography, Fairchild blamed Marlatt for the decline of SPI (*World,* 425). Marlatt was quite upset at this attack and attributed it to Fairchild's jealousy of his success. Marlatt to Fairchild, 30 October 1938, Fairchild Papers; author's conversation with Florence Marlatt.

57. Current concern is evident in U.S. Congress, Office of Technology Assessment, *Harmful Non-indigenous Species in the United States* (Washington, DC: GPO, 1993); Don C. Schmitz and Daniel Simberloff, "Biological Invasions: A Growing Threat," *Issues in Science and Technology* 13 (1997): 33–40.

58. Paul R. Cutwright, *Theodore Roosevelt: The Making of a Conservationist* (Urbana: University of Illinois Press, 1985); Thomas G. Dyer, *Theodore Roosevelt and the Idea of Race* (Baton Rouge: Louisiana State University Press, 1980).

59. *National Cyclopedia of American Biography,* 35:196; William Williams, "Reminiscences of the Bering Sea Arbitration," *American Journal of International Law* 37 (1943): 562–84.

60. Quoted in Prescott F. Hall, "Selection of Immigration," *Annals of the American Academy of Political and Social Science* 24 (1904): 175.

61. Elizabeth Yew, "Medical Inspection of the Immigrant at Ellis Island, 1891–1943," *Bulletin of the New York Academy of Medicine* 56 (1980): 488–510; Kraut, *Silent Travellers*, 50–77; "Second Report of the Committee on Immigration of the Eugenics Section of the American Genetic Association," *Journal of Heredity* 5 (1914): 297–300. The impact of inspection on immigration is uncertain. Williams's procedures resulted in the exclusion of, at most, 2.5 percent of entrants, a much smaller percentage than he envisioned. Yet awareness of the inspection process certainly deterred many individuals with detectable illnesses or disabilities from making the trip.

62. For provisions of the laws, see Marion T. Bennett, *American Immigration Policies: A History* (Washington, DC: Public Affairs Press, 1963); for the change in perspective, see, e.g., Robert De C. Ward, "The Immigration Problem Today," *Journal of Heredity* 11 (1920): 323–28.

63. Stephen J. Gould, *The Mismeasure of Man* (New York: Norton, 1981), 231–32; Daniel J. Kevles, *In the Name of Eugenics* (New York: Knopf, 1985), 96–97; Elazar Barkan, "Reevaluating Progressive Eugenics: Herbert Spencer Jennings and the 1924 Immigration Legislation," *Journal of the History of Biology* 24 (1991): 91–112; Carl N. Degler, *In Search of Human Nature: The Decline and Revival of Darwinism in American Social Thought* (New York: Oxford University Press, 1991); Mae M. Ngai, "The Architecture of Race in American Immigration Law: A Reexamination of the Immigration Act of 1924," *Journal of American History* 86 (1999): 67–92.

64. The issue of links between ecological and political nativism has reemerged in the last few years: see Gert Groening and Joachim Wolschke-Bulmahn, "Some Notes on the Mania for Native Plants in Germany," *Landscape Journal* 11 (1992): 116–26, and Michael Pollan, "Against Nativism," *New York Times*, 15 May 1994, 52–55. More broadly, see William Cronon, ed., *Uncommon Ground: Rethinking the Human Place in Nature* (New York: Norton, 1995). My research on the history of American ecological independence is continuing.

PART II
SPECIALIZATION AND ORGANIZATION

1. MBL, *Annual Report*, 1888, 1889, 1890. On the spelling of the village's name, see chapter 6, note 8.

2. Charles O. Whitman, "Specialization and Organization, Companion Principles of All Progress—The Most Important Need of American Biology," in *Biological Lectures Delivered at the Marine Biological Laboratory of Wood's Holl in the Summer Session of 1890* (Boston: Ginn, 1891), 1–26.

3. Ibid., 1–2, 6, 11.

4. Ibid., 21–25.

5. Senate Committee on Fisheries, *Report of the Subcommittee of the Committee on Fisheries, United States Senate, in the Matter of the Investigation of the United States Fish Commission*, 51st Cong., 2nd Sess., Senate Report 2361, 494–504.

CHAPTER FOUR
LIFE SCIENCE INITIATIVES IN THE LATE NINETEENTH CENTURY

1. On the importance of bureau organization, see A. Hunter Dupree, *Science in the Federal Government: A History of Policies and Activities* (Cambridge, MA: Belknap Press of Harvard University Press, 1957; reprint, Baltimore: Johns Hopkins University Press, 1986).

2. David H. Guston, "Congressmen and Scientists in the Making of Science Policy: The Allison Commission, 1884–1886," *Minerva* 23 (1994): 25–52.

3. Theodore Lyman, "Address on General Phases of Fish Culture and the Fisheries," *Transactions of the American Fisheries Society*, 1884, pp. 84–87; Henry P. Bowditch, "Biographical Memoir of Theodore Lyman," *National Academy of Sciences Biographical Memoirs* 5 (1903): 143–53; Dean Conrad Allard, Jr., *Spencer Fullerton Baird and the U.S. Fish Commission* (New York: Arno Press, 1978), 99. Lyman suffered progressive paralysis, probably from amyotrophic lateral sclerosis; he died in 1897.

4. Agassiz's attacks are discussed in Daniel J. Kevles, *The Physicists*, 2nd ed. (Cambridge, MA: Harvard University Press, 1995), 51–59; Alexander Agassiz to Spencer F. Baird, 26 June 1886, box 13, Spencer F. Baird Collection, RU 7002, Smithsonian Institution Archives (hereafter Baird Collection). Powell had explicitly modeled his work on that of Baird: see Powell's statement, December 1884, in *Testimony before the Joint Commission to Consider the Present Organization of the Signal Service . . .* , 49th Cong., 1st Sess., 1886, Senate Misc. Doc. 82, p. 178.

5. Allard, *Baird*, 325–28; "Prof. Baird's Accounts," *Washington Post*, 2 October 1885; see the voluminous material in folders 1–3, box 67, Baird Collection.

6. John Wesley Powell, "The Personal Characteristics of Professor Baird," *Bulletin of the Philosophical Society of Washington* 10 (1887): 76–77.

7. Smithsonian Institution, *Annual Report*, 1887–88, pp. xiv–xv.

8. "Prof. Baird's Successor," *New York World*, 27 August 1887. Baird had made the commission position anomalous in that it carried no salary.

9. *Investigation of the Fish Commission*, 65–70, 492–93, 498, 514.

10. Lester F. Ward, *Glimpses of the Cosmos*, 6 vols. (New York: G. P. Putnam, 1913–18), 5:74–80.

11. Michael James Lacey, "The Mysteries of Earth-Making Dissolve: A Study of Washington's Intellectual Community and the Origins of American Environmentalism in the Late Nineteenth Century" (Ph.D. diss., George Washington University, 1979), 220; William Healey Dall, *Spencer Fullerton Baird: A Biography* (Philadelphia: Lippincott, 1915).

12. Helen Lefkowitz Horowitz, "The National Zoological Park: 'City of Refuge' or Zoo?" in *New Worlds, New Animals*, ed. R. J. Hoage and William A. Deiss (Baltimore: Johns Hopkins University Press, 1996), 132–33; Leland O. Howard, *History of Applied Entomology (Somewhat Anecdotal)*, Smithsonian Miscellaneous Collections, vol. 84, 1930, p. 93; William Culp Darrah, *Powell of the Colorado* (Princeton: Princeton University Press, 1951), 348; George Brown Goode, *The Origins of Natu-*

ral Science in America, ed. Sally Gregory Kohlstedt (Washington, DC: Smithsonian Institution Press, 1991), 6.

13. On Walcott as a scientific bureaucrat, see Stephen J. Gould, *Wonderful Life: The Burgess Shale and the Nature of History* (New York: Norton, 1989), 241–53.

14. Alfred Russel Wallace, *My Life*, 2 vols. (London: Chapman and Hall, 1905) 2:107–99. In the notes below I will mention Wallace's itinerary and contacts.

15. Frederick C. Jaher, *The Urban Establishment* (Urbana: University of Illinois Press, 1981); Robert F. Dalzell, Jr., *Enterprising Elite: The Boston Associates and the World They Made* (Cambridge, MA: Harvard University Press, 1987).

16. Mona Domosh, *Invented Cities: The Creation of Landscape in Nineteenth-Century New York and Boston* (New Haven: Yale University Press, 1996), 118–19; Edward Lurie, *Louis Agassiz: A Life in Science* (Chicago: University of Chicago Press, 1960), 232–33. Wallace sailed into New York in October 1886, but left almost immediately for Boston; due to lack of money he stayed with acquaintances and in inexpensive hotels. Wallace, *My Life*, 2:107–8.

17. On the construction history of the MCZ, see Mary P. Winsor, *Reading the Shape of Nature: Comparative Zoology at the Agassiz Museum* (Chicago: University of Chicago Press, 1991), 175–92. Wallace spent more than a month in Boston, delivering the Lowell Lectures; he saw all the museums and university facilities he could, and met both famous writers (such as Oliver Wendell Holmes and James Russell Lowell) and working scientists. He discussed the MCZ in depth in "American Museums," *Fortnightly Review* 42 (1887): 342–69.

18. Winsor, *Reading the Shape of Nature*, 174–86; Robert E. Kohler, "The Ph.D. Machine: Building on the Collegiate Base," *Isis* 81 (1990): 638–62.

19. Samuel Eliot Morison, *Three Centuries of Harvard, 1636–1936* (Cambridge, MA: Harvard University Press, 1936), 356.

20. Benjamin L. Robinson, "Botany, 1869–1929," in *The Development of Harvard University since the Inauguration of President Eliot, 1869–1929*, ed. Samuel Eliot Morison (Cambridge, MA: Harvard University Press, 1930), 370; Harvard University, *Annual Report of the President*, 1887–88, p. 21; Richard Evans Schultes and William A. Davis, with Hillel Burger, *The Glass Flowers at Harvard* (New York: E. P. Dutton, 1982).

21. William Albert Setchell, "Biographical Memoir of William Gilson Farlow, 1844–1919," *Scientific Memoirs of the National Academy of Sciences*, 2, no. 4 (1926).

22. Asa Gray, *Darwiniana: Essays and Reviews Pertaining to Darwinism* (New York: D. Appleton, 1876); A. Hunter Dupree, *Asa Gray* (Cambridge, MA: Belknap Press of Harvard University Press, 1959); James R. Moore, *The Post-Darwinian Controversies* (Cambridge: Cambridge University Press, 1979).

23. On James's evolutionary thinking, see Robert J. Richards, *Darwin and the Emergence of Evolutionary Theories of Mind and Behavior* (Chicago: University of Chicago Press, 1987), 409–50.

24. Harvard University, *Catalogue*, 1887–88, p. 107; "Annual Return, 1887–88. Philosophy 13," UA III 15.28, Harvard University Archives; Josiah Royce, *The Spirit of Modern Philosophy* (Boston: Houghton Mifflin, 1892); on Royce, and Harvard philosophy more generally, see Bruce J. Kuklick, *The Rise of American Philosophy, Cambridge, Massachusetts, 1860–1930* (New Haven: Yale University Press, 1977).

25. Nathaniel Southgate Shaler, *Nature and Man in America* (New York: Charles Scribner's Sons, 1891); David N. Livingstone, *Nathaniel Southgate Shaler and the Culture of American Science* (Tuscaloosa: University of Alabama Press, 1987), 121–57.

26. John Fiske, *The Destiny of Man Viewed in the Light of His Origin* (Boston: Houghton Mifflin, 1884); Fiske, *The Idea of God as Affected by Modern Knowledge* (Boston: Houghton Mifflin, 1885); Fiske, *The Critical Period of American History, 1783–1789* (Boston: Houghton Mifflin, 1888); Milton Berman, *John Fiske: The Evolution of a Popularizer* (Cambridge, MA: Harvard University Press, 1961), 157–221; Kuklick, *Rise of American Philosophy*, 80–90.

27. Wallace did not worry. At this time he was pursuing at least four interests: biogeography, evolutionary theory, spiritualism, and social reform. Wallace, *My Life*, 2:107–99.

28. Cynthia Zaitzevsky, *Frederick Law Olmsted and the Boston Park System* (Cambridge, MA: Belknap Press of Harvard University Press, 1982).

29. William M. Wheeler, "The Bussey Institution," in *Development of Harvard*, ed. Morison, 508–17; Ida Hay, *Science in the Pleasure Ground: A History of the Arnold Arboretum* (Boston: Northeastern University Press, 1995), 38–43, 60–64; Harvard University, *Report of the President*, 1887–88, p. 25.

30. S. B. Sutton, *Charles Sprague Sargent and the Arnold Arboretum* (Cambridge, MA: Harvard University Press, 1970); Hay, *Science in the Pleasure Ground*. Wallace did not mention the arboretum, but this may have been because his Boston stay was during the winter.

31. Sally Gregory Kohlstedt, "From Learned Society to Public Museum: The Boston Society of Natural History," in *The Organization of Knowledge in Modern America, 1860–1920*, ed. Alexandra Oleson and John Voss (Baltimore: Johns Hopkins University Press, 1979), 386–406.

32. Bruce Sinclair, "Harvard, MIT, and the Ideal Technical Education," in *Science at Harvard University: Historical Perspectives*, ed. Clark A. Elliott and Margaret W. Rossiter (Bethlehem, PA: Lehigh University Press, 1992), 76–95; E. O. Jordan, G. C. Whipple, and C.-E. A. Winslow, *A Pioneer of Public Health: William Thompson Sedgwick* (New Haven: Yale University Press, 1924), 22–40.

33. Kenneth Ludmerer, "Reform at Harvard Medical School," *Bulletin of the History of Medicine* 55 (1981): 343–70; Henry K. Beecher and Mark D. Altschule, *Medicine at Harvard: The First Three Hundred Years* (Hanover, NH: University Press of New England, 1977), 95–97, 165–67.

34. Walter Cannon, "Henry Pickering Bowditch," *Biographical Memoirs of the National Academy of Sciences* 17 (1924): 183–96; W. Bruce Fye, *The Development of American Physiology* (Baltimore: Johns Hopkins University Press, 1987), 92–128.

35. Frederick T. Lewis, "Charles Sedgwick Minot," *Anatomical Record* 10 (1916): 132–64; Philip J. Pauly, "The Appearance of Academic Biology in Late Nineteenth-Century America," *Journal of the History of Biology* 17 (1984): 369–97.

36. Dorothy G. Wayman, *Edward Sylvester Morse: A Biography* (Cambridge, MA: Harvard University Press, 1942); Mark V. Barrow, Jr., "Birds and Boundaries: Community, Practice, and Conservation in North American Ornithology, 1865–1935" (Ph.D. diss., Havard University, 1992), 74–75; Wallace, a former commercial collector, visited Morse in Salem and presumably met Cassino. *My Life*, 2:110.

37. R. W. Dexter, "The Annisquam Laboratory of Alpheus Hyatt," *Scientific Monthly* 74 (1952): 112–16; Philip J. Pauly, "Summer Resort and Scientific Discipline," in *The American Development of Biology*, ed. Ronald Rainger, Keith R. Benson, and Jane Maienschein (Philadelphia: University of Pennsylvania Press, 1988), 121–50.

38. Winsor, *Reading the Shape of Nature*, 200–207.

39. George W. Pierson, *Yale College: An Educational History, 1871–1921* (New Haven: Yale University Press, 1952); Brooks Mather Kelley, *Yale: A History* (New Haven: Yale University Press, 1974), 235–72; Russell H. Chittenden, *History of the Sheffield Scientific School of Yale University, 1846–1922*, 2 vols. (New Haven: Yale University Press, 1928), 1:95–101, 238–47; Charles Schuchert and Clara Mae Le Vene, *O. C. Marsh: Pioneer in Paleontology* (New Haven: Yale University Press, 1940; reprint, Arno Press, 1978). Wallace spent a day in New Haven, staying with Marsh. *My Life*, 2:112–13.

40. Douglas Sloan, "Science in New York City," *Isis* 71 (1980): 35–76; Simon Baatz, *Knowledge, Culture, and Science in the Metropolis* (Annals of the New York Academy of Sciences, vol. 584, 1990), 95–138; George M. Beard, *American Nervousness; Its Causes and Consequences, a Supplement to Nervous Exhaustion (Neurasthenia)* (1881; reprint, New York: Arno Press, 1972); on Beard, see Charles Rosenberg, *No Other Gods* (Baltimore: Johns Hopkins University, 1976), 98–108.

41. Douglas Preston, *Dinosaurs in the Attic* (New York: St. Martin's Press, 1986); Ronald Rainger, *An Agenda for Antiquity* (Tuscaloosa: University of Alabama Press, 1991), 57–61. Wallace passed through New York three times, but mentioned only a meeting with Clarence King and a brief visit to the American Museum. *My Life*, 2:107, 113, 125.

42. Steven Conn, *Museums and American Intellectual Life, 1876–1926* (Chicago: University of Chicago Press, 1998), 62–67.

43. Pauly, "Appearance of Academic Biology," 390–91. The Wistar Institute became more active academically following its reorganization in 1905.

44. Henry Fairfield Osborn, *Cope: Master Naturalist* (Princeton: Princeton University Press, 1931); Url Lanham, *The Bone Hunters* (New York: Columbia University Press, 1973), 243–61; David Rains Wallace, *The Bonehunters' Revenge* (Boston: Houghton Mifflin, 1999).

45. Wallace apparently passed through Philadelphia a number of times without stopping. He spoke at Vassar but not Bryn Mawr. He lectured in Baltimore at Johns Hopkins and then stayed for about two months in Washington, where he became

particularly "intimate" with Elliott Coues and Lester Ward. Coues and Wallace shared an interest in spiritualism; Ward, described by Wallace as "an absolute agnostic," criticized this belief, but impressed Wallace with both his botanical and sociological work. *My Life*, 2: 113–26; Bernhard J. Stern, ed., "Letters of Alfred Russel Wallace to Lester F. Ward," *Scientific Monthly* 40 (1935): 375–79.

46. Edward J. Renehan, Jr., *John Burroughs: An American Naturalist* (Post Mills, VT: Chelsea Green Publishing Company, 1992), 151–81; John Burroughs, *Signs and Seasons* (Boston: Houghton Mifflin, 1886), 232–35; Burroughs, "Science and Theology," *Popular Science Monthly* 30 (1886): 145–62. On egg collectors, see Mark V. Barrow, Jr., *A Passion for Birds: American Ornithology after Audubon* (Princeton: Princeton University Press, 1998), 27–42.

47. Lurie, *Agassiz*, 360–61; Morris Bishop, *A History of Cornell* (Ithaca: Cornell University Press, 1962), 99–142; Gould Colman, *Education and Agriculture* (Ithaca: Cornell University Press, 1963), 70–90.

48. Colman, *Education and Agriculture*, 99–100; Andrew Denny Rodgers III, *Liberty Hyde Bailey: A Story of American Plant Sciences* (Princeton: Princeton University Press, 1949).

49. Roswell Ward, *Henry A. Ward: Museum Builder to America*, Rochester Historical Society Publications #24, 1948, pp. 54–48, 199–230; Sally Gregory Kohlstedt, "Henry A. Ward: The Merchant Naturalist and American Museum Development," *Journal of the Society for the Bibliography of Natural History* 9 (1980): 647–61.

50. Anne Whiston Spirn, "Constructing Nature: The Legacy of Frederick Law Olmsted," in *Uncommon Ground: Rethinking the Human Place in Nature*, ed. William Cronon (New York: Norton, 1996), 95–96; William Cronon, *Nature's Metropolis: Chicago and the Great West* (New York: Norton, 1991), 148–206. Wallace probably visited Ithaca and mentioned Rochester; with an interest in glaciation, he spent four days at Niagara in mid-winter. He cruised Lake Ontario and the St. Lawrence River at the end of his trip, in August. *My Life*, 2:125–27, 187.

51. Wallace stayed only six hours in Chicago, waiting to change trains, and then visited Michigan Agricultural College, not Milwaukee. *My Life*, 2:185–86.

52. E. P. Allis in E. J. Dornfield, "The Allis Lake Laboratory, 1886–1893," *Marquette Medical Review* 21 (1956): 115–44; Charles O. Whitman, *Methods of Research in Microscopical Anatomy and Embryology* (Boston: S. E. Cassino, 1885).

53. Dornfield, "Allis Lake Laboratory."

54. Andrew D. Rodgers III, *John Merle Coulter: Missionary in Science* (Princeton: Princeton University Press, 1944), 33–52.

55. Thomas D. Clark, *Indiana University: Midwestern Pioneer*, 4 vols. (Bloomington: Indiana University Press, 1973), 1:202–83; David Starr Jordan, "The Story of a Salmon," *Popular Science Monthly* 19 (1881): 1–6. Wallace was there! His lecture schedule took him by train from Washington to Cincinnati, and then St. Louis, so Bloomington was a convenient stop. *My Life*, 2:136, 145.

56. William Barnaby Faherty, S. J., *Henry Shaw: His Life and Legacies* (Columbia: University of Missouri Press, 1987); Kim J. Kleinman, "The Museum in the Gar-

den: Research, Display, and Education at the Missouri Botanical Garden since 1859" (Ph.D. diss., Union Institute, 1997); Carol Grove, "Aesthetics, Horticulture, and the Gardenesque: Victorian Sensibilities at Tower Grove Park," *Journal of the New England Garden History Society* 6 (1998): 32–41. Wallace visited. He liked the plant houses and benches but, as a foreigner, was disappointed that most of the plants were exotic rather than American. *My Life,* 2:146.

57. Daniel Goldstein, "Midwestern Naturalists: Academies of Science in the Mississippi Valley, 1850–1900" (Ph.D. diss., Yale University, 1989). Wallace was on the other side of the state, lecturing to the Sioux City Natural History Society, and touring the slaughterhouse and a local amateur's "zoological farm." *My Life,* 2:147–49.

58. James C. Carey, *Kansas State University: The Quest for Identity* (Lawrence: Regents Press of Kansas, 1977); Earle D. Ross, *A History of the Iowa State College of Agriculture and Mechanic Arts* (Ames: Iowa State College Press, 1942). Wallace spoke in both Lawrence and Manhattan. In Manhattan he met the fathers of both David Fairchild and Charles Marlatt. Sixty miles west, near Salina, he visited a French-American farmer and bryologist who made Wallace feel "quite small" for his inability to converse in French. *My Life,* 2:151–54.

59. Richard A. Overfield, *Science with Practice: Charles E. Bessey and the Maturing of American Botany* (Ames: Iowa State University Press, 1993).

60. Schuchert and LeVene, *Marsh,* 211–12.

61. U.S. National Museum, *Annual Report,* 1888, p. 40.

62. Wallace traveled from San Francisco through Stockton, to Yosemite, and then up to the Calaveras Grove; he remarked a number of times on the devastation left by the miners. *My Life,* 2:160, 165.

63. Lester D. Stephens, *Joseph LeConte: Gentle Prophet of Evolution* (Baton Rouge: Louisiana State University Press, 1982). Wallace met LeConte immediately upon arriving in California. *My Life,* 2:158.

64. Linnie Marsh Wolfe, *Son of the Wilderness: The Life of John Muir* (New York: Knopf, 1945); Thurman Wilkins, *John Muir: Apostle of Nature* (Norman: University of Oklahoma Press, 1997), 159–68; John Muir, ed., *Picturesque California and the Region West of the Rocky Mountains, from Alaska to Mexico,* 5 vols. (San Francisco: J. Dewing, 1888). Muir could give Wallace only one day for a trip into the Oakland foothills to see the stumps of redwood trees burned forty years earlier. Wallace's brother, who lived in Stockton, took the naturalist for a brief trip to Yosemite, but they had neither the money nor the time to climb out of the valley. *My Life,* 2:158–60.

65. Peter Dreyer, *A Gardener Touched with Genius: The Life of Luther Burbank* (Berkeley: University of California Press, 1985).

66. Wallace, on his return trip, botanized in the Sierras and the Rockies, lectured at Michigan Agricultural College, visited friends on the St. Lawrence River, and sailed from Montreal. *My Life,* 2:171–91.

67. Howard Mumford Jones, *The Age of Energy: Varieties of American Experience, 1865–1915* (New York: Viking Press, 1970).

CHAPTER FIVE

ACADEMIC BIOLOGY: SEARCHING FOR ORDER IN LIFE

1. "Introductory," *American Naturalist* 1 (1867): 1–4; the early years of the magazine are discussed in Lynn Nyhart, "The American Naturalist, 1867–1886" (B.A. thesis, Princeton University, 1979). Putnam became involved with the magazine and other ventures and never reissued the directory. As chapter 4 notes, local collector Samuel Cassino took over the directory in 1877 and published it annually after that.

2. H. C. Wood, "A Botanical Excursion in My Office," *American Naturalist* 1 (1867): 517–29; J. S. Newberry, "Modern Scientific Investigation: Its Methods and Tendencies," *American Naturalist* 1 (1867): 449–69.

3. "The Second Decennary of the American Naturalist," *American Naturalist* 11 (1877): 1–3.

4. "Editors' Table," *American Naturalist* 17 (1883): 58.

5. "Second Decennary," 1; "Editors' Table," *American Naturalist* 15 (1881): 987–89; "Editors' Table," *American Naturalist* 17 (1883): 58.

6. "Editors' Table," *American Naturalist* 18 (1884): 393; "Editors' Table," *American Naturalist* 21 (1887): 59.

7. "Editors' Table," *American Naturalist* 31 (1897): 800–801; Michael Sokal, *The Man of Science in Modern America: James McKeen Cattell, 1860–1944*, in press.

8. This section draws on Toby Appel, "Organizing Biology: The American Society of Naturalists and Its 'Affiliated Societies,' 1883–1923," in *The American Development of Biology*, ed. Ronald Rainger, Keith R. Benson, and Jane Maienschein (Philadelphia: University of Pennsylvania Press, 1988), 87–120.

9. AAAS, *Proceedings* 32 (1883). On the AAAS, see Sally Gregory Kohlstedt, Michael M. Sokal, and Bruce V. Lewenstein, *The Establishment of Science in America: 150 Years of the American Association for the Advancement of Science* (New Brunswick, NJ: Rutgers University Press, 1999).

10. ASN, *Records* 1 (1884–95): 22.

11. Ibid., 23; "Editors' Table," *American Naturalist* 18 (1884): 160–61; Appel, "Organizing Biology," 90–99.

12. ASN, *Records* 1 (1884–95): 134, 281, 306.

13. On these developments, see Lawrence Veysey's classic, *The Emergence of the American University* (Chicago: University of Chicago Press, 1963); Roger L. Geiger, *To Advance Knowledge: The Growth of American Research Universities, 1900–1940* (New York: Oxford University Press, 1986); Julie A. Reuben, *The Making of the Modern University* (Chicago: University of Chicago Press, 1996). This section uses some material from my earlier publication, "The Appearance of Academic Biology in Late Nineteenth-Century America," *Journal of the History of Biology* 17 (1984): 369–97, with kind permission of Kluwer Academic Publishers.

14. D. C. Gilman, "Inaugural Address," 22 February 1876, quoted in Allan Chesney, *The Johns Hopkins Hospital and the Johns Hopkins University School of Medicine*, 3 vols. (Baltimore: Johns Hopkins University Press, 1943), 1:43; Hugh Hawkins,

Pioneer: A History of the Johns Hopkins University, 1874–1889 (Ithaca: Cornell University Press, 1960), 48; *Johns Hopkins University Circular* 1 (1881): 104. On Huxley and biology in England, see Joseph Caron, " 'Biology' in the Life Sciences: A Historiographical Contribution," *History of Science* 26 (1988): 250–56; Adrian Desmond, *T. H. Huxley*, 2 vols. (London: Michael Joseph, 1997), 2:37–39, 273.

15. Hawkins, *Pioneer*, 142–45; Martin, "Modern Physiological Laboratories: What They Are and Why They Are," *Johns Hopkins University Circular* 3 (1884): 87; Martin to Gilman, 29 May 1876, 2 January 1877, Gilman Papers, Special Collections, Eisenhower Library, Johns Hopkins University (hereafter Gilman Papers); Keith R. Benson, "William Keith Brooks (1845–1908): A Case Study in Morphology and the Development of American Biology" (Ph.D. diss., Oregon State University, 1979); "Editors' Table," *American Naturalist* 18 (1884): 393. On Martin's work at Hopkins, see Larry Owens, "Pure and Sound Government: Laboratories, Gymnasia, and Playing Fields in Nineteenth Century America," *Isis* 76 (1985): 182–94.

16. Fabian Franklin, *The Life of Daniel Coit Gilman* (New York: Dodd, Mead, 1910), 252; Chesney, *Hopkins Hospital*, 49, 98; Johns Hopkins University, *Half-Century Directory, 1876–1926* (Baltimore: Johns Hopkins University Press, 1926), 428–29; Pauly, "Appearance of Academic Biology," 379–81.

17. Benson, "William Keith Brooks," 73–76, 314; Garland E. Allen, *Thomas Hunt Morgan: The Man and His Science* (Princeton: Princeton University Press, 1978), 45–46; Jane Maienschein, *Transforming Traditions in American Biology, 1880–1915* (Baltimore: Johns Hopkins University Press, 1991), 42.

18. E. O. Jordan, G. C. Whipple, and C.-E. A. Winslow, *A Pioneer of Public Health: William Thompson Sedgwick* (New Haven: Yale University Press, 1924), 19–55; Patricia Peck Gossel, "The Emergence of American Bacteriology, 1875–1900" (Ph.D. diss., Johns Hopkins University, 1988), 332–44; Henry N. Martin to Daniel C. Gilman, 11 July 1883, Gilman Papers.

19. Bryn Mawr College, *President's Report to the Board of Trustees, 1884–1885*, 15–16; M. Carey Thomas, "Conversations about College Organization in 1884," manuscript notebook, College Archives, Canaday Library, Bryn Mawr College, Pennsylvania; E. B. Wilson to M. Carey Thomas, 23 January 1888, M. Carey Thomas Official Correspondence, reel 152, Bryn Mawr Archives, Pennsylvania; Bryn Mawr College, *Program*, 1889, pp. 46–50; Bryn Mawr College, *President's Report*, 1898, p. 70. Wilson and Sedgwick also collaborated on the first original American biology textbook, a detailed anatomical and physiological treatment of one plant and one animal designed to provide an overview of "the properties of matter in the living state" as part of general education: W. T. Sedgwick and E. B. Wilson, *General Biology* (New York: Henry Holt, 1886).

20. Columbia University, *Annual Report*, 1891, p. 6; "Report of the Special Committee on the Disposition of the Legacy of Charles M. Da Costa, Recommending the Establishment of the Department of Biology," Columbia College Papers, January–May 1891, Manuscript Department, Butler Library, Columbia University (hereafter CUM); Columbia University, *Catalog, 1892–93*; J. S. McLane to Seth Low, 1

March 1891, Trustees' Papers, Butler Library, Columbia University; Low to J. G. Curtis, 4 November 1892, Central Files, Low Library, Columbia University; Andrew S. Dolkart, *Morningside Heights: A History of Its Architecture and Development* (New York: Columbia University Press, 1998).

21. E. P. Cheyney, *History of the University of Pennsylvania, 1740–1940* (Philadelphia: University of Pennsylvania Press, 1940), 300–303; University of Pennsylvania, *Handbook of Information Concerning the School of Biology* (Philadelphia: University of Pennsylvania Press, 1889); [C. S. Dolley] to "Provost and Trustees of the University of Pennsylvania," n.d., file 1891 Biological Station, University Archives, University of Pennsylvania.

22. Dorothy Ross, G. *Stanley Hall: The Psychologist as Prophet* (Chicago: University of Chicago Press, 1972), 186–202; Clark University, *Third Annual Report of the President to the Board of Trustees, April 1893* (Worcester, MA: Clark University Press, 1893), 90–106; Clark University, *Decennial Celebration* (Worcester, MA: Clark University Press, 1899), 99–107.

23. Charles O. Whitman to Alexander Agassiz, 14 July 1886, Alexander Agassiz Papers, Museum of Comparative Zoology, Harvard University.

24. Charles O. Whitman, "Biological Instruction in Universities," *American Naturalist* 21 (1887): 507–19; ASN, *Records* 1 (1884–95): 110, 135.

25. Charles O. Whitman, "Report of the Director of the Marine Biological Laboratory for the Fourth Session, 1891," MBL, *Annual Report*, 1891, pp. 15–16.

26. Ross, G. *Stanley Hall*, 218–28; Richard J. Storr, *Harper's University: The Beginnings* (Chicago: University of Chicago Press, 1966), 141; F. P. Mall to W. H. Welch, 15 January 1893, in Florence R. Sabin, *Franklin Paine Mall: The Story of a Mind* (Baltimore: Johns Hopkins University Press, 1934), 112.

27. Whitman to Harper, 19 December 1891, 15 and 26 January 1892, 13 March 1896, Presidents' Papers, University Archives, University of Chicago; on Whitman's plans, see esp. Jane Maienschein, "Whitman at Chicago: Establishing a Chicago Style of Biology?" in *American Development of Biology*, ed. Rainger et al., 151–84.

28. J. P. Campbell, *Biological Teaching in the Colleges of the United States* (U.S. Bureau of Education, Circular of Information, 1891, # 9), 138.

29. W. Bruce Fye, "H. Newell Martin: A Remarkable Career Destroyed by Neurasthenia and Alcoholism," *Journal of the History of Medicine and Allied Sciences* 40 (1985): 133–66.

30. Typed excerpt from Minutes of the Faculty of Pure Science, 20 October 1893, in F. S. Lee Papers, box 1, file "Columbia University—President Butler and Secretary," CUM; G. S. Huntington, "Comparative Anatomy in the Medical Course," *American Journal of Medical Science* 116 (1898): 629–46; Low to Osborn, 12 , 21, and 28 March 1895; Osborn to Low, 16 and 22 March 1895, 6 March 1896; Huntington to Low, 5 December 1894, 2 and 7 March 1895, 10 May 1895; Low to Huntington, 28 February and 10 May 1895; Seth Low to "Faculty of the College of Physicians and Surgeons," n.d., J. W. McLane folder (all this correspondence in Columbia Central Files); G. S. Huntington, printed statement, 5 December 1894, Records of the College of Physicians and Surgeons, file "anatomy," CUM.

31. F. S. Lee, "The Scope of Modern Physiology," *American Naturalist* 28 (1894): 388; Columbia University, *Annual Report*, 1898–99, p. 163, and 1911, p. 83.

32. Sabin, *Mall*, 115–17; Whitman to Harper, 9 April 1895, 3 May, and 4 and 12 September 1899, and Harper to Martin Ryerson, 24 December 1895, all in University of Chicago Presidents' Papers; Howard S. Miller, Dollars for Research (Seattle: University of Washington Press, 1970), 159–62; "Exercises in Connection with the Laying of the Corner Stones of the Hull Biological Laboratories," *University [of Chicago] Record*, 31 July 1896, p. 286; William H. Welch, "Biology and Medicine," *Papers and Addresses*, 3 vols. (Baltimore: Johns Hopkins University Press, 1920), 3:240; Storr, *Harper's University*, 141–44, 286–91, 360–62.

33. Ernest E. Irons, *The Story of Rush Medical College* (Chicago: Rush Medical College, 1953), 32–44; Robert E. Kohler, *From Medical Chemistry to Biochemistry* (New York: Cambridge University Press, 1982), 145–48.

34. William G. Farlow, "The Change from the Old to the New Botany in the United States," *Science* 37 (1913): 85.

35. "Editorial," *Botanical Gazette* 15 (1890): 236–37. Cope's *American Naturalist* responded in kind, arguing that botanists were losing ground in colleges because they did not actually teach about living organisms: "Editors' Table," *American Naturalist* 24 (1890): 1050. It is a puzzle whether the "iniquity" of zoology teaching lay in vivisection, lack of intellectual rigor, or Louis Agassiz's anti-evolutionism.

36. Byron Halsted, "A Section of Botany in the American Association," *Botanical Gazette* 17 (1892): 25–26. The imbalance of fields was evident at the first meeting of the separate sections, in 1893, which included thirty-four botanical papers, but only eleven zoological ones. The officers elected for the zoological section the following year all resigned prior to the next meeting: AAAS, *Proceedings* 43 (1894): xvi.

37. "Editorial," *Botanical Gazette* 17 (1892): 94; Conway Macmillan, "The Emergence of a Sham Biology in America," *Science* 21 (1893): 184–86; see also Macmillan, "Dr. J. P. Campbell's 'Biological Instruction,' " *Botanical Gazette* 17 (1892): 301–2. Federal naturalist C. Hart Merriam, supported privately by Theodore Roosevelt, added his criticism: a number of responses, and Macmillan's reply, can be found in the same volume of *Science*. On these disputes, see Mary P. Winsor, *Reading the Shape of Nature: Comparative Zoology at the Agassiz Museum* (Chicago: University of Chicago Press, 1991), 187–96.

38. Andrew Denny Rodgers III, *John Merle Coulter: Missionary in Science* (Princeton: Princeton University Press, 1944), 111–12, 131–32, 152–59; "Address of Head Professor Coulter," *University [of Chicago] Record*, 31 July 1896, 287–88.

CHAPTER SIX
A PLACE OF THEIR OWN: THE SIGNIFICANCE OF WOODS HOLE

1. Charles R. Crane to John D. Rockefeller, Jr., 22 December 1923, quoted in Frank R. Lillie, *The Woods Hole Marine Biological Laboratory* (Chicago: University of Chicago Press, 1944), 73. On the history of the MBL, see Lillie, *MBL*; Jane Maienschein, *100 Years Exploring Life* (Boston: Jones and Bartlett, 1989). Part of

this chapter first appeared as "Summer Resort and Scientific Discipline," in *The American Development of Biology*, ed. Ronald Rainger, Keith R. Benson, and Jane Maienschein, 121–50. © 1988 University of Pennsylvania Press. Reprinted by permission of the publisher.

2. On conditions in American cities in the nineteenth century, see, e.g., Jacob Riis's classic *How the Other Half Lives* (New York: C. Scribner's Sons, 1890); Harold M. Mayer and Richard C. Wade, *Chicago: Growth of a Metropolis* (Chicago: University of Chicago Press, 1969); Edwin G. Burrows and Mike Wallace, *Gotham: A History of New York City to 1898* (New York: Oxford University Press, 1999), 1089–1185.

3. On middle-class vacations, see Dona Brown, *Inventing New England: Regional Tourism in the Nineteenth Century* (Washington, DC: Smithsonian Institution Press, 1995); for lower-class leisure, Charles E. Funnell, *By the Beautiful Sea: The Rise and High Times of That Great American Resort, Atlantic City* (New York: Knopf, 1975). On the Massachusetts seashore and its meanings, see John R. Stilgoe, *Alongshore* (New Haven: Yale University Press, 1994); more broadly, Lena Lencek and Gideon Bosker, *The Beach: The History of Paradise on Earth* (New York: Viking Press, 1998); Cindy S. Aron, *Working at Play: A History of Vacations in the United States* (New York: Oxford University Press, 1999).

4. Barbara Novak, *Nature and Culture: American Landscape and Painting, 1825–1875* (New York: Oxford University Press, 1980); *Resorts of the Catskills* (New York: St. Martin's Press, 1979); David Schuyler, *Apostle of Taste: Andrew Jackson Downing, 1815–1852* (Baltimore: Johns Hopkins University Press, 1996).

5. Cleveland Amory, *The Last Resorts* (New York: Harper, 1951); Brown, *Inventing New England*. Art colonies such as Cornish, New Hampshire, have been the subject of some study: see Hugh Mason Wade, *A Brief History of Cornish, 1763–1974* (Hanover, NH: University Press of New England, 1976).

6. "The New Biological Laboratory of the Johns Hopkins University," *Science* 3 (1884): 350–54; E. A. Andrews, "The Old Laboratory," in "Hopkins Biology News-Letter," June 1948, mimeograph, Records of the Department of Biology, Ferdinand Hamburger Jr. Archives, Johns Hopkins University.

7. The summer expeditions continued until 1897, when a professor and a postdoctoral fellow died of yellow fever. See William K. Brooks, "Notes from the Biological Laboratory," *Johns Hopkins University Circular* 17 (1897): 1–2; H. L. Clark to D. C. Gilman, 14 September 1897, Daniel Coit Gilman Papers, Department of Special Collections, Eisenhower Library, Johns Hopkins University.

8. On Woods Hole's history, see Theodate Geoffrey [pseud. Dorothy Wayman], *Suckanesset: Wherein May Be Read a History of Falmouth Massachusetts* (Falmouth: Falmouth Publishing Company, 1930), 144–51; Mary Lou Smith, ed., *Woods Hole Reflections* (Woods Hole: Woods Hole Historical Collection, 1983); Mary Lou Smith, ed., *The Book of Falmouth: A Tercentennial Celebration: 1686–1986* (Falmouth: Falmouth Historical Commission, 1986). More specifically, see Prince S. Crowell, "The Pacific Guano Company," typescript, Woods Hole Historical Collec-

tion and Museum, Woods Hole, Massachusetts; Richard A. Wines, *Fertilizer in America* (Philadelphia: Temple University Press, 1985); Dean Conrad Allard, Jr., *Spencer Fullerton Baird and the U.S. Fish Commission* (New York: Arno Press, 1978), 322–41; Paul C. Galtsoff, *The Story of the Bureau of Commercial Fisheries Biological Laboratory, Woods Hole, Massachusetts* (U.S. Department of the Interior, circular #145, 1962). The variations in the spelling of the village name occurred because of Fay's aspirations and nostalgia. He conceived the idea that the Norsemen had visited the area and taught the Indians the term "holl," Norse for "hill," and petitioned the post office to change the official spelling from "Wood's Hole" to "Wood's Holl." After complaints that "Holl" was "the meaningless corruption of Wood's Hole by finical summer visitors," the Federal Board of Geographic Names decreed that the spelling should be "Woods Hole." See Lillie, *MBL*, 3–5; Joseph Story Fay, "The Track of the Norseman," *Magazine of American History* 8 (1882): 431–34; Frederik A. Fernald, "Hole or Holl?" *Popular Science Monthly* 42 (1892): 123.

9. Crowell, "Pacific Guano Company"; Wines, *Fertilizer*, 91; Elizabeth Spooner Fay, "Before You Were Born" (1930), *Book of Falmouth*, 491; MBL, *Annual Report*, 1888, pp. 18–19. On the development of Woods Hole as a resort, see Millard C. Faught, *Falmouth Massachusetts: Problems of a Resort Community* (New York: Columbia University Press, 1945), 5–37.

10. "A Big Firm Goes Under," *New York Times*, 9 February 1889. In "Summer Resort and Scientific Discipline," I erroneously dated the closing of the fertilizer factory to 1880 or thereabouts. On redevelopment efforts, see Candace Jenkins, "The Development of Falmouth as a Summer Resort 1850–1900," *Spritsail: A Journal of the History of Falmouth and Vicinity* 6 (1992): 24–29; Winslow Carleton, "Bankers' Row," *Woods Hole Reflections*, 140–47.

11. Quotation in Edward S. Morse, "Charles Otis Whitman," *National Academy of Sciences Biographical Memoirs* 7 (1912): 269–88 (quotation, 270); on Whitman, see also F. R. Lillie, "Charles Otis Whitman," *Journal of Morphology* 22 (1911): xv–lxxiii; C. B. Davenport, "The Personality, Heredity and Work of Charles Otis Whitman, 1843–1910," *American Naturalist* 51 (1917): 5–30; for a fuller account of Whitman's origins, see Philip J. Pauly, "From Adventism to Biology: The Development of Charles Otis Whitman," *Perspectives in Biology and Medicine* 37 (1994): 395–408.

12. For Whitman's religious anxieties, see Charles O. Whitman, "Death," *The Radical* 4 (1868): 372; Whitman, "Free Inquiry," *The Radical* 5 (1869): 394–96; Whitman, "The Idea of Immortality," *The Radical* (1871): 53–63; Whitman, "Progress Has No Goal," *The Radical* 9 (1871): 201–4. *The Radical* represented the Free Religious Association, a Universalist group.

13. Lillie, "Whitman," xxiii.

14. Charles O. Whitman, preface to *Biological Lectures Delivered at the Marine Biological Laboratory of Woods Holl in the Summer Session of 1890* (Boston: Ginn, 1891).

15. "Annual Circular for 1895," MBL, *Annual Report*, 1894–95, pp. 98–99.

16. Lillie, *MBL*, 176; Maienschein, *100 Years Exploring Life*, 30.

17. MBL, *Annual Report*, 1894–95.

18. Ibid., 14.

19. For an early discussion of life at the MBL, see J. S. Kingsley, "The Marine Biological Laboratory," *Popular Science Monthly* 42 (1892): 605–15; also E. G. Conklin, "Early Days at Woods Hole," *American Scientist* 56 (1968): 112–20; Conklin, "M.B.L. Stories," *American Scientist* 56 (1968): 121–28. On the dining club/"mess," see especially "Report of the Trustees," MBL, *Annual Report*, 1894–95, pp. 10–11.

20. Whitman, "Report of the Director," MBL, *Annual Report*, 1894–95, p. 49.

21. Hyatt, "Report of the Curator," *Proceedings of the Boston Society for Natural History* 23 (1887): 362; Lillie, *MBL*, 34–36, 204–6; Maienschein, *100 Years Exploring Life*, 19–26. Among the seven original trustees, four were the young MIT biologists W. T. Sedgwick and E. G. Gardiner, the Harvard botanist Farlow, and the struggling Harvard Medical School embryologist C. S. Minot.

22. Samuel H. Scudder et al., "A Statement Concerning the Marine Biological Laboratory at Wood's Holl, Massachusetts," *Science* 6 (1897): 529–34; S. F. Clarke, E. G. Gardiner, and J. P. MacMurrich, *A Reply to the Statement of the Former Trustees of the Marine Biological Laboratory* (Boston: Alfred Mudge and Sons, 1897); Whitman, "Report of the Director," MBL, *Annual Report*, 1894–95, p. 23; Lillie, *MBL*, 40–46, 169; Maienschein, *100 Years Exploring Life*, 67–69.

23. Jeffrey Werdinger, "Embryology at Woods Hole: The Emergence of a New American Biology" (Ph.D. diss., Indiana University, 1980); Lillie, *MBL*, 115–56; John M. Farley, *Gametes and Spores: Ideas about Sexual Reproduction, 1750–1914* (Baltimore: Johns Hopkins University Press, 1982), 235–51; Scott F. Gilbert, "The Embryological Origins of the Gene Theory," *Journal of the History of Biology* 11 (1978): 307–51; Philip J. Pauly, *Controlling Life: Jacques Loeb and the Engineering Ideal in Biology* (New York: Oxford University Press, 1987), 109.

24. See, e.g., John M. O'Donnell, *The Origins of Behaviorism: American Psychology, 1870–1920* (New York: New York University Press, 1985), 141–45.

25. Jane Maienschein, ed., *Defining Biology: Lectures from the 1890s* (Cambridge, MA: Harvard University Press, 1986), 21–26; Franklin P. Mall, "What Is Biology?" *Chautauquan* 18 (1894): 411–14; Herbert S. Jennings, *Behavior of the Lower Organisms* (New York: Columbia University Press, 1906), 338–50. While Europeans such as Hans Driesch found Jennings's work profound, his American reviewer, G. H. Parker, treated the book solely on the empirical level. See Parker, "The Behavior of the Lower Organisms," *Science* 26 (1907): 548–49.

26. Charles O. Whitman, "Specialization and Organization, Companion Principles of All Progress—The Most Important Need of American Biology," in *Biological Lectures Delivered . . . 1890*, 22–23.

27. Data compiled from MBL, *Annual Report*, 1888–1910; for a graph of attendance, see Pauly, "Summer Resort and Scientific Discipline," 133. Data on the Fish Commission Laboratory is only rarely available, probably because directors sought to obscure from congressional investigators the fact that private individuals were using the laboratory. U.S. Commission of Fish and Fisheries, *Report of the Commissioner*

for 1888, 513, reported eighteen workers; H. C. Bumpus, "The Work of the Biological Laboratory of the U.S. Fish Commission at Woods Hole," *Science* 8 (1898): 96, referred to twenty-four; F. B. Sumner, "The Summer's Work at the Woods Hole Laboratory of the Bureau of Fisheries," *Science* 19 (1904): 242, listed thirty. For the summers of 1896 and 1897 the laboratory was closed to investigators.

28. Information compiled from Lillie, *MBL*, 252–54; "Leading Men of Science in the United States in 1903," *American Men of Science*, 5th ed. (New York: Science Press, 1933), 1269–78; ASN, *Records*, 1930; "Proceedings of the American Society of Zoologists," *Anatomical Record* 41 (1929): 123–25.

29. On clubs, see E. Digby Baltzell, *Philadelphia Gentlemen* (Chicago: Quadrangle Books, 1971), 335–63; G. William Domhoff, *The Bohemian Grove and Other Retreats: A Study in Ruling-Class Cohesiveness* (New York: Harper and Row, 1974).

30. Edwin Linton, "Reminiscences of the Woods Hole Laboratory of the Bureau of Fisheries," *Science* 41 (1915): 752–53.

31. David A. Dary, *The Buffalo Book: The Full Saga of the American Animal* (Chicago: Swallow Press, 1974), 222–40, 279–87; Andrew C. Isenberg, "Indians, Whites, and the Buffalo: An Ecological History of the Great Plains, 1750–1900" (Ph.D. diss., Northwestern University, 1993).

32. A. W. Schorger, *The Passenger Pigeon: Its Natural History and Extinction* (1955; reprint, Norman: University of Oklahoma Press, 1973), 200; John F. Lacey, "Interstate Commerce in Game and Birds in Violation of State Law; Let Us Save the Birds," Address to the House of Representatives, 30 April 1900," reprinted in *Major John F. Lacey Memorial Volume*, ed. Iowa Park and Forestry Association (Ames: Iowa Park and Forestry Association 1915), 141.

33. Whitman to George E. Atkinson, n.d. (ca. 1904), quoted in W. B. Mershon, *The Passenger Pigeon* (New York: Outing Publishing Company, 1907), 199.

34. Lillie, "Whitman."

35. Whitman to W. B. Mershon, 30 May 1904, quoted in Mershon, *Passenger Pigeon*, 207.

36. Schorger, *Passenger Pigeon*, 28, 105.

37. Ibid., 223.

38. The zoo's director declared that Martha had been born in Cincinnati, but Schorger concluded (*Passenger Pigeon*, 30) that she had belonged to Whitman.

PART III
THE AGE OF BIOLOGY

1. Clarence King, "The Education of the Future," *Forum* 13 (1892): 20–33 (quotations, 20, 27). To my knowledge, this was the first use of the term "age of energy."

2. Ibid., 27.

3. Thurman Wilkins, *Clarence King: A Biography*, rev. ed. (Albuquerque: University of New Mexico Press, 1988), 362–64, 389–91.

4. The Bloomingdale trustees used the proceeds to build a new facility in suburban Westchester County. See Andrew S. Dolkart, *Morningside Heights: A History of Its Architecture and Development* (New York: Columbia University Press, 1998), 31–35, 103–55.

CHAPTER SEVEN
THE DEVELOPMENT OF HIGH SCHOOL BIOLOGY

1. E.g., Stephen Jay Gould, *Hen's Teeth and Horse's Toes* (New York: Norton, 1983), 280–90; Edward J. Larson, *Trial and Error: The American Controversy over Creation and Evolution* (New York: Oxford University Press, 1985), 8–24, 72–89. An earlier version of this chapter was published as "The Development of High School Biology: New York City, 1900–1925," *Isis* 82 (1991): 662–88, © 1991 by the History of Science Society. All rights reserved.

2. Writings on the history of high school biology include C. W. Finley, *Biology in Secondary Schools and the Training of Biology Teachers*, Teachers College Contributions to Education, #199 (New York: Teachers College, 1926); Otto B. Christy, *The Development of the Teaching of General Biology in the Secondary Schools*, Peabody Contributions to Education, #201, 1936; Sidney Rosen, "The Origins of High School Biology," *School Science and Mathematics* (hereafter *SSM*) 59 (1959): 473–89; Paul DeHart Hurd, *Biological Education in American Secondary Schools, 1890–1960* (Washington, DC: AIBS, 1961); William V. Mayer, "Biology Education in the United States during the Twentieth Century," *Quarterly Review of Biology* 61 (1986): 481–507; Eric W. Engles, "Biology Education in the Public High Schools of the United States from the Progressive Era to the Second World War: A Discursive History" (Ph.D. diss., University of California, Santa Cruz, 1991).

3. Finley, *Biology in Secondary Schools*, 1–15; Hurd, *Biological Education*, 10–13.

4. ASN, *Records* 1 (1884–95): 242–43; NEA, *Report of the Committee on Secondary School Studies* (Washington, DC: GPO, 1893; reprint, New York: Arno Press, 1969), 138–42, 158–61.

5. Charles B. Davenport, "Zoology as a Condition for Admission to College," University of the State of New York, High School Department, *Annual Report*, 1898, pp. 462–63; see also discussion by Edwin G. Conklin on pp. 465–68. See also Clifton F. Hodge, "Dynamic Biology and Its Relations to High School Courses," *Pedagogical Seminary* 11 (1904): 395–96.

6. John H. Wadland, *Ernest Thompson Seton: Man in Nature and the Progressive Era, 1880–1915* (New York: Arno Press, 1978), 369–79; on youth organizations, see David I. Macleod, *Building Character in the American Boy: The Boy Scouts, the YMCA, and Their Forerunners, 1870–1920* (Madison: University of Wisconsin Press, 1983), 130–45.

7. U.S. Department of Commerce, Bureau of the Census, *Historical Statistics of the United States* (Washington, DC: GPO, 1975), 1:16, 368–69.

8. Edward A. Krug, *The Shaping of the American High School*, 2 vols. (New York: Harper and Row, 1964), 1:169–89.

9. Statistics from New York City Department of Education, *Annual Report*, 1899, 1911. On education in New York City, see, e.g., Diane Ravitch, *The Great School Wars* (New York: Basic Books, 1974), 107–229.

10. John Tebbel, A History of Book Publishing in the United States, 4 vols. (New York: R. R. Bowker, 1975), 2:565–576.

11. Philip J. Pauly, "The Struggle for Ignorance about Alcohol: American Physiologists, Wilbur Olin Atwater, and the Woman's Christian Temperance Union," *Bulletin of the History of Medicine* 64 (1990): 375.

12. New York City Department of Education, *Course of Study* (1898, 1902); University of the State of New York, *Syllabus for Secondary Schools* (1905); "The Secondary School Course in Zoology," University of the State of New York, High School Department, *Annual Report*, 1899, pp. 528–48, 743–77, esp. James E. Peabody, "Minority Report," 775–77 and remarks by Henry R. Linville, 529.

13. Robert A. M. Stern, Gregory Gilmartin, and J. M. Massengale, *New York 1900: Metropolitan Architecture and Urbanism 1890–1915* (New York: Rizzoli, 1983), 80–83; Edward L. Bernays, *Biography of an Idea: Memoirs of Public Relations Counsel Edward L. Bernays* (New York: Simon and Schuster, 1965), 26–27; author's conversation with Bernays, 6 June 1990; *Clintonian* [student yearbook], 1906, p. 13. A collection of yearbooks can be found in the Alumni Office, DeWitt Clinton High School, Bronx, New York.

14. *Clinton Review* [*Clintonian*], 1901, p. 3; Bernays, *Biography*, 24; *Clintonian*, 1906, p. 13, noted that over 80 percent of graduates continued at universities, primarily Columbia, Cornell, and New York University.

15. Dennis East II, "Linville, Henry Richardson," *Biographical Dictionary of American Educators*, 3 vols., ed. John F. Ohles (Westport, CT: Greenwood Press, 1978), 2:803–4; William E. Eaton, *The American Federation of Teachers, 1916–1961: A History of the Movement* (Carbondale: Southern Illinois University Press, 1975), 12–13, 96–97.

16. Jo Ann Kauffman, "Gruenberg, Benjamin Charles," *Biographical Dictionary of American Educators*, 2:561–62; *New York Times*, 2 July 1965, p. 29; *Clintonian*, 1909, p. 43; Benjamin C. Gruenberg, "Efficiency versus Democracy," *American Teacher* 1 (1912): 79–81; Gruenberg, "Some Economic Obstacles to Educational Progress," *American Teacher* 1 (1912): 89–92.

17. J. M. Cattell, ed., *American Men of Science* 1st–5th eds. (New York: Science Press, 1906–33); MBL, *Annual Report*, 1895–1902; for Hunter's general views, see his *Science Teaching at the Junior and Senior High School Levels* (New York: American Book Co., 1934), 3–12.

18. Francis Lloyd and Maurice A. Bigelow, *The Teaching of Biology in Secondary Schools* (New York: Longmans, Green, 1904). This was in fact two essentially independent books, on botany (by Lloyd) and zoology (by Bigelow), bound together.

19. E.g., Benjamin C. Gruenberg, "Teaching Biology in the Schools," *Atlantic Monthly* 103 (1909): 796–800; Henry R. Linville et al., "The Practical Use of Biology," *SSM* 9 (1909): 121–30; George W. Hunter, "The Methods, Content and Purpose of Biological Science in the Secondary Schools of the United States," *SSM* 10 (1910): 1–10, 103–11; James E. Peabody et al., "Revised Report of the Biology Committee of the National Education Association . . .," *SSM* 16 (1916): 501–17.

20. H. R. Linville and H. A. Kelly, *A Text-book in General Zoology* (New York: Ginn, 1906).

21. Some of Hunter's texts, many of which appeared in multiple editions, were Hunter and M. C. Valentine, *Laboratory Manual of Biology* (New York: Henry Holt, 1903); Hunter, *Elements of Biology* (New York: American Book Co., 1907); Hunter, *Essentials of Biology* (New York: American Book Co., 1911); Hunter, *A Civic Biology: Presented in Problems* (New York: American Book Co., 1914); Hunter, *Problems in Biology* (New York: American Book Co., 1931); Hunter, H. E. Walter, and G. W. Hunter III, *Biology: The Story of Living Things* (New York: American Book Co., 1937); Hunter, *Life Science: A Social Biology* (New York: American Book Co., 1941); Hunter and F. R. Hunter, *Biology in Our Lives* (New York: American Book Co., 1949).

22. Benjamin C. Gruenberg, *Elementary Biology* (New York: Ginn, 1919); H. R. Linville, *The Biology of Man and Other Organisms* (New York: Harcourt Brace, 1923); another Clinton teacher, R. W. Sharpe, prepared *A Laboratory Manual for the Solution of Problems in Biology* (New York: American Book Co., 1911).

23. J. E. Peabody and A. E. Hunt, *Elementary Biology: Plant, Animal, Human*, 3 vols. bound as one (New York: Macmillan, 1913); Peabody and Hunt, *Biology and Human Welfare* (New York: Macmillan, 1924); M. A. Bigelow and A. N. Bigelow, *Introduction to Biology* (New York: Macmillan, 1913); W. M. Smallwood, I. L. Reveley, and G. A. Bailey, *Practical Biology* (Boston: Allyn and Bacon, 1916); Smallwood, Reveley, and Bailey, *New Biology* (Boston: Allyn and Bacon, 1924).

24. Christy, *Development*, 274–75, 321. Between 1910 and 1928, in a student population that continued to double each decade, the percentage enrolled annually in botany dropped from 16.3 to 1.6; zoology from 7.8 to 0.8; and physiology from 15.8 to 2.7. Biology was not included in national statistics until 1915; by 1928, 13.3 percent of high school students were enrolled in biology each year. This was nearly double the numbers in either chemistry (7.3) or physics (7.1).

25. C. N. W., "A Letter Concerning the Biology Symposium," *SSM* 8 (1908): 699. Although New Yorkers led the movement, others, especially in Chicago, shared many of their aims. Midwestern biology educators included Oscar Riddle, Elliott Downing, T. W. Galloway, Otis W. Caldwell (after 1917 at Teachers College), and Herbert E. Walter (after 1906 at Brown University). Clifton F. Hodge, professor of biology at Clark University, was also prominent.

26. On evolutionism, see esp. Peter J. Bowler, *The Non-Darwinian Revolution* (Baltimore: Johns Hopkins University Press, 1988); for its manifestations among educational theorists, see Merle Curti, *The Social Ideas of American Educators* (New

York: Charles Scribner's Sons, 1935), 396–541; Richard Hofstadter, *Anti-Intellectu-alism in American Life* (New York: Knopf, 1963), 362–90; Lawrence A. Cremin, *The Transformation of the School* (New York: Basic Books, 1968), 90–126.

27. Nicholas M. Butler, *The Meaning of Education* (New York: Macmillan, 1898), 31–32; Charles W. Eliot, "Educational Changes and Tendencies," *Journal of Education* 34 (1891): 393.

28. H. R. Linville, "Framing a Course in Biology for Untrained Minds: A Discussion of Principles," University of the State of New York, High School Department, *Annual Report*, 1900, pp. 940–41; Linville, "Biology as Method," 268–69; Linville et al., "Practical Use of Biology," 129–30; T. W. Galloway et al., "A Consideration of the Principles That Should Determine the Courses in Biology in the Secondary Schools," *SSM* 9 (1909): 241–47. Advocates of nature study and woodcraft, by contrast, sought to maintain boys' identification with the natural world explicitly in order to retard maturation; their hope was that slowing down development would lessen its difficulty.

29. C. W. Dodge, discussion following C. B. Davenport, "Zoology as a Condition," 473; Lloyd and Bigelow, *Teaching of Biology*, 94–96; Peter Schmidt, *Back to Nature: The Arcadian Myth in Urban America* (New York: Oxford University Press, 1969), 86–91; quotation, G. W. Hunter, "Pedagogical Experiments from the Biological Laboratory of the DeWitt Clinton High School," *SSM* 18 (1918): 730.

30. Jane Maienschein, "Physiology, Biology, and the Advent of Physiological Morphology," in *Physiology in the American Context, 1850–1940*, ed. Gerald L. Geison (Bethesda, MD: American Physiological Society, 1987), 177–94; Philip J. Pauly, "General Physiology and the Discipline of Physiology," in *Physiology in the American Context, 1850–1940*, ed. Gerald L. Geison (Bethesda, MD: American Physiological Society, 1987), 195–208.

31. University of the State of New York, *Syllabus for Secondary Schools, 1910*, bulletin #607 (1916), 140; Hodge, "Dynamic Biology," 382. See also Davenport, "Zoology as a Condition," 465; E. G. Conklin, "Advances in Methods of Teaching: Zoology," *Science* 9 (1899): 83; Oscar Riddle, "What and How Much Can Be Done in Ecological and Physiological Zoology in Secondary Schools?" *SSM* 6 (1906): 250–51; Hunter, *Essentials of Biology*, 5; Peabody and Hunt, *Elementary Biology*, vii–viii; "Reorganization of Science in Secondary Schools," U.S. Bureau of Education, *Bulletin*, 1920, #26, 31.

32. Peabody and Hunt, *Elementary Biology*, x; Bigelow and Bigelow, *Introduction to Biology*; Hunter, *Civic Biology*; Gruenberg, *Elementary Biology*.

33. G. W. Hunter, *Laboratory Problems in Civic Biology* (New York: American Book Company, 1916), 32; Linville, "Biology as Method," 269.

34. J. K. van Denburg, "The Subject-Matter or the Year of High School Biology Should Be Changed," High School Teachers Association of New York City, *Yearbook*, 1906–7, p. 43; Gruenberg, "The Practical, Pedagogical and Scientific Bases for the Study of Biology," *SSM* 8 (1908): 540–41; Gruenberg, *Elementary Biology*, iv;

Linville, *Biology of Man*, 4–5; T. W. Galloway, "The Function of the Biological Sciences in Education," *SSM* 8 (1908): 546.

35. Linville, *Biology and Man*, 5; Hunter, *Civic Biology*, 7; G. C. Wood, "Essentials of a Practical Course in Biology," *SSM* 14 (1914): 8–9; Linville, "Framing a Course," 941.

36. Lloyd and Bigelow, *Teaching of Biology*, 64; Linville, "Biology as Method," 268, 266, also claimed that biology was a better introduction to science than physics or chemistry because of its inductive nature; although the latter disciplines dealt with simpler entities, their authoritarian assertions about forces and atoms reinforced dogmatic credulity.

37. F. S. Lee, "Teaching Physiology in the Schools," University of the State of New York, High School Department, *Annual Report*, 1900, p. 821; B. C. Gruenberg and F. M. Wheat, *Student's Manual of Exercises in Elementary Biology* (Boston: Ginn, 1921), iii.

38. Hunter, "Methods," 1; Linville et al., "Practical Use of Biology."

39. Wood, "Essentials of a Practical Course," 6; Peabody and Hunt, *Elementary Biology*, ix; Hunter, *Laboratory Problems*, 23.

40. J. E. Peabody, "The Relation of Biology to Human Welfare," *SSM*, 1914, 14:380; Peabody and Hunt, *Elementary Biology*, 44–81; Hunter, *Laboratory Problems*, 25–28, 194–220, 244–46, 262–72; on vitamins, see Gruenberg, *Elementary Biology*, 89–142; Linville, *Biology of Man*, 348–50.

41. Hunter, *Civic Biology*, 69–270, 374; Peabody and Hunt, *Elementary Biology*, 129–38; Gruenberg, *Elementary Biology*, 154–73.

42. Peabody and Hunt, *Elementary Biology*, 23–43; Hunter, *Civic Biology*, 376–96; Gruenberg, *Elementary Biology*, 386–411; also Clifton F. Hodge and Jean Dawson, *Civic Biology: A Textbook of Problems, Local and National, That Can Be Solved Only by Civic Cooperation* (Boston: Ginn, 1918), 231–70.

43. Lloyd and Bigelow, *Teaching of Biology*, 79, 285. As an example of this approach, see the botanical argument against masturbation in Linville et al., "Practical Use of Biology," 126–27.

44. M. A. Bigelow, *Teachers' Manual of Biology* (New York: Macmillan, 1912), 75.

45. M. A. Bigelow, "Biology in Relation to Sex Instruction in Schools and Colleges," *Social Diseases* 2 (1911): 10–15, and succeeding comments by Gruenberg, 23–26, and Peabody, 26–31. J. E. Peabody, "What a New York City High School Has Done," *Journal of the Society for Sanitary and Moral Prophylaxis* 6 (1915): 45–50; "The Problem of Sex Instruction as Viewed by Boards of Education" (including comments by G. W. Hunter and George Donaldson), *Journal of the Society for Sanitary and Moral Prophylaxis* 6 (1915): 149–59; W. H. Eddy, "An Experiment in Teaching Sex Hygiene," *Journal of Educational Psychology* 2 (1911): 451–58.

46. B. C. Gruenberg, "Sex Education in the Secondary Schools," *National Education Association Journal of Addresses and Proceedings* 59 (1921): 675–76; Gruenberg,

The Teacher and Sex Hygiene (American Social Hygiene Association Publication #426, 1923); Sidonie M. Gruenberg, *Sons and Daughters* (New York: Henry Holt, 1916).

47. See David W. Noble, *The Paradox of Progressive Thought* (Minneapolis: University of Minnesota Press, 1958); Hofstadter, *Anti-Intellectualism in American Life*, 359–90.

48. Bernays, *Biography*, 25; H. J. Van Cleave, "The Field Excursion in High School Biological Courses," *SSM* 19 (1919): 7–10.

49. G. W. Hunter, *Laboratory Problems*, 32; Henry Fairfield Osborn, "The Museum of the Future," *American Museum Journal* 11 (1913): 223–25; J. E. Peabody, "How One Crowded High School Uses the Museum," *American Museum Journal* 11 (1913): 240–41; Hunter, "The American Museum's Reptile Groups in Relation to High School Biology," *American Museum Journal* 15 (1915): 405–7. On dioramas, see Donna Haraway, *Primate Visions* (New York: Routledge, 1989), 26–58.

50. American Humane Association, *Report of the American Humane Association on Vivisection and Dissection in Schools* (Chicago: N.p., 1895); Susan E. Lederer, *Subjected to Science: Human Experimentation in America before the Second World War* (Baltimore: Johns Hopkins University Press, 1995).

51. Lee, "Teaching Physiology in the Schools," 819–20; C. W. Dodge, comments following Davenport, "Zoology as a Condition," 472; J. E. Peabody, "Laboratory Work in High-School Physiology," *NEA Journal*, 42 (1903): 867–71; Lloyd and Bigelow, *Teaching of Biology*, 273; Riddle, "How Much Can Be Attempted," 215; Clifford Crosby, "Physiology, How and How Much?" *SSM* 7 (1907): 733–44 (quotation, 735). A number of states, including Massachusetts, Illinois, and Washington, outlawed vivisection in public schools: see N. E. Adams, "The Legal Restrictions Concerning the Teaching of Biology" (M.Ed. thesis, Indiana University, 1930), 13–14.

52. *Humane Association on Vivisection*, 3; "Humane Education and Nature Studies in Public Schools," in American Humane Association, *Twenty-third Annual Report*, 1899, pp. 65–66; Lloyd and Bigelow, *Teaching of Biology*, 359; Lee, "Teaching Physiology," 821; discussion in "The Secondary Course in Zoology," 537, 541–42, 546; Hunter, *Laboratory Problems in Civic Biology*, 161–62; Gruenberg and Wheat, *Student's Manual*. On promotional efforts of the General Biological Supply House, see "Ten Years of Progress: Turtox Service Did It," *Turtox News* 3 (1925): 14–15.

53. Naomi Rogers, "Germs with Legs: Flies, Disease, and the New Public Health," *Bulletin of the History of Medicine* 63 (1989): 610. On the "typhoid fly," see, e.g., Hunter, *Laboratory Problems*, 153; more broadly, see Hunter, *Civic Biology*, 219; Linville, *Biology of Man*, 39. Hodge and Dawson's *Civic Biology* represented the extreme position that generated awareness of a problem: a reviewer complained that "practically the entire book is given over to the harmful forces at work. There is comparatively little of the beautiful and really wonderful operations of nature" (*SSM* 19 [1919]: 286).

54. Peabody, "Relation of Biology to Human Welfare," 380–81; Sharp, *Laboratory Manual*, 226–33.

55. More generally, see Wallace H. Maw, "Fifty Years of Sex Education in the Public Schools of the United States (1900–1950): A History of Ideas" (Ph.D. diss., University of Cincinnati, 1953); Brian Strong, "Ideas of the Early Sex Education Movement in America, 1890–1920," *History of Education Quarterly* 12 (1972): 129–61; Jeffrey P. Moran, " 'A Wholesome Fear': The Evolution of Sex Education in the United States, 1905–1995 (Public Education)" (Ph.D. diss., Harvard University, 1996).

56. M. A. Bigelow, "Selection and Training of Teachers for Sex Instruction," *Journal of the Society for Sanitary and Moral Prophylaxis* 6 (1915): 134–35; Bigelow, *Sex Education* (New York: Macmillan, 1916), 115–20, 146–55; Bigelow, *Manual*, 75–77; Peabody et al., "Revised Report of the Biology Committee," 515; Eddy, "Experiment in Teaching Sex Hygiene," 451–52, 457–58.

57. Bigelow, *Sex Education*, passim; Gruenberg, "[Comment]," *Social Diseases* 2 (1911): 24.

58. Edgar F. Van Buskirk, "How Can Sex Education Be Made a Part of Biology?" *SSM* 19 (1919): 336; American Federation for Sex Hygiene, *Report of the Special Committee on the Matter and Methods of Sex Education* (New York: The Federation, 1913), 3; B. C. Gruenberg, *Manual of Suggestions for Teachers to Accompany "Elementary Biology"* (Boston: Ginn, 1919), 74; W. H. Eddy, *Reproduction and Sex Hygiene: A Text and a Method* (Boston: Ginn, 1916), 57.

59. "Sex Hygiene," *Journal of Education* 79 (1914): 268; Jeffrey P. Moran, " 'Modernism Gone Mad': Sex Education Comes to Chicago," *Journal of American History* 83 (1996): 481–513.

60. Maw, "Fifty Years of Sex Education," 88; Eddy, "Experiment in Teaching Sex Hygiene," 451–52; Peabody, "Relation of Biology to Human Welfare," 382; Peabody, "What a New York City High School Has Done," 45–50; Ira S. Wile, "The Problem of Sex Instruction as Viewed by Boards of Education," *Journal of the Society for Sanitary and Moral Prophylaxis* 6 (1915): 141–48; T. W. Galloway, "Instruction of Young People in Respect to Sex," *SSM* 14 (1914): 676; Newell W. Edson, "Status of Sex Education in High Schools," U.S. Bureau of Education, *Bulletin*, 1922, #14.

61. Executive Committee Minutes, 30 March, 11 May, 8 June 1925, ACLU Papers, vol. 281, Seeley G. Mudd Manuscripts Library, Princeton University.

62. The 1910 *New York Syllabus for Secondary Schools* included, in addition to biology, electives in advanced botany and advanced zoology. The last of these did include (p. 141) the ambiguous directive that "the prominent evidences of relationship, suggesting evolution, within such groups as the decapods, the insects, and the vertebrates, should be demonstrated," as well as "a few facts" indicating adaptation, variation, and the struggle for existence; "but the factors of evolution and the discussion of its theories should not be attempted." The College Entrance Examination Board used similar language for an "optional" unit in biological theory when it first instituted an examination in biology: see College Entrance Examination Board, Doc. No. 72, 1915, p. 60.

Hunter's four pages (out of 432) were considerable compared to both Peabody and Hunt's *Elementary Biology* and the Bigelows' *Introduction to Biology*, the latter of which noted (p. v) that it gave "limited attention to facts and ideas whose applications are aesthetic and intellectual (i.e., evolution)." Gruenberg's *Elementary Biology* devoted 10 pages of 500 to evolution (but see below); Linville's *Biology of Man and Other Organisms* gave the subject about 25 pages out of 470. In contrast to these figures, Linville and Kelly's 1906 *General Zoology* included more than 50 pages, out of 459, of explicit discussion of evolution, including an outline of the various mechanisms; 76 pages (of 329) in D. S. Jordan and V. L. Kellogg's *Animal Life* (New York: Appleton, 1900) dealt with struggle, phyletic adaptation, degeneration, and mimicry; and J. Y. Bergen's 395-page *Foundations of Botany* (Boston: Ginn, 1901) included 19 pages on evolutionary history and the struggle for existence. Gerald Skoog, "The Topic of Evolution in Secondary School Textbooks: 1900–1977," *Science Education* 63 (1979): 621–40, is a frequently cited quantitative analysis of these issues; but since his selection of books before 1920 was arbitrary it is of little value in this context. Larson, *Trial and Error*, 15–24, notes that biology texts supported evolution rather than creation, but his stress on this point leads him to slight the changing importance of the subject prior to 1925. T. T. Martin's *Hell and the High Schools: Christ or Evolution, Which?* (Kansas City, MO: Western Baptist Publishing Co., 1923), 55–69, focused on geography and botany books, and included only one brief reference to a biology text, in its polemical review of the extent of evolutionary teaching in schools.

63. Lloyd and Bigelow, *Teaching of Biology*, 286–89; Hargitt, "Place and Function of Biology," 138; Frank Smith, "The Chief Aims in Zoology Work in High Schools," *SSM* 5 (1905): 344; "Report of Committee on Zoology and Botany," *SSM* 5 (1905): 740. B. C. Gruenberg, "Scientific Education as a Defense against Propaganda and Dogma," *NEA Journal* 63 (1925): 598–607 is the exception, but this address was given before the Scopes trial and was general in content. Linville's views were not recorded in the ACLU minutes. He was probably among the majority of the executive committee that resisted the shift in strategy from a quiet legal challenge to public "circus" implicit in the decision to hire Clarence Darrow as a counterweight to William Jennings Bryan: see John T. Scopes and James Presley, *Center of the Storm: Memoirs of John T. Scopes* (New York: Holt, 1967), 71–73; Edward J. Larson, *Summer for the Gods: The Scopes Trial and America's Continuing Debate over Science and Religion* (New York: Basic Books, 1997).

64. See Gruenberg, *Manual for Teachers*, 90–91; Gruenberg, *Elementary Biology*, iv. Eugenics also played a small part in the thinking of New York biology educators. The subject entered the textbooks about 1914, but in contrast to college texts, treatment was perfunctory, narrowly focused on medically specific defects, and subordinated to—rather than distinguished from—issues of hygiene. Both Gruenberg's and Peabody and Hunt's *Elementary Biology* essentially ignored eugenics. Linville, *Biology of Man*, 168–78, and Hunter, *Civic Biology*, 261–65, embedded the subject within exhortations about personal hygiene and environmental improvement. For a

different perspective, see Steven Selden, *Inheriting Shame: The Story of Eugenics and Racism in America* (New York: Teachers College Press, 1999).

65. Bernays, *Biography*, esp. 287–95, 590–600, 775–804; Bernays, ed., *The Engineering of Consent* (Norman: University of Oklahoma Press, 1955), 5–7; Larry Tye, *The Father of Spin: Edward L. Bernays & the Birth of Public Relations* (New York: Crown Publishers, 1998). The inventors of both mutations (Hermann Muller) and the birth control pill (Gregory Pincus) were early graduates of Morris High School in the Bronx.

66. Emphasis on natural history and dissection was strongly encouraged after 1920 by the General Biological Supply House: see esp. M. M. Wells, "Laboratory Work in High School Biology," *Turtox News* 6 (1928): 1–4.

67. Alfred C. Kinsey, *Methods in Biology* (Philadelphia: Lippincott, 1937), 5.

CHAPTER EIGHT
BIG QUESTIONS

1. Watson Davis to Executive Committee, 25 May 1925; Executive Committee Minutes, 2 June 1925; both in box 1, folder "Executive Committee, 1925," Science Service Papers, RU 7091, Smithsonian Institution Archives.

2. Watson Davis to G. W. Rappleyea, telegram, 7 July 1925; telegrams in box 32, folder "Scopes Trial, 1925, Evolution Witnesses for," Science Service Papers; Watson Davis, "Davis Goes Back Three Centuries: Sees Galileo on Trial before Judge Raulston," *Chattanooga Daily Times*, 14 July 1925; Davis, "Great Picture of Man's Source. Watson Davis in Condensed Version of Day in Court. Sees Lesson in Testimony of Zoology Expert Metcalf, First Scientific Witness," *Chattanooga Daily Times*, 16 July 1925.

3. Gerald Skoog, "The Topic of Evolution in Secondary School Textbooks: 1900–1977," *Science Education* 63 (1979): 621–40; Edward J. Larson, *Summer for the Gods: The Scopes Trial and America's Continuing Debate over Science and Religion* (New York: Basic Books, 1997).

4. Benjamin C. Gruenberg to Maynard Shipley, 16 August 1924, box 10, William E. Ritter Papers, Bancroft Library, University of California, Berkeley (hereafter Ritter-UC).

5. Miscellaneous manuscripts and publications were gathered together in Charles Otis Whitman, *Posthumous Works of Charles Otis Whitman*, 3 vols., ed. Oscar Riddle (Washington, DC: Carnegie Institution of Washington, 1919).

6. Frederic T. Lewis, "Charles Sedgwick Minot," *Anatomical Record* 10 (1916): 133–64.

7. William K. Brooks, *Foundations of Zoology* (New York: Macmillan, 1899); Keith R. Benson, "William Keith Brooks (1845–1908): A Case Study in Morphology and the Development of American Biology" (Ph.D. diss., Oregon State University, 1979).

8. Carl Snyder, "Bordering the Mysteries of Life and Mind," *McClure's Magazine* 19 (2 March 1902): 386–96; Jacques Loeb, "The Limitations of Biological Research," *University of California Publications in Physiology* 1 (1903): 33–37; C. B. Davenport, "Cooperation in Science," *Science* 25 (1907): 361–66; Frank R. Lillie, William Trelease, Henry H. Donaldson, William H. Howell, and James R. Angell, "Cooperation in Biological Research" (American Society of Naturalists Symposium) *Science* 27 (1908): 369–86; D. P. Penhallow, "The Functions and Organization of the American Society of Naturalists," *Science* 29 (1909): 679–90; Toby A. Appel, "Organizing Biology: The American Society of Naturalists and Its 'Affiliated Societies,' 1883–1923," in *The American Development of Biology*, ed. Ronald Rainger, Keith R. Benson, and Jane Maienschein (Philadelphia: University of Pennsylvania Press, 1988), 104.

9. Theodore Roosevelt, T. S. Van Dyke, D. G. Elliot, and A. J. Stone, *The Deer Family* (New York: Macmillan, 1902); Roosevelt, *Biological Analogies in History* (Oxford: Clarendon, 1910); Edward B. Clark, "Roosevelt on the Nature Fakirs," *Everybody's Magazine* 16 (1907): 770–76; Ralph H. Lutts, *The Nature Fakers: Wildlife, Science, and Sentiment* (Golden, CO: Fulcrum Publishing, 1990); Paul R. Cutright, *Theodore Roosevelt, the Naturalist* (New York: Harper, 1956).

10. Daniel J. Wilson, *Arthur O. Lovejoy and the Quest for Intelligibility* (Chapel Hill: University of North Carolina Press, 1980), 36; see, e.g., David Starr Jordan, *The Call of the Nation* (Boston: Beacon Press, 1910).

11. Mary P. Winsor, *Reading the Shape of Nature: Comparative Zoology at the Agassiz Museum* (Chicago: University of Chicago Press, 1991), 187–94; Theodore Roosevelt, "Twisted Eugenics," *The Outlook* 105 (3 January 1914): 30–34.

12. Pauly, *Controlling Life*, 140; A. E. Hamilton, "Eugenics," *Pedagogical Seminary* 21 (1914): 35; Julie Reuben, *The Making of the Modern University* (Chicago: University of Chicago Press, 1996), 163–73.

13. Henri Bergson, *Creative Evolution* (New York: Holt, 1911); *New York Times*, 3, 4, 5, 7, 11, and 20 February 1913; Edwin E. Slosson, *Major Prophets of To-day* (Boston: Little Brown, 1914). The *Reader's Guide to Periodical Literature* listed seventy-five references to Bergson between 1910 and 1914 (three times as many as, for example, either Sarah Bernhardt or Henry James).

14. Pauly, *Controlling Life*, 140; William E. Ritter, "The Controversy between Materialism and Vitalism: Can It Be Ended?" *Science* 33 (1911): 437–41; Arthur O. Lovejoy, "The Meaning of Vitalism," *Science* 33 (1911): 610–11; Herbert S. Jennings, "Vitalism and Experimental Investigation," *Science* 33 (1911): 927–32.

15. Herbert S. Jennings, "Heredity and Personality," *Science* 34 (1911): 902–10. Cf. the preceding years' addresses: T. H. Morgan, "Chance or Purpose in the Origin and Evolution of Adaptation," *Science* 31 (1910): 201–10; Daniel T. MacDougal, "Organic Response," *Science* 33 (1911): 94–101.

16. Edwin G. Conklin, "Heredity and Responsibility," *Science* 37 (1913): 46–54.

17. Edwin G. Conklin, *Heredity and Environment in the Development of Men* (Princeton: Princeton University Press, 1915); Fanny H. Clark and W. Evans Clark,

introduction to George Howard Parker, *Biology and Social Problems* (The William Brewster Clark Memorial Lectures, 1914) (Boston: Houghton Mifflin, 1914), ix–x. See also Henry E. Crampton, *The Doctrine of Evolution* (New York: Columbia University Press, 1911).

18. David Starr Jordan and Harvey Ernest Jordan, *War's Aftermath: A Preliminary Study of the Eugenics of War as Illustrated by the Civil War of the United States and the Late Wars in the Balkans* (Boston: Houghton Mifflin, 1914); Raymond Pearl, "On the Effect of Continued Administration of Certain Poisons to the Domestic Fowl, with Special Reference to the Progeny," *Proceedings of the American Philosophical Society* 55 (1916): 258; William E. Ritter, *War, Science and Civilization* (Boston: Richard G. Badger, 1915).

19. Vernon Kellogg, *Headquarters Nights* (Boston: Atlantic Monthly Press, 1917), 22; Gregg Mitman, *The State of Nature: Ecology, Community, and American Social Thought, 1900–1950* (Chicago: University of Chicago Press, 1992), 58–71.

20. Edwin Grant Conklin, *The Direction of Human Evolution*, 2nd ed. (New York: Charles Scribner's Sons, 1922), xvi, 15–19, 53, 74–75; Herbert S. Jennings, *Prometheus; or, Biology and the Advancement of Man* (New York: E. P. Dutton, 1925); Jennings, *The Biological Basis of Human Nature* (New York: W. W. Norton, 1930); Vernon L. Kellogg, *Mind and Heredity* (Princeton: Princeton University Press, 1923); George H. Parker, "Some Implications of the Evolutionary Hypothesis," *Philosophical Review* 33 (1924): 593–603; Ralph S. Lillie, *Protoplasmic Action and Nervous Action* (Chicago: University of Chicago Press, 1923); Charles M. Child, *Physiological Foundations of Behavior* (New York: Henry Holt, 1924); Charles J. Herrick, *Fatalism or Freedom: A Biologist's Answer* (New York: W. W. Norton, 1926); Winterton C. Curtis, *Science and Human Affairs from the Viewpoint of Biology* (New York: Harcourt Brace, 1922).

21. For an insightful overview, see Ruth P. Pearce, "American Biology in a Mechanistic World: The Search for a Progressive Synthesis" (Ph.D. diss., University of Manitoba, 1986); also Sharon E. Kingsland, "Toward a Natural History of the Human Psyche," in *The Expansion of American Biology*, ed. Keith R. Benson, Jane Maienschein, and Ronald Ranger (New Brunswick, NJ: Rutgers University Press, 1991), 195–230; Kathy Jane Cooke, "A Gospel of Social Evolution: Religion, Biology, and Education in the Thought of Edwin Grant Conklin" (Ph.D. diss., University of Chicago, 1994); Nadine Weidman, "Psychobiology, Progressivism, and the Anti-Progressive Tradition," *Journal of the History of Biology* 29 (1996): 271–88; Mark A. Largent, " 'These Are Times of Scientific Ideals': Vernon Lyman Kellogg and Scientific Activism" (Ph.D. thesis, University of Minnesota, 1999).

22. William E. Ritter to Nelson Ritter, 1 April [1878]; 12 August [1878?], 16 October and 4 December 1881, all in box 1, folder 32, Ritter Family Papers, Scripps Institution of Oceanography Archives, La Jolla, California (hereafter Ritter-SIO); William E. Ritter, *Charles Darwin and the Golden Rule* (Washington, DC: Science Service, 1954), 4.

23. William E. Ritter, "On the Eyes, the Integumentary Sense Papillae, and the Integument of the San Diego Blind Fish (*Typhlogobius Californiensis*, Steindachner)," *Bulletin of the Museum of Comparative Zoology* 24 (1893): 41–98; [William E. Ritter], "Biological Science in the University," January 1889, carton 9, Ritter-UC (quotations 1, 11).

24. William E. Ritter, "The Marine Biological Station of San Diego: Its History, Present Conditions, Achievements, and Aims," *University of California Publications in Zoology* 9 (1912): 137–248 (quotation, 148).

25. Helen Raitt and Beatrice Moulton, *Scripps Institution of Oceanography: First Fifty Years* (Ward Ritchie Press, 1967), 6–22; Ritter, "Marine Biological Station of San Diego," 151–58.

26. Ritter to Wheeler, 12 December 1900, box 1, Ritter-SIO.

27. William E. Ritter, "Preliminary Report on the Marine Biological Survey Work Carried on by the Zoological Department of the University of California at San Diego," *Science* 18 (1903): 361; "Report of Professor Ritter to the Marine Biological Association of San Diego," [1903?], box 3, folder 69, Ritter-SIO.

28. William E. Ritter, "A General Statement of the Ideas and the Present Aims and Status of the Marine Biological Association of San Diego," *University of California Publications in Zoology* 2 (1905): ii, xvi; also Ritter, "Organization in Scientific Research," *Popular Science Monthly* 67 (1905): 50.

29. Ritter, "General Statement," 155–58. In 1903 the population of Los Angeles exceeded 110,000, while that of San Diego was only 17,000. On the early Scripps Institution, see Raitt and Moulton, *Scripps Institution*; Eric L. Mills, *The Scripps Institution: Origin of a Habitat for Ocean Science* (La Jolla, CA: Scripps Institution of Oceanography, University of California, San Diego, 1993).

30. Ritter, "Marine Biological Station of San Diego," 163–66.

31. On Scripps, see Vance H. Trimble, *The Astonishing Mr. Scripps: The Turbulent Life of America's Penny Press Lord* (Ames: Iowa State University Press, 1992); Gerald J. Baldasty, *E. W. Scripps and the Business of Newspapers* (Urbana: University of Illinois Press, 1999); also Charles O. Preece, *Edward Willis and Ellen Browing Scripps: An Unmatched Pair* (Chelsea, MI: Bookcrafters, 1990); Oliver Knight, ed., *I Protest: Selected Disquisitions of E. W. Scripps* (Madison: University of Wisconsin Press, 1966).

32. Ritter, "A Business Man's Appraisement of Biology," *Science* 44 (1916): 820.

33. Quotation in Scripps to Ritter, 31 May 1914, box 19, Ritter-UC; Mary Bennett Ritter, *More Than Gold in California, 1849–1933* (Berkeley: Professional Press, 1933), 306; Knight, *I Protest*, 726; Ritter, "Marine Biological Station of San Diego," 160–62.

34. E. W. Scripps, "What Is the Use of Biology," typescript, 7 March 1909, ser. 4, box 1, vol. 2, Scripps Family Papers, Ohio University, Athens, Ohio (hereafter Scripps Family Papers), 54–64.

35. Scripps to Ritter, 2 and 31 May 1914 (the latter with enclosure, Scripps to George B. Foster, 31 May 1914), box 19, Ritter-UC; Scripps, 1909, quoted in Knight, *I Protest*, 727.

36. Ritter, "A General Statement," vii–ix, xiv–xvii.

37. Mills, *Scripps Institution*, 24–26.

38. "To the Members of the Board of Directors, Marine Biological Station of San Diego," 15 August 1908, Scripps Family Papers.

39. Ritter, "Marine Biological Station of San Diego," 140.

40. Ibid., 225.

41. Ibid., 228, 225.

42. Ibid., 228–31.

43. See, e.g., Ritter, "The Workableness of Natural Religion," address to San Diego Unitarian Church, 10 November 1920, carton 7, Ritter-UC.

44. William E. Ritter, *The Unity of the Organism: The Organismal Conception of Life*, 2 vols. (Boston: Richard G. Badger, 1919); W.[illiam] E. R[itter], "A Rough-draft Report of a Conference on the Work of the Scripps Institution," typescript, 12 pp. (ca. 1917), box 2, Ritter-UC.

45. Quoted in Mills, *Scripps Institution*, 21.

46. William E. Ritter, *War, Science and Civilization* (Boston: Richard G. Badger, 1915), 124; Ritter, "The Place of Description, Definition and Classification in Philosophical Biology" (1915), in Ritter, *The Higher Usefulness of Science* (Boston: Richard G. Badger, 1918); Ritter, "Philosophical Brutism from the Standpoint of Zoology," (1918), carton 5, Ritter-UC; Ritter to Woodrow Wilson, telegram, 25 January 1917, and Ritter to Franklin K. Lane, 3 February 1917, both in box 1, Ritter-UC; Ritter, *Unity of the Organism*, 2:357.

47. Ritter, *Unity of the Organism*, 1:xix.

48. William E. Ritter, "The Philosophy of E. W. Scripps," 1943–44, box 4, folder 95, Ritter-SIO, pp. 78–83.

49. E. W. Scripps, "The American Society for the Dissemination of Science," 5 March 1919, box 1, Science Service Papers; David J. Rhees, "A New Voice for Science: Science Service under Edwin E. Slosson, 1921–1929" (M.A. thesis, University of North Carolina at Chapel Hill, 1979).

50. E. W. Scripps to E. E. Slosson, 28 August 1920, box 19, Ritter-UC; Ritter to Slosson, 16 July 1923, box 15, folder W. E. Ritter, Science Service Papers; Knight, *I Protest*, 728.

51. George E. Webb, *The Evolution Controversy in America* (Lexington: University Press of Kentucky, 1994), 66–78; Larson, *Summer*, 31–59; Benjamin Gruenberg to Maynard Shipley, 16 August 1924, box 10, Ritter-UC.

52. Luther Burbank, "Science and Civilization," 14 November 1924, reprinted in Maynard Shipley, *The War on Modern Science* (New York: Knopf, 1927), 389 (erroneously dated 14 November 1925); Webb, *Evolution Controversy*, 77.

53. Ritter, "What Is at Stake in the Effort to Prevent People from Studying Evolution?" in *War on Modern Science*, ed. Shipley, 398–99.

54. Ritter to Charles T. Fagnani [Union Theological Seminary], 28 January 1925, box 3, Ritter-UC; see notes 1 and 2 above.

55. William E. Ritter, "Science and the Newspapers," *Science* 67 (1928): 282–86.

56. Larson, *Summer*; Ronald L. Numbers, *Darwinism Comes to America* (Cambridge, MA: Harvard University Press, 1998), 76–91; William E. Ritter, *The Natural History of Our Conduct* (New York: Harcourt Brace, 1927), v.

CHAPTER NINE
GOOD BREEDING IN MODERN AMERICA

1. E. G. Conklin, "The Value of Zoology to Humanity: The Cultural Value of Zoology," *Science* 41 (1915): 334; E. G. Conklin, S. A. Forbes, C. A. Kofoid, F. R. Lillie, T. H. Morgan, G. H. Parker, J. Reighard, and H. M. Smith, "Some Suggestions for National Service on the Part of Zoologists and Zoological Laboratories," *Science* 45 (1917): 627–30; "First Report of the Committee on Zoology," *Proceedings of the National Academy of Sciences* 3 (1917): 725–31; National Academy of Sciences, *Annual Report*, 1918, pp. 95–97, 108–9, and 1919, pp. 100–101.

2. Stephen J. Gould, *The Mismeasure of Man* (New York: Norton, 1981); Daniel J. Kevles, *In the Name of Eugenics* (New York: Knopf, 1985). The most recent overview is Diane Paul, *Controlling Human Heredity: 1865 to the Present* (Atlantic Highlands, NJ: Humanities Press, 1996). The diversity of people and aims in eugenics is emphasized by Paul, and in Martin Pernick, "Eugenics and Public Health in American History," *American Journal of Public Health* 87 (1997): 1767–72.

3. Daniel Pick, *Faces of Degeneration* (Cambridge: Cambridge University Press, 1989), surveys these tendencies.

4. Richard L. Dugdale, *"The Jukes": A Study in Crime, Pauperism, Disease and Heredity* (New York: G. P. Putnam's Sons, 1877); Oscar C. McCulloch, "The Tribe of Ishmael: A Study in Social Degradation" (1888), reprinted in *White Trash: The Eugenic Family Studies 1877–1919*, ed. Nicole Hahn Rafter (Boston: Northeastern University Press, 1988), 49–54; David Starr Jordan, *Days of a Man*, 2 vols. (Yonkers: World Book Company, 1922), 1:133. On popular eugenics literature, see Dorothy Nelkin and M. Susan Lindee, *The DNA Mystique* (San Francisco: Freeman, 1995), 19–33.

5. David N. Livingstone, *Nathaniel Southgate Shaler and the Culture of American Science* (Tuscaloosa: University of Alabama Press, 1987), 150–53; John S. Haller, Jr., *Outcasts from Evolution: Scientific Attitudes of Racial Inferiority, 1859–1900* (Urbana: University of Illinois Press, 1971).

6. Philip R. Reilly, *The Surgical Solution: A History of Involuntary Sterilization in the United States* (Baltimore: Johns Hopkins University Press, 1991); James W. Trent, Jr., *Inventing the Feeble Mind: A History of Mental Retardation in the United States* (Berkeley: University of California Press, 1994), 192–202.

7. Thomas G. Dyer, *Theodore Roosevelt and the Idea of Race* (Baton Rouge: Louisiana State University Press, 1980), 154–57.

8. The committee was announced in late 1906 but was not organized until a year later. Other original members were army physician Charles Woodruff, breeder Roswell Johnson, and young biologists Vernon Kellogg, William Castle, and William Tower. American Breeders' Association, *Annual Report* 3 (1907): 137–38; ibid., 4 (1908): 201–8.

9. Mark H. Haller, *Eugenics: Hereditarian Attitudes in American Thought* (New Brunswick, NJ: Rutgers University Press, 1963); Pauline Mazumdar, *Eugenics, Human Genetics and Human Failings* (London: Routledge, 1992); Barbara A. Kimmelman, "The American Breeders' Association: Genetics and Eugenics in an Agricultural Context, 1903–13," *Social Studies of Science* 13 (1983): 163–204; Luther Burbank, *The Training of the Human Plant* (New York: Century, 1906).

10. David Starr Jordan, *The Factors in Organic Evolution: A Syllabus of a Course of Elementary Lectures Delivered in Leland Stanford Junior University* (Boston: Ginn, 1895).

11. Charles B. Davenport, *Eugenics* (New York: Henry Holt, 1910); Charles B. Davenport, *Heredity in Relation to Eugenics* (New York: Henry Holt, 1911); William E. Kellicott, *The Social Direction of Human Evolution* (New York: D. Appleton, 1911); William E. Castle, John M. Coulter, Charles B. Davenport, Edward M. East, and William L. Tower, *Heredity and Eugenics* (Chicago: University of Chicago Press, 1912); Morton A. Aldrich et al., *Eugenics: Twelve University Lectures* (New York: Dodd, Mead, 1914); George H. Parker, *Biology and Social Problems* (Boston: Houghton Mifflin, 1914); Edwin G. Conklin, *Heredity and Environment in the Development of Men* (Princeton: Princeton University Press, 1915); Michael F. Guyer, *Being Well-Born: An Introduction to Eugenics* (Indianapolis: Bobbs-Merrill, 1916); William E. Castle, *Genetics and Eugenics: A Text-book for Students of Biology and a Reference Book for Animal and Plant Breeders* (Cambridge, MA: Harvard University Press, 1916).

12. First International Eugenics Congress, *Problems in Eugenics* (London: Eugenics Education Society, 1912; reprint, New York: Garland, 1984); Race Betterment Foundation, *Proceedings of the First National Conference on Race Betterment* (Battle Creek, MI: Gage Printing Co., 1914).

13. Kenneth M. Ludmerer, *Genetics and American Society* (Baltimore: Johns Hopkins University Press, 1972), 35, notes a number of expressions of enthusiasm.

14. Guyer, *Being Well-Born*, 289–339.

15. William E. Ritter to Charles Kofoid, 10 October 1912, Scripps Family Papers, box 1, folder "Biological Station," Scripps Institution of Oceanography; Pauly, *Controlling Life*, 144–45; Morgan to Edward M. East, 6 April and 6 May 1919, box 9, Jacques Loeb Papers, Manuscripts Division, Library of Congress, Washington, D.C.; Herbert W. Conn, *Social Heredity and Social Evolution: The Other Side of Eugenics* (New York: Abingdon Press, 1914); and Maynard Metcalfe, "Eugenics and Euthenics," *Popular Science Monthly* 84 (1914): 383–89; see also Lester Ward's last essay, "Eugenics, Euthenics, and Eudemics," *American Journal of Sociology* 18 (1913): 737–54. On this theme, see Kathy J. Cooke, "The Limits of Heredity: Nature and

Nurture in American Eugenics before 1915," *Journal of the History of Biology* 31 (1998): 263–78.

16. Charles B. Davenport, "Eugenics, a Subject for Investigation Rather than Instruction," *American Breeders' Magazine* 1 (1910): 68–69; Guyer, *Being Well-Born*.

17. Paul Weindling, *Health, Race and German Politics between National Unification and Nazism, 1870–1945* (Cambridge: Cambridge University Press, 1989); Robert N. Proctor, *Racial Hygiene: Medicine under the Nazis* (Cambridge, MA: Harvard University Press, 1988); Peter Weingart, Jürgen Kroll, and Kurt Bayertz, *Rasse, Blut und Gene: Geschichte der Eugenik und Rassenhygiene in Deutschland* (Frankfurt am Main: Suhrkamp Verlag, 1988); Sheila Faith Weiss, *Race, Hygiene, and National Efficiency* (Berkeley: University of California Press, 1987); Mazumdar, *Eugenics, Human Genetics and Human Failings*.

18. On Davenport, see E. Carleton MacDowell, "Charles B. Davenport, 1866–1944: A Study of Conflicting Influences," *Bios* 17 (1946): 3–50; Oscar Riddle, "Charles Benedict Davenport," *National Academy of Sciences Biographical Memoirs* 25 (1949): 75–110; Charles E. Rosenberg, *No Other Gods* (Baltimore: Johns Hopkins University Press, 1976), 89–97; Kevles, *In the Name of Eugenics*, 44–56.

19. Charles B. Davenport, "*Cristatella*: The Origin and Development of the Individual in the Colony," *Bulletin of the Museum of Comparative Zoology* 20 (1890): 101–51, esp. 122; Davenport, "Observations on Budding in *Paludicella* and Some Other Bryozoa," *Bulletin of the Museum of Comparative Zoology* 22 (1891): 1–114, esp. 72; Davenport, *Experimental Morphology*, 2 vols. (New York: Macmillan, 1897–99.

20. Charles B. Davenport, "Zoology of the Twentieth Century," *Science* 14 (1901): 317; Davenport, "The Animal Ecology of the Cold Spring Harbor Sand Spit, with Remarks on the Theory of Adaptation," *University of Chicago Decennial Publications*, 1st ser., multivolume (Chicago: University of Chicago Press, 1903), 10:157–76; Lee Richard Hiltzik, "The Brooklyn Institute of Arts and Sciences' Biological Laboratory, 1890–1924: A History" (Ph.D. diss., State University of New York at Stony Brook, 1993).

21. Charles B. Davenport to John S. Billings, 18 August 1902, 3 May 1903, file Cold Spring Harbor beginnings, Charles B. Davenport Papers, American Philosophical Society Library, Philadelphia (hereafter Davenport Papers); on Davenport's entrepreneurship, see MacDowell, "Davenport"; more broadly, see Nathaniel Comfort and Bentley Glass, *Building Arcadia: A History of Cold Spring Harbor Laboratory* (Cold Spring Harbor, NY: Cold Spring Harbor Laboratory Press, in press); Elizabeth L. Watson, *Houses for Science: A Pictorial History of Cold Spring Harbor Laboratory* (Plainview, NY: Cold Spring Harbor Laboratory Press, 1991).

22. Carnegie Institution of Washington, *Year-Book*, 1904, p. 33, and 1906, p. 94; American Breeders' Association, *Proceedings* 2 (1906): 11.

23. Jordan invited Davenport to join the Eugenics Committee only in late October 1908. In January 1909, after the ABA meeting, Davenport sent Jordan proposed questions on heredity for inclusion in the 1910 census and volunteered himself for

the position of secretary of the committee. By late March Jordan had given him carte blanche: see Jordan to Davenport, 29 October 1908, 23 March 1909; Davenport to Jordan, 27 January 1909, and other correspondence in folder "Jordan, David S.," Davenport Papers.

24. Charles B. Davenport, "Heredity in Man," *Harvey Society Lectures* 4 (1908–9): 280; Paul R. Cutright, *Theodore Roosevelt: The Making of a Conservationist* (Urbana: University of Illinois Press, 1985), 231. The Harvey Society was an elite New York City medical organization; the lecture was delivered 6 March 1909. Sagamore Hill, Roosevelt's home, was about two miles north of the Station for Experimental Evolution, along the harbor. Davenport had frequently invited Roosevelt to visit the station during his presidency, but apparently without success: see letters prior to 1909 in Roosevelt's file in the Davenport Papers. On the ERO, see Garland E. Allen, "The Eugenics Record Office at Cold Spring Harbor, 1910–1940: An Essay in Institutional History," *Osiris* 2 (1986): 225–64.

25. Davenport to Vernon Kellogg, 30 October 1912, Davenport Papers.

26. MacDowell, "Davenport," 32–33; Riddle, "Davenport," 88–89.

27. David S. Jordan to Davenport, 20 July 1910, Davenport Papers; Allen, "Eugenics Record Office"; William H. Goetzmann and Kay Sloan, *Looking Far North: The Harriman Expedition to Alaska, 1899* (Princeton: Princeton University Press, 1982).

28. Charles B. Davenport, *The Trait Book* (ERO, bulletin #6, 1912); Davenport, *The Family-History Book* (ERO, bulletin #7, 1912); Paul, *Controlling Human Heredity*, 54–58; Allen, "Eugenics Record Office," 251.

29. Charles B. Davenport, *Huntington's Chorea in Relation to Heredity and Eugenics* (ERO, bulletin #17, 1916); Henry S. Conard and Charles B. Davenport, *Hereditary Fragility of Bone (Fragilitas Osseus, Osteopsathyrosis)* (ERO, bulletin #14, 1915).

30. Florence H. Danielson and Charles B. Davenport, *The Hill Folk: Report on a Rural Community of Hereditary Defectives* (Memoirs of the ERO, no. 1, 1912); Arthur H. Estabrook and Charles B. Davenport, *The Nam Family: A Study in Cacogenics* (Memoirs of the ERO, no. 2, 1912); Charles B. Davenport, *The Feebly Inhibited* (Carnegie Institution of Washington Publication no. 236, 1915); Charles B. Davenport, assisted by Mary Theresa Scudder, *Naval Officers, Their Heredity and Development* (Carnegie Institution of Washington Publication no. 259, 1919); Kevles, *In the Name of Eugenics*, 49; Rafter, *White Trash*, 82.

31. See Leila Zenderland, *Measuring Minds: Henry Herbert Goddard and the Origins of American Intelligence Testing* (New York: Cambridge University Press, 1998), 153–69, on Davenport's relations with Goddard; for criticisms of Davenport, see J. E. Wallace Wallin, "The Hygiene of Eugenic Generation," *Psychological Clinic* 8 (1914): 121–37, 170–79; L. L. Bernard, review of *Naval Officers, Their Heredity and Development*, by Charles B. Davenport, *American Journal of Sociology* 25 (1919–20): 241–42; George W. Stocking, *Race, Culture, and Evolution* (New York: Free Press, 1968); Cooke, "The Limits of Heredity," 263–78.

32. Estabrook and Davenport, *Nam Family*; Arthur H. Estabrook, *The Jukes in 1915* (Carnegie Institution of Washington Publication no. 240, 1916); Frances J. Hassencahl, "Harry H. Laughlin, 'Expert Eugenics Agent' for the House Committee on Immigration and Naturalization, 1921 to 1931" (Ph.D. diss., Case Western Reserve University, 1970).

33. For more detail on this work, see Philip J. Pauly, "How Did the Effects of Alcohol on Reproduction Become Scientifically Uninteresting in the Early Twentieth Century?" *Journal of the History of Biology* 29 (1996): 1–28; Comfort and Glass, *Building Arcadia*.

34. E. C. MacDowell and E. M. Vicari, "Alcoholism and the Behavior of White Rats, I: The Influence of Alcoholic Grandparents upon Maze-Behavior," *Journal of Experimental Zoology* 33 (1921): 209–91 (quotations, 210); MacDowell, "Alcoholism and the Behavior of White Rats, II: The Maze-Behavior of Treated Rats and Their Offspring," *Journal of Experimental Zoology* 37 (1923): 418–56.

35. Charles R. Stockard and collaborators, *The Genetic and Endocrinic Basis for Differences in Form and Behavior, as Elucidated by Studies of Contrasted Pure-Line Dog Breeds and Their Hybrids* (Philadelphia: Wistar Institute of Anatomy and Biology, 1941); Staff minutes, 12 February 1930, folder 975, and "Cornell University—Stockard Dog Farm," 8 pp., [1940], folder 978, both in box 81, series 200, Rockefeller Foundation record group 1.1, Rockefeller Archive Center, Pocantico Hills, New York.

36. Martin Pernick, *The Black Stork* (New York: Oxford University Press, 1995).

37. Garland E. Allen, "Old Wine in New Bottles: From Eugenics to Population Control in the Work of Raymond Pearl," in *The Expansion of American Biology*, ed. Keith R. Benson, Jane Maienschein, and Ronald Rainger (New Brunswick, NJ: Rutgers University Press, 1991), 231–61; Diane B. Paul, *The Politics of Heredity* (Albany: State University of New York Press, 1998), 117–32; Allen, "Eugenics Record Office."

38. Kevles, *In the Name of Eugenics*; Mazumdar, *Eugenics, Human Genetics and Human Failings*.

39. Leslie Clarence Dunn and Theodosius Dobzhansky, *Heredity, Race, and Society* (New York: Penguin Books, 1946), endpage and 114–15. An enlarged second edition was published by New American Library in 1952. More broadly, see Elazar Barkan, *The Retreat of Scientific Racism* (New York: Cambridge University Press, 1992).

40. Paul Robinson, *The Modernization of Sex* (New York: Harper, 1970); Vern L. Bullough, *Science in the Bedroom* (New York: Basic Books, 1994); Linda Gordon, *Woman's Body, Woman's Right* (New York: Viking, 1976); James Reed, *From Private Vice to Public Virtue: The Birth Control Movement and American Society since 1830* (New York: Basic Books, 1978); Allan M. Brandt, *No Magic Bullet: A Social History of Venereal Disease in the United States since 1880* (New York: Oxford University Press, 1985); Nelly Oudshoorn, *Beyond the Natural Body* (London: Routledge, 1994); Adele E. Clarke, *Disciplining Reproduction: Modernity, American Life Sciences, and the Problems of Sex* (Berkeley: University of California Press, 1998); James H. Jones,

Alfred C. Kinsey: A Public/Private Life (New York: W. W. Norton, 1997); Julia A. Ericksen, *Kiss and Tell: Surveying Sex in the Twentieth Century* (Cambridge, MA: Harvard University Press, 1999). See also Jonathan Gathorne–Hardy, *Sex the Measure of All Things* (Bloomington: Indiana University Press, 2000).

41. Kenneth Manning, *Black Apollo of Science* (New York: Oxford University Press, 1983), 81; Jane Maienschein, *100 Years Exploring Life, 1888–1988* (Boston: Jones and Bartlett, 1989), 117–50.

42. Pauly, *Controlling Life*, 101.

43. Timothy J. Gilfoyle, *City of Eros: New York City, Prostitution, and the Commercialization of Sex, 1790–1920* (New York: Norton, 1992); Brandt, *No Magic Bullet*; Edward L. Bernays, *Biography of an Idea* (New York: Simon and Schuster, 1965), 53–65; Nathan G. Hale, Jr., *Freud and the Americans* (New York: Oxford University Press, 1971).

44. Earl Zinn to Macfie Campbell, 28 March and 4 April 1921, Beginning of program: presentation of project to NRC divisions, NRC Program Files, Medical Sciences Division, Committee for Research in Problems of Sex, National Academy of Sciences Archives, Washington, D.C. (hereafter NRC-CRPS). Within a few years the Public Health Service withdrew from sex education, but by that time the Rockefeller organizations were set in their new directions. The basic history of the CRPS is Sophie D. Aberle and George W. Corner, *Twenty-five Years of Sex Research* (Philadelphia: W. B. Saunders, 1953). My account draws on Bullough, *Science in the Bedroom*, and Clarke, *Disciplining Reproduction*, but diverges from their argument that the committee's biological approach was a retreat from an initial more desirable commitment to human sexology.

45. Zinn to Campbell, 28 March and 4 April 1921, Walter Miles to Zinn, 20 April 1921, Zinn to Miles, 22 April 1921, all in Beginning of program: presentation of project to NRC divisions, 1921, NRC-CRPS.

46. "Conference on Sex Problems," stenographic report, 28 October 1921, Beginning of program: Conference on Sex Problems, 1921, NRC-CRPS.

47. "Preliminary Report of the Committee for Research on Sex Problems to the Division of Medical Sciences of the National Research Council, 15 March 1922, Exhibit 'C': Report on Sex Project for the period, July 1, 1921–January 1, 1922," Preliminary Report, 1922, NRC-CRPS. They were joined by psychiatrist T. W. Salmon, and, after some maneuvering, sociologist Katherine Bemont Davis, head of the Bureau of Social Hygiene (the committee's sponsor). See Minutes, 12 December 1921, 17 January, 3 April, and 9 May 1922, NRC-CRPS.

48. "Preliminary Report of the Committee for Research on Sex Problems to the Division of Medical Sciences of the National Research Council, 15 March 1922, Exhibit 'D': Preliminary Report of Sex Research in Progress and Proposed," Preliminary Report, 1922, NRC-CRPS.

49. "First Annual Report of Committee for Research on Sex Problems of the Division of Medical Sciences, National Research Council," 1 March 1923, NRC-CRPS. Conklin resigned in June 1922, apparently when he recognized that the committee

would have little to do with the invertebrates on which he worked. Lillie, by contrast, had been working on mammals and birds for nearly a decade. Minutes, 23 June 1922, in Appointments: Members 1921–1938, NRC-CRPS.

50. "Preliminary Report of the Committee for Research on Sex Problems to the Division of Medical Sciences of the National Research Council, 15 March 1922, Exhibit 'D': Preliminary Report of Sex Research in Progress and Proposed," Preliminary Report, 1922; Minutes, 14 March 1925, NRC-CRPS.

51. Ibid.

52. Minutes, 1927–28, passim, NRC-CRPS; Donald Fisher, *Fundamental Development of the Social Sciences* (Ann Arbor: University of Michigan Press, 1993), 36–42; Hamilton Cravens, *The Triumph of Evolution* (Philadelphia: University of Pennsylvania Press, 1978), 180–88.

53. Lillie to Wycliffe Rose, 17 June 1924, General Education Board subject 905, box 357, f. 3680 (Woods Hole Marine Biological Laboratory 1924–27), Rockefeller Archive Center; Lillie to Earl Zinn, 8 July 1927, folder CRPS 1922–34, Grantees: Lillie, F.R., NRC-CRPS; Lillie to Robert Maynard Hutchins, 23 June 1930, Presidents' Papers 109:2, University Archives, Regenstein Library, University of Chicago; Gregg Mitman, *The State of Nature: Ecology, Community, and American Social Thought, 1900–1950* (Chicago: University of Chicago Press, 1992), 97–109, discusses Lillie's outlook and situates most of these approaches within it.

54. D. Erskine Carmichael, *The Pap Smear: Life of George N. Papanicolaou* (Springfield, IL: Charles C. Thomas, 1973), 44–53.

55. Lists of expenditures, grantees, projects, and publications are in Aberle and Corner, *Twenty-five Years of Sex Research.*

56. This mixed record is emphasized in both Oudshoorn, *Beyond the Natural Body,* 24–41, and Clarke, *Disciplining Reproduction,* 122–39. On Pincus, see Reed, *Private Vice,* 320–26; Lillie assessed his field in "Zoological Sciences in the Future," *Science* 88 (1938): 65–72.

57. George W. Corner, *Attaining Manhood: A Doctor Talks to Boys about Sex* (New York: Harper, 1938), v. Corner described both male and female anatomy and development, and explained that masturbation was "simply a kind of substitute for the normal thing, caused by the postponement of normal mating in our modern civilization." On the other hand, he cautioned against "necking" and discussed intercourse only in general terms.

58. Jones, *Kinsey*; Stephen J. Gould, *The Flamingo's Smile* (New York: W. W. Norton, 1985), 155–66.

59. Jones, *Kinsey,* 417–41; George W. Corner, *The Seven Ages of a Medical Scientist* (Philadelphia: University of Pennsylvania Press, 1981), 269.

60. Alfred C. Kinsey, Wardell B. Pomeroy, and Clyde E. Martin, *Sexual Behavior in the Human Male* (Philadelphia: W. B. Saunders, 1948); Jones, *Kinsey,* 501–33.

61. Kinsey, *Sexual Behavior in the Human Male,* 195.

62. Jones, *Kinsey*, 577–93, 638–48; Margaret Mead, "An Anthropologist Looks at the Report," American Social Hygiene Association, *Problems of Sexual Behavior*, 1948, p. 64.

63. Reed, *Private Vice*.

64. Lester Ward to Emily Palmer Cape, 9 June 1910, quoted in Cape, *Lester F. Ward: A Personal Sketch* (New York: G. P. Putnam, 1922), 93; Lionel Trilling, "The Kinsey Report" (1948), reprinted in Trilling, *The Liberal Imagination* (New York: Viking Press, 1950), 234.

EPILOGUE

1. Carol Gruber, "The Overhead System in Government-Sponsored Academic Science: Origins and Early Development," *Historical Studies in the Physical and Biological Sciences* 25 (1995): 243.

2. M. Susan Lindee, *Suffering Made Real: American Science and the Survivors at Hiroshima* (Chicago: University of Chicago Press, 1994); John Beatty, "Genetics in the Atomic Age: The Atomic Bomb Casualty Commission, 1947–1956," in *The Expansion of American Biology*, ed. Keith R. Benson, Jane Maienschein, and Ronald Rainger (New Brunswick, NJ: Rutgers University Press, 1991), 284–324; Toby A. Appel, *Shaping Biology: The National Science Foundation and Federal Support of Biology in the Cold War Era* (Baltimore: Johns Hopkins University Press, in press); Barbara B. Clowse, *Brainpower for the Cold War: The Sputnik Crisis and National Defense Education Act of 1958* (Westport, CT: Greenwood Press, 1981).

3. National Research Council, *A Century of Doctorates* (Washington, DC: National Academy of Sciences, 1978), 12; *Biological Abstracts*; NIH, *Almanac*, 1967, p. 73; NSF, *Federal Funds for Science*, 1954, p. 196.

4. Lindsey R. Harmon and Herbert Soldz, *Doctorate Production in United States Universities 1920–1962* (Washington, DC: National Academy of Sciences, 1963), 15.

5. Robert E. Kohler, *Partners in Science: Foundation Managers and Natural Scientists, 1900–1945* (Chicago: University of Chicago Press, 1991); Lily E. Kay, *The Molecular Vision of Life: Caltech, the Rockefeller Foundation, and the Rise of the New Biology* (New York: Oxford University Press, 1993).

6. Robert E. Kohler, *Lords of the Fly: Drosophila Genetics and the Experimental Life* (Chicago: University of Chicago Press, 1994).

7. James D. Watson, *The Double Helix*, ed. Gunter Stent (London: Weidenfield and Nicolson, 1981), 17, 27, 46; Horace Freeland Judson, *The Eighth Day of Creation: Makers of the Revolution in Biology* (New York: Simon and Schuster, 1979).

8. Nathaniel Comfort and Bentley Glass, *Building Arcadia: A History of Cold Spring Harbor Laboratory* (Cold Spring Harbor, NY: Cold Spring Harbor Laboratory Press, in press).

9. John C. Burnham, *How Science Lost and Superstition Won: Popularizing Science and Health in the United States* (New Brunswick, NJ: Rutgers University Press, 1987); Vannevar Bush, *Science—the Endless Frontier: A Report to the President on a*

Program for Postwar Scientific Research (1945; reprinted Washington, DC: National Science Foundation, 1980); James H. Jones, *Alfred C. Kinsey: A Public/Private Life* (New York: W. W. Norton, 1997); Thomas Hager, *Force of Nature: The Life of Linus Pauling* (New York: Simon and Schuster, 1995).

10. For an impersonal statement about these anxieties and their linkage to the tradition of NIH patronage, see Martin Kenney, *Biotechnology: The University-Industrial Complex* (New Haven: Yale University Press, 1986).

11. Paul Rabinow, *Making PCR: A Story of Biotechnology* (Chicago: University of Chicago Press, 1996).

12. Albert R. Jonsen, *The Birth of Bioethics* (New York: Oxford University Press, 1998).

13. Barbara A. Mikulski, "Science in the National Interest," *Science* 264 (1994): 221–22, and the accompanying interview with Mikulski: "The Hand on Your Purse Strings," *Science* 264 (1994): 192–94.